THE AUSTERE LANDSCAPE: The Paintings of Hung-jen

THE AUSTERE LANDSCAPE
The Paintings of Hung-jen

by

Jason C. Kuo

SMC Publishing Inc.

Taipei • New York

in cooperation with

University of Washington Press

Seattle • London

SMC Publishing Inc.
P.O. Box 13-342, Taipei 10764
Taiwan, R. O. C.

University of Washington Press
P. O. Box 50096
Seattle, Washington 98145-5096
U. S. A.

ISBN 957-9482-11-X
LCCN 89-83323

Printed in Taiwan, R.O.C.

to Edie

TABLE OF CONTENTS

ACKNOWLEDGEMENTS

This book grows out of my Ph.D. dissertation written at the University of Michigan in Ann Arbor. Since it was written in 1980, a few previously unpublished paintings by Hung-jen have come to light. I have attempted to incorporate some of them into the book.

I am most indebted to my principal adviser Richard Edwards for his many years of teaching. His encouragement and advice have been invaluable. Most importantly, he has taught me how to see. Thanks are also due to other members of my dissertation committee: Thomas Lawton, formerly Director of the Freer Gallery of Art, Washington, D.C.; Shuen-fu Lin, the University of Michigan; Walter Spink, the University of Michigan.

James Cahill, the University of California, Berkeley, has generously shared his ideas and rare material on Hung-jen with me. In the spring of 1980, I participated in his seminar on the Anhui School. Many ideas presented here took final shape during my sojourn in Berkeley. I am grateful to him for his suggestions.

For financial support, I am grateful to the Department of the History of Art, the Center for Chinese Studies, and the Horace H. Rackham School of Graduate Studies, all at the University of Michigan and the JDR 3rd Fund.

The publication of this book is made possible through the support of grants from the Division I Faculty Research Fund of Williams College and the Office of Graduate Studies and Research, the University of Maryland at College Park. The support of my colleagues in the Department of Art, Williams College and the Department of Art History and Archaeology, the University of Maryland at College Park is gratefully acknowledged. I am grateful to James Douglas Farquhar for his constant encouragement and support. To Ellen Johnston Laing, Jerome Silbergeld, Louise Yuhas, and many other colleagues, I offer my sincere thanks.

J.C.K.

LIST OF PLATES

17. Yang Ming-shih: "Old Tree and Bamboo." Hanging scroll, ink on paper, 136×32 cm. Palace Museum. Peking.

18. Hou Mou-kung: "Landscape." Hanging scroll, ink and light colors on paper, 112×34.9 cm. National Palace Museum, Taipei.

19. Li Yung-ch'ang: "In the Manner of Yüan Masters," 1634. Hanging scroll. Whereabouts unknown.

20. Ch'eng Chia-sui: "Horse-Riders on the Willow Bank,"1639. Hanging scroll, ink and colors on paper, 108×32 cm. T'ien-tsin Museum.

21. Hung-jen: "Landscape," 1663. Hanging scroll. Liaoning Provincial Museum.

22. Wang Chih-jui: "Landscape." Hanging scroll. Whereabouts unknown.

23. Liu Shang-yen and others: "Joint Handscroll of Landscape for Li Sheng-pai," 1639. Shanghai Museum.

24. Wang Chih-jui: "Landscape," 1649. Album leaf. Whereabouts unknown.

25. Wang Chih-jui: "Landscape." Album leaf. Whereabouts unknown.

26. Hsiao Yün-ts'ung: 'Walking with a Staff in a Sparse Forest," 1648. Hanging scroll, ink on paper, 107.2×31.5 cm. T'ien-tsin Museum.

27. Hsiao Yün-ts'ung: "Travellers in Autumn Mountains," 1657 . Handscroll, ink and colors on paper, 25.4×250.4 cm. Tokyo National Museum. Section 2

28. Hsiao Yün-ts'ung: "Travellers in Autumn Mountains," Section 3.

29. Hsiao Yün-ts'ung: "Going Home and Living Abroad Are the Same Thing," 1656. Handscroll, ink and colors on paper, 23.5×1,302 cm. Drenowatz collection, Zürich. Section 1.

30. Hsiao Yün-ts'ung: "Going Home and Living Abroad Are the Same Thing." Section 4.

31. "Joint Handscroll of Landscape for Ko-an," 1650-51. Whereabouts unknown. Section 1.

32. "Joint Handscroll of Landscape for Ko-an." Section 2.

33. "Joint Handscroll of Landscape for Ko-an." Section 3.

34. "Joint Handscroll of Landscape for Ko-an." Section 4.

35. "Joint Handscroll of Landscape for Ko-an." Section 5.

36. "Joint Handscroll of Landscape for Ko-an." Section 6.

37. "Joint Handscroll of Landscape for Ko-an." Section 7.

168. Hung-jen: "The Ten-Bamboo Studio." Detail.

169. Hung-jen: "Plum-Blossom Pavilion," 1658. Short handscroll, ink on paper, 52.5×84.5 cm. Whereabouts unknown (formerly Shen Chien-fu collection).

170. Hung-jen: "Frosted Trees," 1658. Hanging scroll, ink on paper, 83.5×30.2 cm. Musée Guimet, Paris.

171. Hung-jen: "Plum House and Pine Spring," 1658. Hanging scroll, ink and colors on paper, 185.5×86.5 cm. Shanghai Museum.

172. Hung-jen: "Landscape in the Style of Ni Tsan." Hanging scroll, ink on paper, 23×18½ in. Mrs. Jackson Burke collection, New York.

173. Hung-jen: "The Unending Songs of Birds," 1659. Hanging scroll, ink and colors on paper, 218×67 cm. C. C. Wang collection, New York.

174. Hung-jen: "Landscape," 1659. Hanging scroll, ink on paper, 89.4×41.8 cm. Shanghai Museum.

175. Hung-jen: "Landscape." Detail.

176. Hung-jen: "Landscape." Detail.

177. Hung-jen: "Landscape." Detail.

178. Hung-jen: "Landscape after Lu Kuang," 1658. Hanging scroll. Ching Yüan Chai Collection, Berkeley.

179. Hung-jen: "Quiet House amid Bamboo and Rocks," 1659. Hanging scroll, ink on paper, 58.5×34 cm. T'ien-tsin Museum.

180. Hung-jen: "Landscape." Hanging scroll, 103×46.5 cm. Li Ch'u-li collection.

181. Hung-jen: "Heavenly Capital Peak," 1660. Hanging scroll , height about 280 cm. Nanking Museum.

182. Hung-jen: "Landscape in the Style of Ni Yün-lin," 1660. Hanging scroll. Whereabouts unknown (formerly Meng-shui shan-fang collection).

183. Hung-jen: "Pines and Rocks at Mt. Huang," 1660. Hanging scroll, ink and colors on paper, 198.7×81 cm. Shanghai Museum.

184. Hung-jen: "Pines and Rocks at Mt. Huang." Detail.

185. Hung-jen: "Pines and Rocks at Mt. Huang." Detail.

186. Hung-jen: "Pines and Rocks at Mt. Huang." Detail.

187. Hung-jen: "Pines and Rocks at Mt. Huang." Detail.

188. Pines and rocks at Mt. Huang. From Huang Yen-p'ei and Lü I-shou, *Huang-*

254. Chu Ch'ang: "Streams and Mountains without End." Section 11 and Section 12.

255. Chu Ch'ang: "Streams and Mountains without End." Section 13 and Section 14.

256. Cha Shih-piao: "Landscape," 1659. Fan, ink on paper. Steve Addiss collection, Lawrence, Kansas.

257. Cha Shih-piao: "Landscape." Fan. K. C. Ho collection, New York.

258. Cha Shih-piao: "Landscape," 1662. Album, leaf no. 1 (right) and Leaf No. 2 (left), ink on silk, 30×25.5 cm. each. T'ien-yin-t'ang collection, Tokyo.

259. Cha Shih-piao: "Landscape." Leaf no. 7 (right) and Leaf no. 8 (left), ink and colors on silk, 30×25.5 cm. each.

260. Cha Shih-piao: "Landscape." Leaf no. 9 (right), ink and colors on silk, and Leaf no. 10 (left), ink on silk, 30×25.5 cm. each.

261. Cha Shih-piao: "Pavilion in Autumn Morning," 1678. Fan, ink on paper. Shanghai Museum.

262. Chiang Chu: "Pines and Rocks at Mt. Huang." Hanging scroll. Whereabouts unknown.

263. Chiang Chu: "Arranging Rocks and Planting Flowers." Hanging scroll, ink and colors on paper, $22\frac{1}{4}\times10\frac{5}{8}$ in. Whereabouts unknown (formerly Richard Bryant Hobart collection).

264. Yao Sung: "Landscape," 1707. Hanging scroll, ink on paper, $58\times23\frac{1}{2}$ in. Drenowatz collection, Zürich.

265. Yao Sung: "High Dwelling in a Quiet Valley." Hanging scroll, ink on silk, 147×50.5 cm. Hashimoto Sueyoshi collection, Japan.

266. Yao Sung: "Leisurely View in the Autumn Mountain," 1717 . Hanging scroll, ink on paper, 202×101.5 cm. Nanking Museum .

267. Yao Sung: "Landscape." Album, 21.3×24.5 cm. each. Franco Vannotti collecton, Lugano. Leaf no. 1.

268. Yao Sung: "Landscape." Leaf no. 2.

269. Yao Sung: "Landscape." Leaf no. 3.

270. Yao Sung: "Landscape." Leaf no. 4.

271. Yao Sung: "Landscape." Leaf no. 5.

272. Yao Sung: "Landscape." Leaf no. 6.

273. Yao Sung: "Landscape." Leaf no. 7.

INTRODUCTION

Hung-jen 弘仁 (1610–1663) is one of the most important and influential artists in the history of seventeenth century painting. He has been known as one of the "Four Great [Painter]-Monks of the Late Ming Dynasty (*Ming-mo ssu-seng* 明末四僧)" ranking with K'un-ts'an 髡殘 (active 1657–1674), Chu Ta 朱耷 (1625–after 1705), and Tao-chi 道濟 (1641–ca.1720).[1] He is also the leading figure among the "Four Great Masters of Hsin-an (*Hsin-an ssu-ta-chia* 新安四大家)," sometimes called the "Four Great Masters of Hai-yang (*Hai-yang ssu-ta-chia* 海陽四大家)," a group which comprises Sun I 孫逸 (died after 1663), Cha Shih-piao 查士標 (1615–1698), and Wang Chih-jui 汪之瑞 (active ca. 1650), and which, together with other lesser masters, has been recognized as the Hsin-an school or Anhui school.[2] The name refers to the geographical area that produced these artist as well as to the style of their painting. Yet, except for a few brief comments or catalog entries recently written by scholars, no comprehensive study of Hung-jen and the Anhui school, of which he is the central figure, has been undertaken.

Although scholars and connoisseurs have long recognized the importance of the paintings of Hung-jen and the Anhui School, relatively little study has been done. More than twenty years after Huang Pin-hung 黃賓虹 (1864–1955), an artist also from Anhui, published his pioneering modern biography of Hung-jen entitled "Chien-chiang Ta-shih shih-chi i-wen 漸江大師事蹟佚聞" in the periodical *Chung-ho yüeh-k'an* 中和月刊 in Peking in 1940,[3] a booklet entitled *Hung-jen,K'un-ts'an* 弘仁髡殘 was published by Cheng Hsi-chen 鄭錫珍 in Shanghai in 1963, giving a brief account of Hung-jen's life and work[4]. This booklet was supplemented in 1964 by a more extensive anthology called *Chien-chiang tzu-liao-chi* 漸江資料集, which was composed of writings by Hung-jen himself and by contemporary and later critics of his work. The anthology, published in Ho-fei, Anhui, in 1964, was edited by Wang Shih-ch'ing 汪世清 and Wang Ts'ung 汪聰.[5] These publications in Chinese, though useful in many ways, are mostly compilations of literary documents rather than critical studies of Hung-jen's paintings.

In the West, James Cahill's brief treatment of Hung-jen and some of his followers in the Asia House Gallery exhibition catalog *Fantastics and Eccen-*

trics in Chinese Painting contributed to a more critical and broader understanding of the artists.[6] His views were later substantiated by his article "Hung-jen" in *Dictionary of Ming Biography: 1368–1644,* the exhibition catalog *Shadows of Mt. Huang: Chinese Painting and Printing of the Anhui School* edited by him, and his book *The Compelling Image: Nature and Style in Seventeenth-Century Chinese Painting.*[7] In addition, Marilyn and Shen Fu worte a catalog entry on Hung-jen in their *Studies in Connoisseurship: Chinese Paintings from the Arthur M. Sackler Collection in New York and Princeton*[8] which, like Chu-tsing Li's catalog entry on Hung-jen in his *A Thousand Peaks and Myriad Ravines: Chinese Paintings in the Charles A. Drenowatz Collection*[9], provided extensive scholarly research and laid the groundwork for future study. Each of these, however, is somewhat limited to the particular painting of Hung-jen under consideration in the catalogs; together, they do not establish a comprehensive critique of the artist and his followers.

The present study attempts to present a fuller understanding of the art of Hung-jen. It is divided into four chapters. The first chapter provides a biographical sketch of Hung-jen and the geographical, economic, and cultural backgrounds in Hsin-an, Hung-jen's birthplace. In the second chapter, the stylistic sources of Hung-jen's painting are described. After a discussion of the stylistic mode of Ni Tsan 倪瓚 (1301–1374) to whom Hung-jen was greatly indebted, the various stylistic currents in Hsin-an in the late sixteenth and early seventeenth centuries are analyzed in relation to Hung-jen's style. In the third chapter, a survey of Hung-jen's paintings is given. Dated paintings are arranged chronologically, and undated paintings are often placed in relation to the dated ones on stylistic grounds. Because of the relatively short period, from 1651 to 1663, in which we have reliable dated works, no attempt has been made to differentiate stylistic periods within Hung-jen's artistic career. It is hoped, however, that the stylistic analysis offered in this chapter will contribute to a fuller understanding of Hung-jen's art. Among the "Four Great [Painter]-Monks of the Late Ming Dynasty," it should be pointed out, Hung-jen alone attracted a large following and thus established himself as the leader of the Hsin-an school. The fourth chapter will therefore be devoted to Hung-jen' s influence on his followers. Through a comparison of the older master with the lesser masters, the unique style of Hung-jen will stand out more clearly.

CHAPTER I

HUNG-JEN AND HIS WORLD

Hung-jen was born in 1610 and died on the twenty-second day of the twelfth month in 1663.[1] He was a native of She-hsien 歙縣, Hui-chou Prefecture 徽州府, in modern Anhui Province. His secular name was Chiang T'ao 江韜, and his literary name (*tzu* 字) was Liu-ch'i 六奇. When he left Anhui around 1645 and went to Fukien 福建, he changed his name, as many of his contemporaries did upon the fall of the Ming Dynasty (1368–1644), to Chiang Fang 江舫, and adopted the literary name Ou-meng 鷗盟, probably to signify a new personal identity. Soon afterwards, probably in 1646, he was ordained a Buddhist monk by Tao-chou Ku-hang 道舟古航 (1558–1655) at Man-t'ing Peak 幔亭峰 of Mt. Wu-i 武夷 in northwestern Fukien and took the Buddhist name Hung-jen 弘仁 and the literary names Wu-chih 旡智 and Wu-chih 旡執. His style names or sobriquets (*hao* 號) include Chien-chiang 漸江, and Chien-chiang hsüeh-jen 漸江學人. The name Lan-ch'ü 嬾臞, though apparently not recorded, appears in seals on Hung-jen's paintings dated from the 1650s and probably was adopted at the time. The name Mei-hua ku-na 梅花古衲, first used by Chang Keng 張庚 (1685–1760) in his *Kuo-ch'ao hua-cheng-lu* 國朝畫徵錄 (completed in 1735),[2] does not appear in Hung-jen's paintings and is probably a posthumous name, as is the name Mei-hua lao-na 梅花老衲. The name Chien-chiang refers to a river, also called Chien-shui 漸水, that passes through the southwestern part of She-hsien. The last two names, Mei-hua ku-na and Mei-hua lao-na, suggest the identification of Hung-jen by his admirers with the plum blossom, symbol of independence, perserverance, and purity for the Chinese literati. His adoption of a new set of personal names and the radical change in life style he underwent by becoming a monk are, in a sense, a symbolic death or rebirth and attest to Hung-jen's loyalty to the Ming Dynasty.

As a young man, Hung-jen studied the Five Classics (*Wu-ching* 五經) with a local scholar, Wang Wu-ai 汪無涯. After a few years of study he attained the equivalent of a bachelor's degree but did not go on to take other examinations that would lead to a career in the government. Instead, Hung-jen seemed to devote himself to the care of his widowed mother and to the study of literature and art.

Hung-jen was closely associated with Wang Mu-jih 汪沐日 (1605–1679), literary name Fu-kuang 扶光, also from She-hsien. Wang Mu-jih passed his provincial examination (*hsiang-shih* 鄉試) in 1633 and became a licentiate (*chü-jen* 舉人). He later held the positions of Secretary of the Board of Operations (*chih-fang-ssu chu-shih* 職方司主事) and of Vice-President (*shao-ssu-ma* 少司馬) in the Ministry of War and participated in the resistance against the Manchus. After She-hsien fell to the Manchus in late 1645, he went to Fukien together with Hung-jen and was also ordained by Tao-chou Ku-hang. He took the Buddhist name Hung-chi 弘濟 and the literary name I-jan 益然. Later he became an abbot in P'u-ch'eng in Fukien. After he died, the prominent scholar, philosopher, and loyalist Huang Tsung-hsi 黃宗羲 (1610–1695) wrote the epitaph for him.[3]

Hung-jen never served in the government in the Ming Dynasty. With his training in the Confucian classics and his association with Wang Mu-jih, however, it is reasonable to believe that Hung-jen himself probably also participated in the resistance against the Manchus. Although there is no explicit evidence of his political involvement, his escape from Anhui to Fukien, where the Prince of T'ang 唐王 had established his own regime under the support of Huang Tao-chou 黃道周 (1585–1646) and others in Fuchou 福州 in 1644, suggests his allegiance to the Ming Dynasty.

Furthermore, the fact that there is considerable confusion about the dates of Hung-jen's escape to Fukien and his ordination suggests that his early biographers might have chosen to be ambiguous in this particular matter in order to avoid censorship by the Manchus. His biographer Wang T'ai-cheng 王泰徵 (1600–1675), a native from She-hsien and a member of the politico-scholarly society Fu-she 復社 ("Restoration Society"), reported in 1664 that Hung-jen went to Fukien the year after the fall of She-hsien to the Manchus.[4] Since the Manchus took over She-hsien in late 1645, it is likely that Hung-jen left for Fukien in 1646. However, the local gazetteer *She-hsien-chih* 歙縣志 completed in 1690 reported that Hung-jen went to Fukien in 1645.[5] Since the discrepancy would amount to only a few months, it seems likely that Hung-jen left Anhui for Fukien around 1645. It is also likely that he went to Fukien in order to join the resistance regime of the Prince of T'ang in Fuchou. Unfortunately, the regime was plagued with feuding and disloyalty and was soon defeated by the Manchus in 1645.[6] It is perhaps at this time that Hung-jen and Wang Mu-jih became monks.

According to Ch'eng Sui 程邃 (1605–1691), after Hung-jen "escaped from Confucianism" (*t'ao-ju* 逃儒), i.e., became a Buddhist monk, he tried to persuade Hung-jen to return to secular life, but Hung-jen refused.[7] Another friend whose writings were censored by the Manchus, Huang Sheng 黃生 (1622-1696), wrote the following poem to Hung-jen:

After the turbulence of our native land,
You went to Mt. Wu-i.
When you left, you discarded your hat and hair;
When you came back, you brought clouds in your sleeves.
The worldly atmosphere floats behind your straw-hat;
Your predilection is to live among egrets and gulls in the wild.
Your renown in painting eclipses your noble conduct,
But your name shall be heard in other generations.[8]

Hung-jen's association with the politico-scholarly societies is a further indication of his political involvement. One of Hung-jen's closest friends, Hsü Ch'u 許楚 (1605-1676), was closely associated with the Fu-she and was in fact a protégé of one of its co-founders, Chang P'u 張溥 (1602-1641).[9] After the fall of the Ming Dynasty, many members of the Fu-she either committed suicide or were executed by the Manchus for their resistance activities. For example, in 1647, Hsü Ch'u himself was jailed for his involvement in anti-Manchu activities but was soon released.[10] Since much of what we know today about Hung-jen's life comes from Hsü Ch'u's writings, among them his epitaph for Hung-jen, his description of Hung-jen's art collection, and his biographical sketch of Hung-jen, it is likely that Hung-jen shared with Hsü Ch'u ideals and experiences as a member of the Fu-she.[11]

Hung-jen was a member of the Pai-yü-she 白楡社 ("White Elm Society") reestablished by Hsü Ch'u and another close friend of Hung-jen's, Ch'eng Shou 程守 (1619-1689). Previously the membership of the Pai-yü-she had included Chan Ching-feng 詹景鳳 (1528-1602, *chü-jen* 1567), from Hsiu-ning 休寧, Anhui, whose catalogue of paintings and calligraphy *Tung-t'u hsüan-lan-pien* 東圖玄覽編 (preface dated 1591) has attracted art historians' attention for its catholic and discriminating judgment, as well as Chou T'ien-ch'iu 周天球 (1514-1595) from Suchou and Mo Shih-lung 莫是龍 (born ca. 1539, d. 1587) from Sung-chiang.[12]

There is further evidence of Hung-jen's association with the politico-literary societies. In 1639, while still a young student of the Confucian classics, he painted a joint handscroll (now in the Shanghai Museum) together with four artists from She-hsien for Li Sheng-pai 李生白 whom he addressed as a "society brother" (*she-hsiung* 社兄).[13] Later, several Anhui artists including

Hsiao Yün-ts'ung (in 1647), Wang Chih-jui (in 1649), Cha Shih-piao (in 1653), and Hung-jen (around 1652), painted a joint album for Ch'en Ying-ch'i 陳應頎 (literary name Hsiang-shih 香士, from Wu-hu 蕪湖, Anhui) whom they addressed as "society brother" or "society comrade" (*she-meng* 社盟).[14]

Although one must always guard against the tendency of earlier biographers of artists to glorify their subjects, there is no reason to doubt that Hung-jen's character was a model in the Confucian tradition. First of all, he was known for his filial piety. His father had died when Hung-jen was still a child and he subsequently struggled to support his mother by means of various menial jobs. Once, when he was late in carrying rice to his mother, he attempted suicide by drowning himself in a river. When his mother died, Hung-jen did his best to repay the loans for the funeral as soon as possible in order not to "be obliged to others for one's own sake," according to Yin Shu 殷曙 (1623-1692) who was a native of She-hsien.[15]

Hung-jen's decision to become a monk, however, was more politically motivated than simple filial piety would suggest. A few later writers allude to his politicization. For example, Wu Chia-chi 吳嘉紀 (1618-1684), a prominent contemporary poet, says in his colophon on a recorded painting by Hung-jen:

> In the past, when the landscape [of the Ming] changed,
> You felt the turbulent pains inside you.
> You would paint just to buy wine.
> When the way to dress was changed [by the Manchus],
> You began to wear the monk's robes.
> You called yourself the "Mountain Monk";
> Others called you the "Master of Painting."
> But your aspirations,
> How many in this world can understand?[16]

The phrase "the way to dress was changed" refers to the Manchus' notorious edict of 1645 that the people of China should henceforward conform in outward appearance to their conquerors by shaving the fronts of their heads and growing the queue. Indeed, it was this hair-cutting and queue-growing edict, more than any other act, that has been considered the symbol of utter submission to the "barbarian" regime of the Manchus, and to have engendered the desperate action of resistance that occurred in the Chiang-nan area.

Like Chu Ta and Tao-chi, Hung-jen became a monk primarily in order to escape from the foreign rule under the Manchus, not because he was particularly religious. His action, in other words, is symbolic of his loyalty to the fallen Ming Dynasty and his resistance to alien rule. Therefore, like Tao-chi, who would call himself a man "with hair and hat" and disclaim his religious affiliation, Hung-jen, when excited, would rent a boat to fetch fish, shrimp, and wine to share with his friends.[17] Hung-jen thus joined the many who, as the Ming cause collapsed in the southeastern part of China, took Buddhism as a recourse to manifest their determination not to serve the new regime, and to avert retribution from the Ch'ing for participating in the Southern Ming courts.[18] Before he died, he asked his friends to plant plum trees on top of his tomb so that he could have the company of the plum blossoms in his afterlife. He was thus very much concerned with noble character as symbolized by the plum blossoms.[19]

Hung-jen disliked the rich and pompous and always tried to help the poor and humble. For instance, according to T'ang Yen-sheng 湯燕生 (1616-1692), who was a close friend of Hung-jen, the day before he died, Hung-jen painted several paintings for two poor people so that they could trade them for food in the cold winter.[20]

Given the tremendous turbulence during the time of Hung-jen, our knowledge of his life is bound to be incomplete. According to Cheng Min 鄭旼 (1633-1680), a close friend of Hung-jen from She-hsien who collected Hung-jen's poems in a posthumous anthology called *Hua-chi* 畫偈 (postscript dated 1668), "The master [Hung-jen] has completed several poems and paintings. When the army was defeated, they were all captured. How pitiful that the master's noble conduct should therefore be unknown to the world."[21] Elsewhere, however, Cheng Min said tersely of Hung-jen the man: "The master understood the moral principles: filial piety and loyalty to the state." [22] In a poem written for a portrait of Hung-jen, now perhaps lost, T'ang Yen-sheng said:

> Who is this man that, through the springs and rocks in the mountains,
> Exudes the reflections of ice and snow?
> Like the tea plant that is often thin and tall,
> And the plum tree that is solitary and eternal by nature.[23]

Perhaps T'ang Yen-sheng would agree that the austerity in Hung-jen's landscape paintings is an extension of his personality.

She-hsien, Hung-jen's birthplace, was established as a political subdivision under K'uai-chi 會稽 during the Ch'in Dynasty (221-206 B.C.). The name Hsin-an 新安 was adopted in the Tsin Dynasty (265-317) for the division (*chün* 郡) that incorporated, among five other counties, She-hsien. Hsin-an was called She Prefecture *(She-chou* 歙州) during the Sui Dynasty (581–618) and the T'ang Dynasty (618–906). She Prefecture was changed to Hui Prefecture (Hui-chou 徽州) in the Northern Sung (960–1127). She-hsien has been the political and cultural center in the area generally called Hui-chou or Hsin-an in the southeastern part of Anhui; She-hsien, Hui-chou, and Hsin-an are often used interchangeably.[24]

Hui-chou has been famous for its mountainous scenery. For example, the Suchou artist Lu Chih 陸治 (1496–1576) visited Mt. Pai-yüeh 白嶽 near Hsiu-ning in mid-autumn of 1544 and painted an album of sixteen leaves, now in the Fujii Yurinkan Museum in Kyoto.[25] Lu Chih's inscriptions on several of the leaves describe the landscape seen by the artist-traveller:

> Ju-hsiang t'ai 乳香臺:
> Thousand-faceted cliffs descend into the stream.
> On a clear day it is said that [one can see] stalactites hanging down from above. When I passed here it was cloudy, and we could not see.
>
> Heng-shih t'an 橫石灘:
> A large rock lies in the middle of the stream surrounded by a profusion of smaller rocks. The stream cascades around them with frightening force. The boats pass above or below them with great caution and fear.[26]

The scenery of Hui-chou was also depicted in a 1617 album by the late Ming painter Li Liu-fang 李流芳 (1535–1619), a native of She-hsien who lived in Chia-ting 嘉定, Kiangsu. The whereabouts of the album is unknown, but one can sense how the artist marvelled at the beauty of what he saw in this recorded inscription:

> On the land, from Yü-hang [in Chekiang] to Feng-kan 豐干 [in Anhui], the streams and mountains are filled with red trees which now mask the landscape and now reveal it; the landscape is now a wide expanse, now an impenetrable depth; all as if in a picture. Returning by boat from T'un-ch'i 屯溪 [the upper course of the Hsin-an River], I find the water is so clear as to reveal the bottom of the river in which strange peaks and fantastic rocks are randomly intermingled. The river is enclosed by two walls of rocks which would seem to reach the clouds and the sun. This is what has been described as: "The boat seems to have reached a dead end; suddenly there emerges an

> unlimited expanse." People used to mention the Hsin-an River. Now I see why.[27]

The mountainous and scenic landscape in his native place must have made a strong impression on Hung-jen as he grew up. Later, as he travelled more extensively, his art in landscape painting became greatly affected by experience of natural scenery and contacts with artists in other regions.

Without his knowledge of landscape through travel, Hung-jen's art would have been different. As Cha Shih-piao 查士標 (1615–1694), one of the "Four Masters of Hsin-an," observed: "Master Chien's painting style changed with his sojourn in Mt. Wu-i; after his return from Mt. Huang, his style became even more wonderful. The ancients regarded the ten thousand things under the heaven and on the earth as their teachers. This is even truer when one considers that mountains and streams can be empathized within a painting with a resulting harmony of spirits."[28]

Hung-jen's first major journey took place when he went to Mt. Wu-i, around 1645. His modern biographer Huang Pin-hung 黃賓虹 suggested that Hung-jen travelled through Mt. Yen-ling 嚴陵 (near T'ung-lu 桐廬, Chekiang), as well as Mt. San-ch'ü 三衢 (near Ch'ang-chou 常州, Chekiang). He then proceeded through the Hsien-hsia Pass 仙霞關 on the border between Fukien and Chekiang, and finally arrived at the Man-t'ing 幔亭 Peak of Mt. Wu-i, where he became a monk.[29]

Hung-jen was apparently very much impressed by the scenery of this mountain (P1. 195). In an inscription on a recorded hanging scroll entitled "Mountains and Valleys at Mt. Wu-i" (*Wu-i yen-ho* 武夷岩壑), for instance, he wrote: "The mountains and valleys of Mt. Wu-i are high and precipitous, just like what I have painted in this picture. I once stayed there for a number of years and, therefore, I have ventured to paint it according to my own ideas."[30] The undated hanging scroll entitled "Landscape of Mt. Wu-i" (*Wu-i shan-shui* 武夷山水) seems to be Hung-jen's only extant painting on this subject (P1. 194). In the inscription on this painting, stylistically datable to about 1660, Hung-jen described the landscape in this manner: "How could Creation have composed the Nine Bends River? The spindly peaks and bulging cliffs are glassy in appearance. The Taoist's brush lies unused on the boat window sill; he can only stare with wide eyes, too bemused to poetize."[31] The world of the transparent and yet majestic beauty of Mt. Wu-i suggested in Hung-jen's recollection is reminiscent of the manner in which its scenery in a moon-lit night was described in an early seventeenth-century gazetteer of Mt. Wu-i: "Ten thousand mountains are quiet, as if in a world of glass." The gazetteer published in 1619 in Ch'ung-an 崇安, Fukien, incidentally contained several woodcut prints that bear striking similarity to Hung-jen's painting of Mt. Wu-i just mentioned.[32] Although Hung-jen's paintings generally do not have the same intense interest in narrative or illustrative elements and are marked by a stronger sense of concentration and lucidity than the woodcut prints, the similarity does suggest not only Hung-jen's contact with the scenery itself, but also the way Mt. Wu-i was depicted at that time.

From 1646 to 1651, Hung-jen stayed in Fukien. On his way back to Anhui, he probably visited Nanking where he was to return for a longer stay in 1657. The visit of 1651 is suggested by the fact that Hung-jen painted a section of the joint handscroll for a certain Mr. Ko-an 格庵 (unidentified) dated 1650–1651 (Pls. 31–39).[33] Now, in the early seventeenth century, both Fukien and Nanking were centers for the revival of the style of the tenth and eleventh century, and many artists would go to Nanking from Fukien. Wu Pin 吳彬, for instance, was exposed to the interest in the Northern Sung style in Fukien; later he went to Nanking and forged an extremely personal and often dramatic and fantastic style derived basically from the Northern Sung style. The style of Wang Chien-chang 王建章 (active ca. 1628–1644), another artist from Fukien who later, like Wu Pin, went to Nanking, is characterized by angular shapes and abstraction of forms and very often is in a fantastic mode similar to Wu Pin's.[34]

From the summer of 1657 to early 1658, Hung-jen stayed in Nanking. It is likely that Hung-jen met Kung Hsien (1599–1689) there at this time. Kung Hsien was interested in the transformation of the style of the Northern Sung, perhaps not without influence from Wu Pin; his attempt around this time at the mode of Northern Sung can be seen, for example, in a hanging scroll by him of a mountain landscape dated 1655 and now in the Hwa Shu-ho 華叔和 collection in Taipei.[35] In 1657, Hung-jen painted an album which included a leaf in the style of Fan K'uan (P1. 160),[36] indicating that he might be sharing the same interest in the revival of the Northern Sung style as the Nanking artists. It is also likely that Hung-jen might have met Fang I-chih 方以智 (1611–1671), the scholar-monk painter from T'ung-ch'eng 桐城, Anhui. During Hung-jen's visit, Fang I-chih was passing by Nanking. At this time, the majority of Buddhist temples in Nanking were controlled by monks descended from the monk Wu-ming Hui-ching 無明慧經 (1548–1618) of the Ts'ao-tung 曹洞 sect.[37] Both Fang I-chih and Hung-jen belonged to this sect. It is thus conceivable that Hung-jen's sojourn in Fukien and Nanking exposed him not only to the real landscape but also to certain stylistic modes which he would then incorporate into his personal style.

Beside Nanking and Fukien, Hung-jen also travelled to Wu-hu 蕪湖, Anhui, in 1652. In 1657, having returned from Nanking, he visited Hsüan-ch'eng 宣城, Anhui. He visited Yang-chou and Hang-chou in 1661. In 1662, Hung-jen went to Po-yang 鄱陽, Kiangsi, in an attempt to climb up Mt. Lu with Tai Pen-hsiao 戴本孝 and other friends, but their plan was aborted by heavy snow. Next spring, in 1663, Hung-jen finally succeeded in climbing up Mt. Lu.[38]

Among all the places he visited, Hung-jen was most fascinated by Mt. Huang, which he first climbed in 1656. He believed that the scenery of Mt. Wu-i was characterized by what was visible, while that of Mt. Huang created a kind of mirage or wonderland (Cf. P1. 53.), and therefore he thought that the latter far surpassed the former in its marvels of landscape. On one occasion, after returning from Mt. Huang, he painted, surprisingly, not a landscape of Mt. Huang, but of Mt. Wu-i, for a friend. His reason was that he would not "dare to insult the Heavenly Capital 天都 [Peak] " of Mt. Huang.[39] In addition, many of his poems attest to his fascination with Mt. Huang.

The famous traveller and geographer Hsü Hsia-k'o 徐霞客 (1486–1641) gave this account of his ascent to the Heavenly Capital Peak that perhaps explains in part Hung-jen's fascination with Mt. Huang. Hsü Hsia-k'o wrote:

> At this time, dense fog rose and subsided at intervals. When it rose nothing could be seen. Most of the hills on the side of Lotus Flower [Peak]were wrapped in the fog, but it favored me as regards Heavenly Capital, for it moved to its back when I was in its front, and it floated out from the left when I would round to its right. Many of the pines here····were still able to retain their stubborn and proud postures, but the cedars were all clinging flatly to the rocks like moss and lichens although they had trunks as large as men's arms. The wind was boisterous at this great height and the vapoury fog came and went as it pleased, Looking down, the hills below appeared like emerald islets when their summits emerged from the fog; it all became one silvery sea when they were wholly immersed. But looking further down, the world was flooded in glittering sunshine.[40]

And of the Lotus Flower Peak, the highest of all the peaks at Mt. Huang (estimated height: 1,820 meters), Hsü Hsia-k'o wrote:

> On the brow of this rock even Heavenly Capital was bowing under us;

> we were up in the boundless blue of the sky. Situated in the center of Huang Shan, this peak is also far above all the others. Looking down we beheld its enclosing cliffs standing round, flooded with radiant rays of the rising sun, reflecting layers and layers of dazzlingly brilliant hues. It was a most overpowering sight; I cried out in ecstasy and could have danced out of sheer joy and admiration.[41]

Mt. Huang is famous for its pine trees of every imaginable fantastic shape. Hsü Hsia-k'o observed:

> When we reached the promontory, we were rewarded with the splendid view of the two peaks, Lotus Flower and Cloud Gate, as well as other peaks competing in beauty and wonder around the Heavenly Capital, for which they are the sentinels. From there we made our way into the overhanging cliffs and precipices, whose surfaces are strewn with clinging pines of the oddest shapes. The tallest of these pines are a scant ten feet, while the low ones are only a few inches high. They had flat heads, short boughs, gnarled roots, and knotty trunks. The older they are, the shorter they seem to be, and the smaller they are, the queerer they look. It was an unexpected joy to find such wonders on such a wonderful mountain.[42]

The granite rocks at Mt. Huang are rich in joint systems which, after erosion, produce different kinds of strangely shaped peaks. Consequently, the granite would fracture into chutes where vertical joints devloped.[43] Graphic and descriptive names have been given to the bare rocky peaks and crags of every conceivable shape and posture (e.g. "Rocky Bamboo-shoot Promontory" [*Shih shun-ch'iang* 石筍矼] and "Sitting Monk Rock" [*Sheng-tso-shih* 僧坐石]), and to the strangely tortured pines that twist in and out of rocky clefts (e.g. "Protracting Dragon Pine" [*Wo-lung-sung* 臥龍松] and "Unsettling Dragon Pine" [*Jao-lung-sung* 擾龍松], etc.). In addition, there are also the fabulous "Seas of Cloud" (*Yün-hai* 雲海) that would appear in a variety of ever-changing magical illusions (Pls. 53–58).

Thus, the scenery at Mt. Huang has been conducive to graphic description in poetry and prose. Thousands of poems and travelogues have been written about Mt. Huang by poets ranging from Li Po 李白 (701–762) of the T'ang Dynasty to Ch'ien Ch'ien-i 錢謙益 (1582–1644) of the late Ming and early Ch'ing Dynasties.[44] According to the research of Hsü Ch'u, the tradition of illustrating the scenery of Mt. Huang could be traced back to at least the

Ching-yu 景祐 reign (1034-1037) of the Northern Sung Dynasty. Hsü Ch'u apparently had access to local material when he stated that subsequently illustrations were made in 1100, 1156, 1208, 1371, 1462, 1581, 1582, and 1619.[45] These early illustrations, no doubt schematic, are unfortunately lost. The handscroll dated 1376 and attributed to Hsü Pen 徐賁 (1335-1380), depicting the sites at Mt. Huang with the names of sites written in red, probably is a copy after an old composition; the handscroll is now in the Museum of Fine Arts, Boston.[46] It, however, suggests that the pictorial imagination of artists was really attracted to the fantastic, and very often anthropomorphic or animistic forms of trees, rocks, and peaks at Mt. Huang.

It was only in the late Ming, probably after the monk P'u-men 普門 (active ca. 1573-1619) and his colleagues opened up the mountain by building more monasteries and providing access to hitherto isolated areas, that increasing numbers of visitors came to Mt. Huang.[47] Following Ting Yün-p'eng and Hung-jen, many seventeenth-century artists, including Tai Pen-hsiao, Mei Ch'ing, and Tao-chi, came to know Mt. Huang firsthand and painted its fantastic scenery. Thus, Mt. Huang provide these individualists the opportunity to get back to nature for "a reconsideration of the old concept of landscape painting as the outcome of a direct encounter between the artist's mind and the exterior world."[48] On the other hand, paradoxically, these artists' individual transformation, sometimes even violation, of visual truthfulness might have also been inspired by the scenery at Mt. Huang.

We shall discuss Hung-jen's landscape in terms of the artist's use of space and brushwork, and point out his indebtedness to traditions as well as the importance of his own formal inventions. However, it is important to keep in mind the artist's experience of the natural scenery itself. Hung-jen had a special affinity toward his native Hui-chou, where most of the Mt. Huang scenes are located. He was born there, lived there most of his life, and died there. More importantly, Hung-jen produced the greatest number of his works while near Mt. Huang, and he repeatedly utilized Mt. Huang as his subject matter. For Hung-jen, Mt. Huang had become almost an icon, and painting Mt. Huang became almost like a ritual act of worship. This is natural if one considers that, for many of his contemporaries from the same region, the majestic landmark of Mt. Huang represented the survivial and affirmation of a cultural heritage; the depiction of Mt. Huang therefore was perhaps one of the best ways to express cultural regionalism. Many cultural heroes were associated with the area, particularly specific sites at Mt. Huang which, after all, was itself named to commemorate the legend of Huang-ti 黄帝, the Yellow

Emperor who has been known as the legendary initiator of Chinese culture.[49] Furthermore, the area of Mt. Huang, being more secluded and inaccessible, did not suffer the ravages of Manchu troops when Hui-chou Prefecture was captured. Thus, for the loyal "left-over" or "surviving" literati (*i-min* 遺民), Mt. Huang might have remained in a real sense untainted by the alien conquerors, and thus still belonged in spirit to the fallen Ming Dynasty.[50] Loyalists came here to visit, to write poems and paint paintings, and even, like Hung-jen, to stay. Away from the turbulence of the world, Mt. Huang became a spiritual sanctuary for the dislocated.

A comparison of modern photographs taken at Mt. Huang (Pls. 55-58) with Hung-jen's paintings reveals that the artist owed a lot to the real scenery. This can be seen in Hung-jen's undated hanging scroll "Landscape" in the Freer Gallery of Art (Pl. 59). The artist was interested in depicting the rock contours with sharp, angular brushstrokes. He also fractured the rock surfaces with repeated application of overlapping rectangular shapes. Although no explicit reference to Mt. Huang was made in the artist's inscription, there is not doubt that the artist was depicting Mt. Huang where the predominant form is rectilinear, the rocks are sharply cut and deeply fissured, and long and intersecting cracks divide the rook surfaces into flat geometric patterns (Pls. 56-58).

One must, however, guard against a naturalistic approach to the art of Hung-jen. As his younger contemporary Wang Chi-ch'un 汪濟淳, who was a co-compiler of the gazetteer of Mt. Huang (*Huang-shan-chih*) published in 1674, pointed out: "The Protracting Dragon Pine painted by Master Chien does not at all look like the same tree at Mt. Huang. The wonder of his supreme brushwork lies exactly in unlikeness."[51] Comparing the seventeenth-century artists who are most renowned for their paintings about Mt. Huang, the modern critic and artist Ho T'ien-chien 賀天健 (1893-1974) observed: "Hung-jen captured the bones (*ku* 骨)of Mt. Huang, Mei Ch'ing the gestures (*tzu* 姿), and Tao-chi the spirit (*ch'i* 氣)."[52] He also pointed out that several techniques of depicting rocks in Hung-jen's painting are closely related to the topography at Mt. Huang. Another modern artist and connoisseur, Chang Ta-ch'ien 張大千 (1899-1983) comments in this way: "Hung-jen captured the bones and spirit of Mt. Huang, Tao-chi the feelings, Mei Ch'ing the changing appearances, and Tai Pen-hsiao the strange and fantastic shapes."[53] As we shall see in Chapter IV, by first realizing the pictorial possibilities of Mt. Huang (one remembers that Hung-jen was the oldest among these artists), Hung-jen exerted a considerable influence on Tai Pen-hsiao, Mei Ch'ing, and Tao-chi.

In order to further understand the genesis of Hung-jen's landscape art, one must also look into the social, cultural, and artistic environment in Hui-chou during his time. The mountainous terrain of Hui-chou is not suitable for agriculture. Most people consequently took up trade and became successful merchants in many parts of China.[54] The famous scholar Ku Yen-wu 顧炎武 (1613-1682), in his *T'ien-hsia chün-kuo li-ping shu* 天下郡國利病書 (preface dated 1622), an outstanding geographic study with special consideration of the effect of geographical conditions on political and economic development, quoted a contemporary local source about the social and economic situation of She-hsien:

> Our state's material prosperity and moral well-being reached their peak in the Hung-chih reign [1488-1505]. In that time, households and individuals were amply provisioned. There were adequate houses for dwelling. There were sufficient fields for tilling. There were forested mountains for firewood. There were fertile gardens for planting. There was no vexatious pressing for taxes. No banditry arose. Marriages were carried out at the proper time. Villages were peaceful and secure. Spinning was done by the wives. Men concerned themselves with mulberry[wood for bows]and p'eng[bramble for arrows]. Indentured servants bore their burdern. Neighbors were helpful.
>
> Coming down to the end of the Chen-te [1506-1521] and beginning of the Chia-ching [1522-1566] reigns, the situation was different. As merchants and traders became more numerous, farming was not emphasized. Property was exchanged. Prices fluctuated. Those who were able were successful. Those who were a bit slow were ruined. The family on the east might become rich while the family on the west was impoverished. As the equilibrium between those of higher and lower status was lost, everyone struggled over paltry sums. People were mutually exploitative; each sought to further himself. Thereupon, deceptive practices sprouted, contention arose, recreation was off-color, extravagance was everywhere. By the end of the Chia-ching [1522-1566] and Lung-ch'ing [1567-1572] reigns, the situation differed even more. Instances of wealth from the superfluous [i.e., commerce] were numerous, and instances of wealth from the basic [i.e., agriculture] were few. The rich became richer and the poor got poorer. Those who had risen were overbearing. Those who had fallen were skittish. Property thus was oppressive, and possessions did not stay long with anyone. Trade was a tangle. Unchecked desires made people calculating. Powerful

persons who did wrong created turmoil. Men without scruples preyed on others.

Today [i.e.,about 1600], another thirty-some years after that, the differences are virtually incomparable [to the condition of society in 1500]. One out of a hunderd is wealthy, but nine out of ten are impoverished. The impoverished are unable to oppose the wealthy, so that, contrary to what should be, the few control the many. Silver and copper cash seem to dominate even heaven and earth····[55]

Historically, the rise of merchants in Hui-chou can be traced back to the Southern Sung Dynasty (1127-1279). Commerce reached its apex and occupied an important position in the economic life in China after the middle of the Ming Dynasty.[56] By the time of the Wan-li (1573-1619) reign, merchants from Hui-chou became as powerful as those from Shansi. As the well-travelled scholar and official Hsieh Chao-chih 謝肇淛 (1567-1624) pointed out: "In the Chiang-nan [south of the Yang-tze River] area, the richest merchants come from Hsin-an [Hui-chou]; in the Chiang-pei [north of the Yang-tze River] area, the richest merchants come form Shansi."[57] By the end of the Ming Dynasty, there were travelling merchants as well as shopkeepers from Hui-chou all over the country; in some cases, they were even engaged in foreign trade. Hsieh Chao-chih went on to say:

> The great merchants of Hui-chou have made fisheries and salt their occupation and have amassed fortunes amounting to one million taels of silver. Others with a fortune of two or three hundred thousand can only rank as middle merchants. The Shansi merchants are engaged in salt, silk, reselling, or grain. Their wealth even exceeds that of the former. This is because the Hui-chou merchants are extravagant, but those of Shansi are frugal. In fact, people of Hui-chou are also extremely miserly as to food and clothing····but with regard to concubines, prostitutes and lawsuits, they squander gold like dust.[58]

Hsieh Chao-chih also said, however, that people from Hui-chou "liked refinement and culture."[59] This can be seen in their architecture and their production of a variety of paper, brushes, ink-cakes, and ink-stones, the so-called "Four Treasures of the Scholar's Studio" (*wen-fang ssu-pao* 文房四寶).[60]

The mountainous terrain of Hui-chou isolated the area geographically and contributed to closer ties among its clansmen. In addition, because of their economic strength, merchants from Hui-chou helped to forge a stronger regionalism than was evidenced in other centers of commerce. Even when they were doing business in other provinces, they would maintain their close association through the format of a "meeting house" or "provincial lodge" (*hui-kuan* 會館) which would serve as a social club.[61]

Regionalism was reflected in the architecture of Hui-chou.[62] A recent study of extant houses dating from the late Ming Dynasty in Hui-chou, mainly in She-hsien, reveals a regional style which is characterized by a two-story house that is doorless in the back and windowless on the first floor. The merchants who erected these secular buildings were either forced to adopt, or simply preferred, a style that was different from the more pretentious houses occupied by government officials. The plan is, in general, irregular and asymmetrical and yet is marked by clarity and compactness. Regionalism in architecture can also be seen in the absence of a living room and a study so common in the lower Yang-tze River basin, as well as the lack of a garden which was also common in Suchou. Given the fact that the area surrounding She-hsien is famous for its paint, what is more interestingly regional is that a great number of buildings had an unpainted fine wooden surface, showing the marked preference in the area for a natural wood texture.[63]

As in the case of architecture, there was also a strong regionalism in the sphere of the fine arts. Once the uniqueness of the style of a master like Hung-jen was recognized, other artists in the same region would follow, making possible an Anhui School as we know it today.

The rise of a merchant class must have stimulated both artists and collectors. An examination of the local gazetteers of Hui-chou compiled in the early sixteenth century (1505 and 1566) reveals that there were relatively few outstanding painters from that area.[64] It was during the second half of the sixteenth century that a greater number of artists began to distinguish themselves. Artists from this period, such as Wang Chao 汪肇 (active ca. 1500), were mostly professionals who followed the style of the Che school. Even a somewhat later artist such as Cheng Chung 鄭重 (active ca. 1610-1644) still maintained a strong professional bent and specialized in Buddhist figure and landscape paintings. By that time, however, the distinction between literati and professional styles, as in Suchou, had become less distinct than before.[65] Some of Cheng Chung's paintings are close in style to those of the scholar and poet Ch'eng Chia-sui 程嘉燧 (1565-1643).[66]

With their wealth, some merchants went on to acquire paintings in major cities such as Suchou, Hangchou, Yangchou, Nanking, and Peking. In 1585, the well-known and knowledgeable man of letters Wang Shih-chen 王世

貞 (1526–1590) complained that the collectors in Suchou and Hui-chou had inflated the price of antiques, even that of relatively modern art objects. He wrote:

> In regard to painting, one should treasure Sung paintings. But during the past thirty years Yüan paintings have suddenly become very much sought after—so much so that the price of works by Ni Tsan, and even of those by Shen Chou of the Ming Dynasty, has quickly increased tenfold. As for ceramics, one should treasure the Ko and the Ju wares. But during the past fifteen years Hsüan-te wares have suddenly become very much treasured—so much so that the price of the Yung-lo and the Ch'eng-hua wares has also quickly increased tenfold. In general, this trend has been started by people from Suchou and followed by those from Hui-chou. This is indeed hard to explain.[67]

The intensive activities of collecting in Hui-chou naturally resulted in the accumulation of a wide variety of art objects and in higher levels of connoisseurship. Thus, it is not surprising that in 1588 the Hui-chou connoisseur Chan Ching-feng was able to demonstrate his knowledge and perceptiveness to Wang Shih-chen. In his *Tung-t'u hsüan-lan-pien* Chan Ching-feng recounted how he once correctly dated an excavated stone engraving from the early tenth century on stylistic grounds alone. Subsequently, his suggested date was confirmed by inscription. Thereupon, Wang Shih-chen had to agree with a friend from Hui-chou who said: "In the past, people with good eyes were from Suchou, but now they are from Hsin-an."[68] This anecdote suggests that collectors and connoisseurs in Hui-chou had become confident and competent. As an illustration of their collecting activities, Wu Ch'i-chen 吳其貞 (1607–1677), author of *Shu-hua-chi* 書畫記 (completed ca. 1677), mentioned that in his native village, antique dealers from all over sold their goods in front of a temple during the eighth and ninth months of the year.[69]

One of the collectors from Hui-chou was Ch'eng Chi-po 程季白 (d. 1626), whose collection of paintings and calligraphy was, according to Wu Ch'i-chen, of excellent quality.[70] Ch'eng Chi-po's collection included the famous handscroll "Rivers and Mountains after Snow" (*Chiang-shan hsüeh-chi* 江山雪霽) attributed to Wang Wei 王維 (699–759), now in the Ogawa collection in Kyoto.[71]

Another famous painting in the Ch'eng family collection was the handscroll "Water Village" (*Shui-ts'un-t'u* 水村圖) by Chao Meng-fu 趙孟頫 (1254–1322) dated 1302 and now in the Palace Museum, Peking.[72] The painting had been in the collections of such prominent artists and connoisseurs as Wen Cheng-ming 文徵明 (1470-1599), Hsiang Yüan-pien 項元汴 (1525-1590), and Tung Ch'i-ch'ang 董其昌 (1555-1636), among others before coming to the Ch'eng collection. Wu Ch'i-chen saw the Wang Wei handscroll, the Chao Meng-fu handscroll, and a hanging scroll of landscape by Ching Hao 荆浩 (ca. 870-880 to ca. 935-940), among other paintings in the house of Ch'eng Cheng-yen 程正言, son of Ch'eng Chi-po, in Hui-chou in 1637.[73]

An examination of Wu Ch'i-chen's *Shu-hua-chi* does not leave any doubt about the richness and excellence of the private collections of paintings and calligraphy in Hui-chou in the first half of the seventeenth century. The collection of Ch'eng Chi-po and his descendents is one of the best examples. Wu Ch'i-chen, himself an impeccable connoisseur and dealer, who also travelled widely, gave the following historical assessment of the private collections in Hui-chou. He wrote:

> There were no better places that examplified the prosperity of Hui-chou than She-hsien and Hsiu-ning-hsien. The possession of antiquities determined whether one was cultivated or vulgar. Therefore people contended for acquisition at all costs. Hearing that, antique dealers from everywhere came to [Hui-chou], and the merchants travelling in other cities searched and brought back antiques. Consequently, acquisition increased greatly. This trend was first begun by the Vice-Minister of War Wang [Tao-k'un] and his brother, and then was continued by the Wu family from the village Ch'i-nan, and the Wang family from the village Ts'ung-mu-fang. My fellow village folk from Shang-shan, the Wu family; the Chu family from the city of Hsiu-ning; the Huang family from the village Chü-an; and the Ch'eng family from the village Yü-ts'un--their collections contain pieces that are famous in the country.[74]

Another prominent collector from She-hsien was Wu T'ing 吳廷, also called Wu Kuo-t'ing 吳國廷 (active ca. 1575-1625).[75] He was a good friend of Tung Ch'i-ch'ang and Ch'en Chi-ju 陳繼儒 (1558-1639) and shared the same interest in ancient calligraphy and painting. In 1609, for instance, Ch'en Chi-ju, Tung Ch'i-ch'ang, Wu T'ing together viewed an early T'ang tracing copy of the famous *Hsing-jan-t'ieh* 行穰帖 by Wang Hsi-chi 王羲之 (ca. 301-361), then in the collection of Wu T'ing and now in the Art Museum, Princeton University.[76] On another occasion, Tung Ch'i-ch'ang wrote a colophon on Wang Meng's 王蒙 (1310-1385) undated hanging scroll "The Forest Chamber

Grotto at Chü-ch'ü (*Chü-ch'ü lin-wu* 具區林屋), now in the National Palace Museum in Taiwan, in the very presence of Wu T'ing and Ch'en Chi-Ju.[77] Tung Ch'i-ch'ang also authenticated rubbings for Wu T'ing to be published in the series entitled *Yü-ch'ing-chai t'ieh* 餘清齋帖.[78] Extant works of calligraphy and painting from the collection of Wu T'ing, numbering over twenty, indicate the richness and variety of art collections in Hui-chou.[79]

Like Wu T'ing, the She-hsien collector Wu Chen 吳楨 (active 1621–1627) was on good terms with Tung Ch'i-ch'ang and Ch'en Chi-ju and often invited them to stay in his house and to discuss calligraphy and painting. He also published rubbings of calligraphy from his collection under the title *Ch'ing-chien-t'ang chu-t'ieh* 清鑑堂主帖 in volumes which also included calligraphy by Tung Ch'i-ch'ang and Ch'en Chi-ju.[80] Extant works of calligraphy and painting from his collection suggest that Wu Chen was also involved with the accumulation of art collections in Hui-chou.[81]

From Chan Ching-feng's *Tung-t'u hsüan-lan-pien* and Wu Ch'i-chen's *Shu-hua-chi*, it is clear that Ch'eng Chi-po, Wu T'ing, and Wu Chen are but a few examples of collectors who furthered the pursuit of acquisition and connoisseurship in Hui-chou in the late sixteenth century and early seventeenth century. The impact of their activities on the development of painting in the area would soon be felt. One should also remember that among the paintings and calligraphy that Wu Ch'i-chen saw in Suchou, Hangchou, Nanking, and Yangchou, many were owned by merchants from Hui-chou, and thus it is likely that many of them were at one time or another available to artists in Hui-chou.

Wu Ch'i-chen was a friend of Li Yung-ch'ang 李永昌 also from She-hsien, who was one of the artists who painted the 1639 joint handscroll now in Shanghai, a section of which is the earliest dated extant work by Hung-jen. The dates of Li Yung-ch'ang are unknown. Recorded paintings by him show that he was active probably between 1627 and 1650, and thus he can be considered as Hung-jen's older contemporary.[82] As an example of the impact of art collections on style, it should be interesting to note what paintings were available to Li Yung-ch'ang . According to Wu Ch'i-chen, Li Yung-ch'ang once owned a painting entitled "Tall Pines at Level Distance" (*Kao-sung p'ing-yüan* 高松平遠) by Wu Chen 吳鎮 (1280–1354) which he bought from Wu Ch'i-chen as soon as he had laid eyes on it.[83] In 1638, Wu Chi-chen and Li Yung-ch'ang together looked at paintings in the collection of a monk called Yü-shih shang-jen 毓石上人. Among the paintings they saw were the Wu Chen just mentioned, a painting entitled "The Straw-hut at Nan-kang" (*Nan-kang ts'ao-t'ang* 南岡草

堂) by Wang Meng, and a painting entitled "Marketplace in the Mountains under Clear Sky" (*Shan-shih ch'ing-luan* 山市晴巒) attributed to Hsü Tao-ning 許道寧 (ca. 1001-ca. 1066) but judged by Wu Ch'i-chen to be a Yüan work in the style of Kuo Hsi.[84] There is no doubt that these viewing sessions were conducted with a critical spirit which touched on the many facets of style and quality.[85] In the same way, Li Yung-ch'ang also exhibited great discernment both in his own collection and in his attitude toward painting as a creative process. Li Yung-ch'ang owned, besides the Wu Chen just mentioned, a calligraphy handscroll, "One Thousand Characters"(*Ch'ien-tzu-wen* 千字文) by the Emperor Kao-tsung 高宗 (reign 1127-1163) of the Southern Sung Dynasty, now in the Shanghai Museum, a painting entitled "Wintry Forest on a Clear Day" (*Ch'ing-tung han-lin* 晴冬寒林) by Kuan T'ung 關同 (active early tenth century), and a painting entitled "Cloudy Mountains" (*Yün-shan* 雲山), with a colophon by Tung Ch'i-ch'ang. This latter painting was attributed to Mi Yu-jen 米友仁 (1073-1153) but Wu Ch'i-chen considered it to be a Yüan work.[86] Whether or not Li Yung-ch'ang's collection included valid attributions, it still reveals his broad and eclectic interest in a range of past styles. In his inscription on one of his recorded paintings, Li Yung-ch'ang wrote:

> Sun I once copied Huang Kung-wang's [1269-1354] painting "Snow Clearing over Nine Peaks" (*Chiu-feng hsüeh-chi* 九峰雪霽) and almost surpassed it. Being unwilling to be slavish to Huang Kung-wang, I painted this according my own idea. Ha! Ha! Ha ! [87]

It is now difficult to positively identify the paintings in Li Yung-ch'ang's collection, although the range of stylistic choices available to him, as suggested by the titles of paintings, seems quite broad. The Huang Kung-wang painting, however, can be identified with the 1349 hanging scroll of the same title, now in the Palace Museum, Peking.[88] Without going into the history of this hanging scroll, it is likely that Li Yung-ch'ang and Sun I might have had first-hand experience of it. In 1655 when Wu Ch'i-chen saw it, it was still in the collection of a friend of his from Hui-chou who then resided at Nanking.[89] Judging from the paintings in his collection, Li Yung-ch'ang's interests range from the powerful, impressive mountain forms traditionally associated with K'uan T'ung, to the formal construction seen in Huang Kung-wang's "Snow Clearing over Nine Peaks" in Peking or "The Stone Cliff at the Pond of Heaven" (*T'ine-ch'ih shih-pi* 天池石壁), a hanging scroll dated 1341, whose whereabouts is now unknown.[90]

Hung-jen was a close friend of Wu Hsi 吳羲 (died. after 1663), also from

She-hsien. He and Wu Hsi would share the pleasure of looking at ancient masterpieces at the latter's home. In a gesture of reciprocity, Hung-jen painted several paintings for Wu Hsi of which at least two are extant. One handscroll, "Dawn on the River" (*Hsiao-chiang feng-pien* 曉江風便), dated 1661, is now in the Anhui Provincial Museum (Pls. 207–218).[91]

A certain Wu family in She-hsien, which in all probability was related to Wu Hsi, had in their collection several Ni Tsan paintings which would naturally exert a formative influence on the development of Hung-jen's style. According to the late Ming connoisseur and collector Chang Ch'ou 張丑 (1577–1643), the Wu family owned at least four Ni Tsan paintings. These paintings were hanging scrolls entitled "Wintry Pines in a Quiet Valley" (*Yu-chien han-sung* 幽澗寒松), "The Straw-hut at Tung-kang" (*Tung-kang ts'ao-t'ang* 東岡草堂), "Trees on a Bank and the Distant Mountains " (*Ting-shu yao-ch'en* 汀樹遙岑), and "Mountain Colors at Wu-sung" (*Wu-sung shan-she* 吳淞山色),[92] Chang Ch'ou himself saw the first two paintings and has left descriptions of them. On the basis of stylistic comparison, he believed that the first one was an example of Ni Tsan's late work, whereas the second one was from the artist's early period. The first one can be identified with the undated hanging scroll of the same title now in the Palace Museum, Peking (Pl. 3).[93] Later Wu Ch'i-chen, based on his observation of many Ni Tsan paintings, placed it among the artist's best works.[94] The second one can be identified with the now lost hanging scroll dated 1338, which exists only in a copy by Tung Ch'i-ch'ang dated 1629, now in the National Palace Museum, Taiwan.[95] The third painting recently was in the collection of the modern connoisseur and collector Wu Hu-fan 吳湖帆 (ca. 1984–ca. 1965).[96] The fourth painting has not yet been identified.

In addition, according to Wu Ch'i-chen, at least two extant paintings by Ni Tsan were in private collections in Hui-chou. One is the undated hanging scroll "Tzu-chih Studio"(*Tzu-chih shan-fang* 紫芝山房), which Wu Ch'i-chen saw in 1639, a painting now in the National Palace Museum.[97] The other is a 1363 handscroll entitled "Portrait of Yang Chu-hsi" (*Yang Chu-hsi hsiao-hsiang* 楊州西小像) jointly painted with Wang I 王繹 (ca. 1333–?) which Wu Ch'i-chen saw in 1656. It is now in the Palace Museum, Peking.[98]

Hung-jen's personal art collection, according to Hsü Ch'u, included the following 9 important items: (1) one handscroll of a landscape by Ni Tsan; (2) one hanging scroll of a landscape by Wang Meng; (3) rubbings of calligraphy made in the Sh'un-hua 淳化 period (990–995) of the Northern Sung Dynasty (960–1127); (4) one ancient ink-stone from She-hsien; (5) one [ceramic] drinking bowl in the shape of the trunk of a plum tree; (6) one bamboo walking staff; (7) one ceramic washing bowl; (8) one ceramic bottle from Ting-chou 定州, Hopei; (9) one ceramic tea-kettle from Yang-hsien 陽羨, modern I-hsing 宜興, Kiang-su.[99] I-hsing, it should be mentioned, is noted for its ware characterized by simplicity of design and elegance of execution; many of the designs are also distinguished by an archaistic fantasy and angularity.[100] According to the early Ch'ing poet Chu I-tsun 朱彝尊 (1629–1709), Hung-jen also owned a painting attributed to Wu Tao-tzu 吳道子 (ca. 720–760) which Hung-jen had bought from the Wu family in She-hsien, but later sold to Ch'eng Sui 程邃 (1605–1691).[101] Unfortunately, it is not possible to identify the works of art that were once in Hung-jen's personal art collection.

It would seem that Hung-jen himself, being poor when he was young and also when he became a monk, did not own an extensive art collection. But he took advantage of the art collections that had been established by rich merchants in Hui-chou. As Hung-jen's early biographer Wang T'ai-cheng pointed out, Hung-jen since his youth had been interested in the art of painting and would try his best to seek out collectors who owned authentic works from the Tsin, T'ang, Sung, and Yüan Dynasties.[102] It is likely that his involvement with the mode of Ni Tsan was closely related to the availability of that master's paintings in the local collections,

It can be assumed from the *Tung-t'u hsüan-lan-pien* compiled by Chan Ching-feng and the *Shu-hua-chi* complied by Wu Ch'i-chen, that in Hui-chou there were even more Ni Tsan paintings than those mentioned above. It is no surprise, then, to see that Hung-jen's older contemporaries from She-hsien such as Ch'eng Chia-sui tended to paint in a reductionist mode. This mode is characterized by sparseness in the landscape and is ultimately derived from the style of Ni Tsan. In the next chapter, we shall turn to a discussion of the transformation of the mode of Ni Tsan by these Anhui artists.

The fact that Hui-chou was a center of commerce and that private collecting increased in the late sixteenth century helped offset Hui-chou's relative geographic seclusion and consequently inhibited artistic provincialism. The artistic exchange with Suchou and Sung-chiang, two major artistic centers at that time, was active. As mentioned earlier, Chan Ching-feng was a member of the literary society called "The White Elm Society" (*Pai-yü she* 白榆社) whose membership included Mo Shih-lung from Sung-chiang and Chou T'ien-ch'iu from Suchou. Chan Ching-feng was apparently well-acquainted with the collection of Mo Shih-lung. Although the size of his collection appears to have been small and less comprehensive than that of his friend

Tung Ch'i-ch'ang, Mo Shih-lung's paintings were certainly known to the artists from Hui-chou.[103] As mentioned earlier, Tung Ch'i-ch'ang and Ch'en Chi-ju also frequented Hui-chou. Their ideas about painting as well as their style could not help but leave their marks on the development of the Anhui artists. Tung Ch'i-ch'ang himself reported that the Hui-chou artist Ting Yün-p'eng 丁雲鵬 (1547–ca. 1621) visited him at his house in Sung-chiang and painted for him and that he inscribed on Ting Yün-p'eng's paintings.[104] Ting Yün-p'eng's handscroll "Five Forms of Kuan-yin," now in the Nelson-Atkins Gallery, Kansas City, is stylistically datable to about 1580. It bears a colophon by Tung Ch'i-ch'ang stating that the artist was in his thirties when he painted the handscroll. It also says that Ting Yün-p'eng later became pedantic and could no longer return to the "skill and wonder" demonstrated in this painting.[105]

There was no less artistic contact between Hui-chou and Su-chou. We have already mentioned Chou T'ien-ch'iu, who, like Mo Shih-lung, was a member of the "White Elm Society." On a recorded handscroll depicting the landscape of Hui-chou dated 1573 by the Sung-chiang artist Sung Hsü 宋旭 (1525–1605) for a certain patron Wu Chi-ho 吳季和 from Hui-chou, Chou T'ien-ch'iu wrote a colophon in which he recounted his two trips, in 1559 and 1580, to Hui-chou.[106] It is no surprise, then, to find that Chou T'ien-ch'iu's handscroll "Ink Orchid" (*Mo-lan* 墨蘭), dated 1580 and now in the National Palace Museum, Taiwan, bears colophons by such Hui-chou people as Chan Ching-feng, Wang Tao-k'un 汪道昆, Wang Tao-kuan 汪道貫, Wang Yin 王寅, Fang Hung-ching 方宏靜, Hsü Kuo 許國, Pao Ying-ao 鮑應鼇, Wang Hui 汪徽, Ling Kuan 淩琯, Wang Ta-ch'eng 汪大成, Wang Yüan-kung 汪元功, Lo Wen-jui 羅文瑞, Chiang Lai-min 江來岷, and Wu Shou-huai 吳守淮.[107]

There are several additional examples of artistic exchange between Hui-chou and Su-chou. For instance, the Suchou artist Sheng Mao-yeh 盛懋曄 (active 1594–1637) collaborated with Ting Yün-p'eng. There are three hanging scrolls dated from 1594 which are the result of such a collaboration. They may have belonged to a large set of perhaps twenty-five, depicting the five hundred arhats in a landscape setting.[108] On two occasions, Lu Chih also painted for patrons from She-hsien. Sometime during the 1530s, Lu Chih painted a handscroll entitled "Streams and Rocks" (*Ch'i-shih t'u* 溪石圖), now in the Nelson-Atkins Gallery, Kansas City, for a Mr. Wu Yun-sheng (active mid-sixteenth century) from She-hsien.[109] The style of this painting, it is interesting to note, is marked by a rare experimentation with the sparser aspect of the mode of Wen Cheng-ming, which is close to the style of Wen Chia and can be related to the sparse style of the Auhui School; the exact nature of that relationship, however, will be discussed in the next chapter. In 1541, Lu Chih again painted a handscroll entitled "Clearing Sky over Mountains and Rivers" (*Ch'i-shan yü-ai* 溪山餘靄) for a Mr. Wu Ch'i-yang 吳溪陽 (d. ca. 1573) from She-hsien, now in the collection of Kuo-t'ai mei-shu-kuan 國泰美術館, Taipei.[110]

Furthermore, in 1652, Hung-jen's contemporary Sun I, who was called the "reincarnation of Wen Cheng-ming," copied an album entitled *Ho-lin yü-lu* 鶴林玉露 by the Suchou artist T'ang Yin 唐寅 (1470–1523). In his colophon to Sun I's album, Hsiao Yün-ts'ung reported that he had copied the same work when he was very young.[111] Other Suchou artists who would visit She-hsien and execute works for patrons there included Chu Yün-ming 祝允明 (1461–1527).[112] The artistic contact between Hui-chou and Suchou can be best seen perhaps in the late sixteenth century in which artists of high calibre from Suchou, including painters, calligraphers, seal-carvers, and musicians, were led by Wang Shih-chen 王世貞 (1526–1590) as a delegation to She-hsien and a cordial but intense artistic contest between the guests and the local artists took place.[113]

As an example of the convergence of different stylistic currents at Hui-chou, it is interesting to note that Liu Shang-yen 劉上延, an older contemporary of Hung-jen's, who participated in painting the 1639 joint handscroll now in the Shanghai Museum, with Hung-jen and three other artists from the same area, also painted for Wen Chia 文嘉 (1501–1583), Wen Cheng-ming's second son. It is likely that he knew Wen Chia and possibly other Suchou artists as well. Also significant is that Liu Shang-yen's calligraphy and painting followed Tung Ch'i-ch'ang. Later, his style was so close to Tung Ch'i-ch'ang that he became one of Tung Ch'i-ch'ang's "substitute" artists, according to the colophon written on the 1639 handscroll by the scholar Huang Yüeh 黃鉞 (1751–1841).[114] Thus, as we shall see, the emergence of the Auhui School was in part due to the influx of Suchou as well as Sung-chiang artistic experience.

Mainly due to the availability of material, the area around Hui-chou has been known in Chinese history as a major center for the production of the "Four Treasures of the Scholar's Studio" —paper, ink-stone, ink-cake, and brush, the very media for calligraphy and painting.[115] The notorious rivalry between the two great makers of ink-cakes in Hui-chou, Ch'eng Ta-yüeh 程大約 (active ca. 1541–1616) and Fang Yü-lu 方于魯 (active ca. 1570–1619) in the late sixteenth and early seventeenth century highlights the scale of their production.[116] The intensity of the consequent commercial promotional activities can be seen in the publication of *Ch'eng-shih mo-yüan* 程氏墨苑 by Ch'eng

Ta-yüeh in 1606 and *Fang-shih mo-p'u* 方氏墨譜 by Fang Yü-lu in 1589, two voluminous catalogs of ink-cakes with exquisite illustrations of designs on and shapes of the ink-cakes. *Ch'eng-shih mo-yüan* included 500 pieces of different designs and shapes; prefaces were contributed by eminent personalities at that time such as Chiao Hung 焦竑 (1541-1620), the philosopher and tutor to heir apparent to the throne,[117] T'u Lung 屠隆 (1542-1605), the famous scholar and the purported author of the resourceful book on high living and art-collecting *K'ao-p'an yü-shih* 考槃餘事, [118] and Tung Ch'i-ch'ang, among others. The illustrations were designed by artists such as Ting Yün-p'eng. The latter book, equally impressive with 385 different designs and shapes, included prefaces written by such celebrated persons as Wang Tao-k'un 汪道昆 (1525-1593), vice-president of the Ministry of War,[119] and the Suchou scholar Wang Chih-teng 王穉登 (1535-1612), among others; the illustrations were designed by Ting Yün-p'eng, his pupil Wu T'ing-yü 吳廷羽, and Yü Chia-shen 俞嘉諶, all from the area of Hui-chou. In fact, one ink-cake dated 1602 is now in the Metropolitan Museum of Art, New York.[120] One should keep in mind that many of the ink-cakes, like the ink-stones, were regarded as works of art rather than merely as articles of utility, For instance, sometimes they are marked "Not for Grinding" (*pu-k'o-mo* 不可磨).[121]

Interestingly, the production of ink-cakes by Ch'eng Ta-yüeh and Fang Yü-lu in Hui-chou was recognized by their contemporary culture critics as an important aspect of cultural achievement. For example, the scholar Hsü Shih-p'u 徐世溥 (1608-1685) was quoted approvingly by Chou Liang-kung 周亮工 (1612-1672) as saying:

> Since 1573, cultural activities in the nation have flourished. The moral integrity of Chao Nan-hsin 趙南星 [1550-1627], Ku Hsien-ch'eng 顧憲成 [1550-1612], Tsou Yüan-piao 鄒元標 [1551-1624], and Hai Jui 海瑞 [1514-1587]; the investigation of moral principles by Yüan Huang 袁黃 [1586 *chin-shih*]: the broad study of things by Chiao Hung 焦竑 [1541-1620]; the calligraphy and painting of Tung Ch'i-ch'ang; the astronomy of Hsü Kuang-ch'i 徐光啓 [1562-1633] and Matteo Ricci [1552-1610]; the poetry and drama of T'ang Hsien-tsu 湯顯祖 [1550-1617]; the herbal medicine of Li Shih-chen 李時珍 [1518-1593]; the philology of Chao Huan-kuang 趙宧光 [1559-1625]; as well as the pottery of Mr. Shih 時, the metallurgy of Mr. Ku 顧, the ink-cakes of Mr. Fang [Yü-lu] and Mr. Ch'eng [Ta-yüeh]; the jade of Mr. Lu 陸, and the seal-carving of Mr.Ho [Chen 何震] —they are all equals to the ancients.[122]

The publication of the wood-block designs for ink-cakes such as *Ch'eng-shih mo-yüan* and *Fang-shih mo-p'u* not only attests to the high regard they received in the cultural scene, but also points to the sophisticated development of the art of woodcut printing in Hui-chou. The printing and publishing of books in Hui-chou reached its apex, ranking with Nanking and Wu-hsing and replacing Suchou and Hangchou in the late sixteenth and early seventeenth century. Hui-chou is also known to have perfected the technique of multi-color woodcut printing. This development was closely related to the art of wood-carving or wood-engraving as seen in the carved details on extant Hui-chou architecture dating from the late Ming and briefly discussed earlier.[123] The art of multi-color woodcut printing can be best seen in *Shih-chu-chai shu-hua-p'u* 十竹齋書畫譜 and *Shih-chu-chai chien-p'u* 十竹齋箋譜, two manuals on painting published in 1627 and 1644 respectively by Hu Yüeh-ts'ung 胡曰從 (1584–1674), who was a native of Hui-chou, in Nanking.[124] The prints are delicately cut, with refined lines and well articulated motifs; they are of a high quality that would not be surpassed by later followers such as the *Chieh-tzu-yüan hua-chuan* 芥子園畫傳 (Mustard-seed Garden Painting Manual) compiled by Wang Kai 王概 (active ca. 1677–1705), first published in 1679 in Suchou and subsequently reprinted and expanded.[125]

Hsieh Chao-chih, who visited Fang Yü-lu himself in Hui-chou, reported that the *Fang-shih mo-p'u* was so widely circulated and imitated, due to exquisite designs on and shapes of the inkcakes, that "suddenly paper became extremely expensive."[126] The availability of these woodcut designs to the literati, as we shall see, was to have an impact on the style of painting in Hui-chou. It should be mentioned here that Hung-jen painted a hanging scroll (Pl. 167) entitled "Ten-Bamboos Studio" (*Shih-chu-chai-t'u* 十竹齋圖), dated 1657 for Hu Yüeh-ts'ung's son Hu Ch'i-i 胡其毅 (active ca. 1650–1700); it was probably painted in Nanking, where Hung-jen visited in 1657 and probably also earlier in 1651 on his way back from Fukien to Anhui.[127] Hu Ch'i-i also owned a hanging scroll "Plum blossoms," undated, by Hung-jen (Pl. 230).[128] Thus, Hung-jen's contact with the wood-cut prints in Hui-chou was not just accidental.

Besides Ting Yün-p'eng, Wu T'ing-yü, and Yü Chia-shen, several other Hui-chou artists were involved with designing for publication. For example, Sun I is said to have painted twenty-four pictures of the mountains in She-hsien which were later published in woodcut prints.[129] Hsiao Yün-ts'ung's *T'ai-p'ing shan-shui t'u-hua* 太平山水圖畫, consisting of 43 pictures depicting the landscape at T'ai-ping, near Hui-chou, was published in 1648 in Hui-chou (Pls. 13–15).[130] The 60-leaf album "Mt. Huang" (*Huang-shan-t'u* 黃山圖) by Hung-jen might well have been intended for publication in the form of woodcut prints (Pls. 65–124).[131] In fact, several leaves in the album later did find their way into the gazetteer of Mt. Huang (Pl. 231).[132] It is conceivable that the austerity and linearity in Hung-jen's painting may be attributed at least in part to his experience with woodcut prints in which crystalline geometric surface pattern is more important than spatial depth and that certain narrative elements in his paintings were probably dictated by the conventions of illustrating the literary works in woodcut prints. The stylistic affinity between, for example, the hanging scroll entitled "Parting at End of the Year" (*Sui-mu fu-pieh* 歲暮賦別) dated 1585 by Ting Yün-p'eng[133] and the designs on ink-cakes by Ting Yun-p'eng himself and by Yü Chia-shen (Pls. 10–12) suggests that the widely published prints, linear and flat by nature, created an artistic environment in which the angular and austere style of Hung-jen and his followers was to develop.[134]

CHAPTER II

HUNG-JEN AND HIS PREDECESSORS

In discussing the origins of the Anhui School and the genesis of the early style of Hung-jen, one can begin with some comments made by Kung Hsien 龔賢 (1599–1689). He provided a kind of genealogy for what he called the "Tien-tu school" (*T'ien-tu p'ai* 天都派) in an inscription on a handscroll now in the Fogg Museum of Art, Harvard University, Cambridge (Pls. 1–2). Kung Hsien wrote:

> Meng-yang [Ch'eng Chia-sui] begins the T'ien-tu School. By the time of Chou-sheng [Li Yung-ch'ang] the school begins to gather momentum. The style of Meng-yang is like that of Yün-lin [Ni Tsan], and the style of Chou-sheng is like that of Shih-t'ien [Shen Chou] imitating Yün-lin. Meng-yang's family name is Ch'eng and his given name Chia-sui; Chou-sheng's family name is Li and his given name Yung-ch'ang. They are all from T'ien-tu. Later on Fang Shih-yü [Fang Yü-ju], Wang Tsun-su [Wang Hsüan-tu], the monk Chien-chiang [Hung-jen], Wu Tai-kuan [Wu Shan-t'ao], Wang Wu-jui [Wang Chih-jui], Sun Wu-i [Sun I], Ch'eng Mu-ch'ien [Ch'eng Sui], Cha Erh-chan [Cha Shih-piao], all learn from these two people. They are all from T'ien-tu, hence the name "T'ien-tu School".

The name "T'ien-tu" specifically refers to Mt. Huang, located near She-hsien in Anhui. It also encompasses the area know as Hui-chou or Hsin-an of which She-hsien has always been the major city.

Thus Kung Hsien used a name with historical connotation that is evocative of regionalism. The name modern Chinese scholars use is the "Hsin-an School" (*Hsin-an p'ai* 新安派) which also carries with it historical references to the area.[1] In the west, the term "Anhui School" has recently gained currency.[2] It should be noted that Kung Hsien's statement follows a section of his handscroll in which he painted in a style reminiscent of Ni Tsan, as if to indicate the sources of the "T'ien-tu School."

Elsewhere, in his inscription on a hanging scroll of landscape probably from his late period and now in the University of California Art Museum in Berkeley, Kung Hsien wrote that Ch'eng Chia-sui and Li Yung-ch'ang might be said to belong to the "untrammeled class" (*i-p'in* 逸品) of artists, suggesting that, in his opinion, they not only derive their style from the same source but also share a certain stylistic affinity.[3]

Kung Hsien had an intimate knowledge of the art world of Hui-chou. The recent rediscovery of Kung Hsien's literary anthology *Ts'ao-hsiang-t'ang chi* 草香堂集, first published around 1653, and William Wu's study of Kung Hsien's life, have increased our knowledge about the relationship between Kung Hsien and the artists from Anhui. As early as the late 1640s, for instance, Kung Hsien painted an album with an Anhui artist Wu Ting 吳定 (active ca.1640).[4] While staying at Yang-chou in the 1650s and 1660s, Kung Hsien came to know many men of Anhui origin, including Wang Hsüan-tu who was mentioned in Kung Hsien's comment on the "T'ien-tu School" quoted above; Hung-jen visited Yang-chou in 1661.[5] Kung Hsien's connection with people from Anhui can also be seen in an inscription he wrote in 1670 for one of his paintings finished about twenty years earlier, in the 1650s, and now in the Cleveland Museum of Art. That particular hanging scroll had been owned by a certain Wu Chin-ming 吳晉明 from Hsin-an who subsequently brought it back to Nanking where the artist was requested to write an inscription.[6] Kung Hsien's circle included a great number of important artists, scholars, connoisseurs, and collectors of his time, including Hung-jen. He and Hung-jen, for instance, each contributed a leaf of landscape to an album now in the Ching Yüan Chai collection in Berkeley, California, for a Mr. Chung-fu 中符 and dated 1663. Kung Hsien's personal acquaintance of Hung-jen can also be seen in his inscription on an album painted when he was young. He wrote: "Whenever Chang Ta-feng 張大風 [Chang Feng 張飌] from Yeh-ch'eng 冶城 [Nanking] and Chien-chiang [Hung-jen] from Mt. Huang finished their paintings , they would ask for my colophons‥‥We all belong to the same class [of artists]." [7] Although no Hung-jen paintings with Kung Hsien's colophons have survived, the authenticity of that colophon suggests the mutual admiration between Kung Hsien and Hung-jen. Kung Hsien's testimony thus opens the way for an understanding of the origins of the Anhui School and Hung-jen's stylistic sources in a broad context.

Two main points emerging from Kung Hsien's comments are crucial to our understanding of the Anhui School. First of all, these artists all attempted to emulate the mode of Ni Tsan. Secondly, the Wu School, through the master Shen Chou, also considerably influenced the formative Anhui coterie. By extension, Kung Hsien might have meant to suggest that there was a strain of influence from Wen Cheng-ming and his followers in the formation of the Anhui School.

In the following pages, in order to grasp the tradition in which Hung-jen found himself and through which he revealed his inventiveness, the mode of Ni Tsan and that of Wen Cheng-ming will be examined in relation to Hung-jen's predecessors .

In emphasizing Ni Tsan's influence on the art of Hung-jen, one tends to risk the danger of neglecting other important aspects of his development. By looking at the artistic ambience around Hung-jen, one begins to see its complexity. The situation is complex because of the co-existence of different stylistic currents in Hung-jen's time and place. The stylistic sources of Hung-jen, like those of many other artists in the sixteenth and seventeenth centuries, are to be sought not only in contemporary masters with whose work the artist came into contact, but also in the classic tradition of the Sung and Yüan dynasties.

We know very little about the early work of Hung-jen. It is possible that his early paintings have simply disappeared through the ravages of time. The scarcity of his early work might also be due to their conscious elimination by Hung-jen himself. According to a traditional account, after having searched in vain for a genuine Ni Tsan work, Hung-jen finally saw a certain Ni Tsan painting in an unidentified Wu family collection near his hometown. Having studied the Ni Tsan for three months, he destroyed all of his earlier paintings which now appeared unsatisfactory to him.[8] It is also possible that some of his paintings, like his writings, might have been censored by the Manchus because of his connection with the resistance movement.

We have just mentioned Hung-jen's fascination with works by Ni Tsan, a penchant he shared with many of his older contemporaries such as Ch'eng Chia-sui. Indeed, it is almost a cliché among connoisseurs and historians that Hung-jen follows the style of Ni Tsan. It is true that among Hung-jen's extant writings, Ni Tsan is invoked time and again. Although the names of Tung Yüan, Ching Hao, Kuan T'ung, Huang Kung-wang, and Wang Meng are also mentioned, they appear much less frequently.[9] Among the antiquities in his personal collection, there was at least one painting by Ni Tsan.[10]

As Susan Arkush suggests in her recent study of Ni Tsan and the image of Yüan painting in the eye of later critics, Ni Tsan had become almost synonymous with Yüan painting as a whole, and his shortcomings, if any, were also symptomatic of Yüan literati painting. As Arkush correctly points out :

> Modern historians are highly conscious of politics and have brought to light many interesting correspondences between broad social circumstances and artistic movements which may represent responses to them. The austerity, willful lack of seductiveness, intellectuality, and non-public quality of Yüan art are now widely interpreted as a visual expression of the artists' alienation from the hated foreign regime and its policy of exclusion towards themselves.[11]

It is therefore not surprising that Ni Tsan came to be identified with the protest by later literati against what is politically, socially, and intellectually foreign, particularly after the fall of the Ming dynasty to the Manchus. By following Ni Tsan, Hung-jen immediately establishes a frame of reference for the possible political meanings of his paintings. Style then becomes an idea, and an integral part of meaning.

Furthermore, as Arkush ably demonstrates, the main stylistic features of Ni Tsan came to be regarded as those of Yüan literati painting in later Chinese art history. She says:

> Spontaneity, a total and dazzling originality reaching the limits of "unpracticed" art; a consequent inimitability; sparse, open compositions and relaxed, bony brushwork expressive of the artist's unconcern with trivia, his native taste for the essential, a lonely flavor springing from his stern virtue and paradoxically conveying a sense of exhilaration, even richness and "moistness," in its embodiment of nobility and purity of spirit.[12]

On the other hand, it should be pointed out that, when Ni Tsan was criticized, sometimes for paintings not by his own hand, the criticism was also directed toward Yüan literati painting as a whole. "The characteristics that unenthusiastic writers on Yüan painting tended to stress—excessive thinness, shallowness, dissipation, and narrowness—are the very ones that are mentioned on the infrequent occasions when the critic has unkind things to say about Ni Tsan."[13] The mode of Ni Tsan, then, presents itself both as an easy and trite means of conveying the lofty ideals traditionally associated with Ni Tsan, and as a challenge for those artists who wished to explore its formal and expressive possibilities. The mode requires both understanding of Ni Tsan's style and technical ingenuity on the part of those who attempt to emulate him As we shall see, very few artists were able to achieve the transformation that is so characteristic of Hung-jen.

By the seventeenth century, the mode of Ni Tsan has become, so to speak, a part of an artist's interior landscape, no matter how his understanding of that mode came into being. As Ernst H. Gombrich has remarkably demonstrated, the history of representational art is created by a specific tradition of reinterpreted images, by memory-guided responses that are conditioned by the artists' experience in the world. Changes in this tradition of images occur when the artist checks past schema against nature, modifying and redefining them in accord with his perception. Thus the origin of a work of art is to be found in the experience of other works of art as well as in the nature of external reality.[14] It is therefore appropriate to define the artistic conventions Hung-jen adopted in creating his own style, especially those that Hung-jen used to recreate the mode of Ni Tsan.

Hung-jen himself wrote of Ni Tsan: "With sparse trees and cold mountains Ni Tsan created a plain and lofty gesture, even though he had long realized it was not fashionable. This painting by Ni Tsan is a treasure in my family. And every year I burn incense in front of it as my teacher."[15] In another poem, Hung-jen wrote of his experience of the work of Ni Tsan and of the mountains; the two kinds of experience seemed to have merged in the artist's consciousness. He wrote: "Returning from the mountains, I sit down to meditate. Bending my body and suddenly seeing the sparse forest, I remember the painting by Ni Tsan I have seen."[16] In still another poem, Hung-jen revealed how his experience of Ni Tsan came to him in his encounter with the landscape: "The old tree has weathered the ravages of time. The level hillocks extend endlessly. Reading these, I think of Yüan-chen [Ni Tsan]."[17] Perhaps it is through his deep understanding of the work of Ni Tsan that Hung-jen was able to make this perceptive observation on the fourteenth-century master's style: "The legendary Yün-lin-tzu [Ni Tsan], I am afraid, may not be merely sparse and flat. As I understand it, artistic perfection is to be found both in the simple and the complex."[18]

Ni Tsan's formative influence on Hung-jen was not only acknowledged by Hung-jen himself, but also recognized by Hung-jen's contemporaries. For example, the famous art patron and collector Chou Liang-kung 周亮工 (1612-1672) reported in his *Tu-hua-lu* 讀畫錄： " [Hung-jen] likes to imitate Ni Tsan and later becomes perfect. People in the Chiang-nan region regard the possession of Hung-jen's work as the proof of good taste, just as they would a Ni Tsan work. All agree that to own a painting by Hung-jen is the same as having a Ni Tsan."[19] Ni Tsan's influence on Hung-jen has, in our time, led scholars to reinforce and reiterate the traditional view. For instance, Roger

Goepper writes: "The principles of Ni Tsan's techniques have been carried to their ultimate logical conclusion in the exceedingly fine landscape 'The Coming of Autumn' [in Honolulu] painted by Hung-jen." Seeing Hung-jen as the artist who transformed the mode of Ni Tsan, he goes on to say about the Honolulu hanging scroll (Pls. 219-223):

> In such details as the dots of moss in the branches and the division of the rock, on which the few rather darker spots stand out as accents rather than as natural forms, we can see clearly foreshadowed that tendency to schematization and abstraction of the object which had reached a comparatively advanced stage by the second half of the seventeenth century—for example in the landscape by Hung-jen.[20]

In discussing the role of Ni Tsan in the formation of Hung-jen's personal style, other scholars are more guarded. For instance, Cahill agrees that Hung-jen shares with Ni Tsan a "cleanness and clarity" and yet points out that "so little of the Ni Tsan manner remains in his mature style, and that so thoroughly transformed, that it is difficult to distinguish more than a similarity in taste."[21] One is thus confronted with the question: What goes into the transformation of Ni Tsan in the style of Hung-jen? In other words, what are the sources of Hung-jen's style?

To answer that question, one must first look into the historical development of the mode of Ni Tsan. The thousands of paintings in the manner of Ni Tsan range from the slavish and pedantic to the more daring and inventive. Within this broad range, a few examples, however, will be selected to help explore the transformation of Ni Tsan during the nearly three hundred years since his death in 1374. This task, in turn, behooves us to define the master's own style as best and as briefly as possible. A few examples from his oeuvre will, perhaps, suffice to demonstrate its nature and range.

Ni Tsan followed the mode of Tung Yüan in his early years. Compared with the painting entitled "Wintry Groves and Layered Banks" traditionally attributed to Tung Yüan and now in the Kurokawa collection,[22] Ni Tsan's early works such as the 1339 hanging scroll "Enjoying the Wildness in an Autumn Grove" (*Ch'iu-lin yeh-hsing* 秋林野興) in the Crawford collection in New York,[23] the 1345 hanging scroll "Six Gentlemen" (*Liu-chün-tzu* 六君子) formerly in the P'ang Yüan-chi 龐元濟 collection,[24] and the 1354 hanging scroll "Pavilion in a Pine Grove" (*Sung-lin t'ing-tzu* 松林亭子) in the National Palace Museum in Taipei reveal his stylistic affinity to the tradition of Tung Yüan.[25] For instance, in the Crawford hanging scroll, Ni Tsan mainly used the hemp-fiber texture strokes (*p'i-ma ts'un* 披麻皴), that is, long dragging lines applied more or less in parallel to the contour line of the rocks and mountains. The foliage of the trees is varied and luxuriant, and the pavilion and other details are painted meticulously. In general, the brushwork is dense, the textures are richly layered, and the land masses are soft and round. The painting, thus, can be readily related to Tung Yüan.

In his late works, however, Ni Tsan turned to the style of the Ching Hao and Kuan T'ung tradition, both in the overall composition and in the construction of mountain masses and rocks. The style of Ching Hao 荆浩 (active ninth and early tenth centuries) and his follower Kuan T'ung 關同 (active early tenth century) has been traditionally characterized by an awe-inspiring and somewhat somber monumentality. Paintings attributed to them, for example the hanging scroll "The K'uang Lu Mountains" (*K'uang Lu t'u* 匡廬圖) attributed to Ching Hao (P1. 232) and the hanging scroll "Waiting for the Ferry at a Mountain-stream" (*Shan-ch'i tai-tu* 山谿待渡) attributed to Kuan T'ung (P1. 233), both in the National Palace Museum in Taipei, exhibit steep and rugged granitic mountains.[26] The stern and dramatic cliffs confront the viewer and occupy the middle of the painting. The brushwork is mostly sharp and angular and contributes to the crystalline and barren rock formation.

We know from both literary and visual evidence that Ni Tsan admired and transformed the style associated with these two artists. For example, in a colophon dated 1344 on a landscape by Hsia Kuei 夏珪 (active ca. 1190–1225), Ni Tsan wrote:

> In Hsia Kuei's "Thousand Peaks Contending for Elegance," the mountain peaks and valleys go winding again and again. The scenery, with its forest, trees, pavilions, and lookouts, is deep and secluded, clear and distant. It cannot be achieved by the ordinary craftsman. Li T'ang derived his style from Ching Hao and Fan K'uan. Hsia Kuei and Ma Yüan, in turn, derived their style from Li T'ang.[27]

If this statement is indeed from the hand of Ni Tsan, he was probably referring to a painting such as the hanging scroll of winter landscape attributed to Hsia Kuei now in the National Palace Museum in Taipei.[28] Ni Tsan's proclivity towards the tradition of Ching Hao and Kuan T'ung, especially in his late years, can be seen in his inscription on the handscroll "The Lion Grove" (*Shih-tzu-lin t'u* 獅子林圖) he painted in 1373 jointly with Chao Yüan 趙原 (active in the 1370s). The inscription reads in part: "The painting is indeed done in the spirit of Ching Hao and Kuan T'ung. Artists like Wang Meng could not have conceived such a thing even in their dreams."[29]

The rocks in "The Lion Grove" handscroll, with their angular forms built up by layers of brushwork, are indeed reminiscent of the style of Ching Hao and Kuan T'ung. The surface on the rocks and the effect of shading on them are the result of dry brushstrokes applied with a relatively small amount of ink. The method is called the "folded-sash texturing stroke" (*Che-tai-ts'un* 折帶皴). It is achieved by holding the brush obliquely and moving it horizontally, then pushing it sharply downward so that the first part is formed by dragging the tip of the brush, whereas the finish is made by the side of the brush. The calligraphic quality of the individual brush stroke is submerged in the process of applying the stroke in multiple layers. As we shall see, this is also the predominant approach to rock forms in Ni Tsan's late work.

One prime example is the hanging scroll "The Jung-hsi Studio" (*Jung-hsi-chai t'u* 容膝齋圖), now in the National Palace Museum in Taipei (Pl. 4).[30] Painted in 1372, just two years before the artist's death, it has been regarded as Ni Tsan's most representative work. The composition can be divided into three parts. It begins at the bottom with a clearly defined foreground, continues into an extended view in the middle ground, and ends with a strip of distant mountains on top. Within this lucid design, the artist used a subtle method of formal construction. By means of delicate and intricate relationship of ink tonality, shape, and position, the texturing imparts a sense of three-dimensionality to rocks which have pointed edges and irregular facets. The tree branches are convincingly drawn and the leaf-dots are clearly related to the branches. The foreground boulders as well as the background mountains, modelled by repeated layers of light ink washes and dry brushstrokes, are indeed reminiscent of Ching Hao and Kuan T'ung. However, Ni Tsan frequently used a concave arc in the sides of the rocks. Moreover, he used brisk, straight lines with abrupt twists in a neat and discreet motion. His neat, elongated and triangular dots are made by holding the brush at an angle. Ni Tsan also achieved a sense of randomness by means of the dots which act as accents as well as unifying elements, and by means of blurred contour lines.

No single element in the whole composition, simple and schematic at first glance, stands as dominant, nor do elements have any real independent existence. Each form is completed within the limits of the picture plane, and overlapping is reduced to a minimum so that the relationships among forms are clarified on a single plane. The artist achieved a delicate equilibrium between nature and his own rendering of it. The factual appearance of nature, in other words, seems to be balanced in a kind of "controlled disorder" with the more essential structural build-up stripped of all its artistically irrelevant clutter and momentary illusion. As we shall see, it is precisely the seemingly simple and yet complex style of Ni Tsan that was to elude so many later emulators. Unfortunately, many later emulators of Ni Tsan, perhaps failing to comprehend the sources of Ni Tsan's style, tended to overemphasize the simple and dry aspect of his style. Their imitation was consequently far from innovative. Tao-chi strongly criticized these imitators in an inscription on an inventive reinterpretation of Ni Tsan, i. e., his 1697 hanging scroll now in the Princeton University Art Museum. He wrote perceptively:

> Noble Ni Tsan's paintings were like waves and sand, streams and stones. Bending and flowing they grew out of nature. And there was a piece of empty, pure, rich air coldly, coldly compelling us. Later generations have only imitated his withered dryness, his cool economy; and that is why their paintings have no profound divinity.[31]

Anticipating Tao-chi, as we have seen, Hung-jen had earlier pointed out that the style of Ni Tsan was not "merely sparse and flat." His deep understanding of Ni Tsan and his own inventiveness enabled him to achieve the rare transformation of the deceptively simple style of the fourteenth-century master. Chapter III of the present study will be devoted to Hung-jen's major works in order to convey his achievement.

In order to see how Hung-jen's transformation differed from and at the same time may have grown out of others' imitations of Ni Tsan, several artists from the Ming Dynasty should be briefly discussed here. Among the early Ming emulators of Ni Tsan, we should first of all mention Wang Fu 王紱 (1362-1410). His hanging scroll, entitled "Landscape," dated to 1401 and formerly in the Yamamoto collection in Tokyo (Pl. 5), is clearly derived from Ni Tsan's works such as "The Jung-hsi Studio."[32] By means of the wide separation of water between the foreground rocks and distant mountains, the simplicity of its design, and the elongated foreground trees, it somehow succeeds in capturing the style of Ni Tsan. The artist also captured the brush manner of Ni Tsan in his pale and dry brush-strokes and repeats his compositional idioms such as an empty hut and the trees in the foreground, a foreground which is separated by empty water from the mountains in the far distance. However, Wang Fu seems to lack the structural approach of Ni Tsan and to pay more attention to the individual details. For example, he over-emphasizes the almost horizontal pair of twigs at the top of the foreground trees. By failing to come to grips with the complexity of the mode of Ni Tsan, Wang Fu can only produce its formal elements.

By contrast, another early Ming artist Liu Chüeh 劉珏 (1410-1472) is more innovative in his emulation of Ni Tsan. This can be seen in his undated hanging scroll "Landscape" in the Musée Guimet, Paris (P1. 6).[33] Although his landscape is stylistically derived from Ni Tsan, there are several differences. First, the composition as a whole is crowded with trees, boulders, and mountains. By compressing these elements into almost overlapping planes and densely packing them into rising forms, Liu Chüeh achieves an effect of solid mass and dynamic movement. Secondly, the mountains and land forms in the Liu Chüeh tend to either push into the picture plane or surge out of it. By not slavishly following Ni Tsan, Liu Chüeh is able to achieve a more interesting reinterpretation of the mode of Ni Tsan.

Liu Chüeh's transformation of the mode of Ni Tsan is further developed in a painting by one of his young contemporaries, Shen Chou 沈周 (1427-1509). The fact that Shen Chou wrote the colophon on Liu Chüeh's "Landscape" suggests Liu Chüeh must have exerted a specific influence on the younger artist in this mode. Shen Chou regarded Ni Tsan's style as derived from that of Ching Hao and Kuan T'ung. He wrote: "In the time of the last dynasty Yün-lin's [Ni Tsan's] character was lofty and spontaneous;···· his painting followed Kuan T'ung, simple in brush and pure in thought."[34] He also wrote: "Master Yün-lin was pure and lofty···· He was fond of the painting of the two masters Ching [Hao] and Kuan [T'ung], admiring the purity and plainness of their painting style."[35]

In his undated hanging scroll entitled "Walking with a Staff" (*T'se-chang-t'u* 策杖圖), now in the National Palace Museum (P1. 7)[36], Shen Chou sought to reinterpret the style of Ni Tsan, but, like Liu Chüeh, did not attempt to capture merely the simplicity of a typical Ni Tsan landscape. Rather, he seemed to be more interested in Ni Tsan's prototypes in Ching Hao and Kuan T'ung. For instance, the painting is crowded with more trees, boulders, and mountain masses than "The Jung-hsi Studio" by Ni Tsan. In addition, the human element of a scholar is put into the composition, enhancing its narrative , and at the same time, its monumentality through the contrast in scale between the figure and the mountain masses. Furthermore, the brushwork is much denser and broader than that found in Ni Tsan's painting. More important is the way in which the whole composition is organized, as compared with "The Jung-hsi Studio." In the latter, the land masses in the foreground and far distance are connected by a continuous movement of rocks; in the former, the land masses are linked by the foreground trees. In the latter, the vertical and frontal tripartite division of the picture plane predominates; in the former, the trees serve to unify the composition.

Noting the similarities between Ni Tsan's "Jung-hsi Studio" and Shen Chou's "Walking with a Staff," Richard Edwards has written of Shen Chou's "new" landscape:

> The scholar, the man, inevitably becomes a vital core—a seed—from which grows and radiates the whole dynamic life of the painting; and from this focal firmly placed center the shapes which define Shen Chou's natural scene move upward toward the left, back to the right and ultimately are centered so that the furthest hills remain firm. But it is a center that is not an exact center, rather an implied axial line whose sureness rests on the dynamics of shape. Thus there is always the strong subtle quality of movement. This is a mountainscape that grows upward and also expands outward so that the mountain is not just there as a shape but as a force.[37]

Elements that are innovative in the Shen Chou include: forms that tend to move in and out of the picture plane; land forms reduced to basic three-sided shapes; and pronounced diagonal thrusts and turns. Above all, Shen Chou infuses a structural solidity into the painting.[38]

Though Shen Chou is said to have been chided by his teacher because his attempt to emulate the brushwork of Ni Tsan is "over-done," i.e., exaggerated or too assertive, he is successful in exploring and expanding the formal possibilities offered by the mode of Ni Tsan.[39]

By the time of Shen Chou, it would seem, the mode of Ni Tsan had already assumed an existence of its own and became a style to be emulated. It is like a language converted into a code. Furthermore, the style itself assumed a mystic character; it was regarded as in the category of the untrammeled style (*i-p'in* 逸品), the meaning of which had now been associated with something that was elusive and inimitable, something that could only be aspired to.[40] As the mid-sixteenth century critic Ho Liang-chün 何良俊 wrote in his *Ssu-yu-chai hua-lun* 四友齋畫論:

> Shen Chou's studies of Huang Kung-wang, Wu Chen, and Wang Meng all came very close to their models and often even surpassed them. It was only in his emulation of Ni Tsan that he could not get a good likeness. I own a small scroll of Shen Chou's which is a study of Ni Tsan; at the end, he wrote this inscription: "This scroll is painted in the mode of Ni Tsan. But where Ni Tsan is simple, I am complex. Now, though his style is simple, it fully conveys his conception. This is what is hard to achieve in his style."[41]

Once the conception of the mode of Ni Tsan as inimitable and inaccessible (even to such a great master as Shen Chou) was established, the artists were then inevitably confronted with a dilemma. They could follow to the letter the mode of Ni Tsan and regard themselves as belonging to the category of *i-p'in*, in the literati circle, if not in terms of artistic merits. Or they could search for the roots, so to speak, of Ni Tsan's style and integrate that art-historical understanding into their own style. The second path would, of course, be more difficult, but more fruitful. Shen Chou apparently took the second path in his "Walking with a Staff."

In the sixteenth century, it was Wen Cheng-ming and Lu Chih who succeeded in continuing the more innovative exploration and expansion of the formal possibilities suggested by the mode of Ni Tsan. For example, in Wen Cheng-ming's undated handscroll "Autumn Mountains" (Pls. 235-236) in the Art Institute of Chicago, the artist seemed to be interpreting the mode of Ni Tsan in its fuller and more crowded aspect, an approach very likely inspired by Shen Chou.[42] Despite the fullness of the composition, Wen Cheng-ming sought to minimize the effects of volume and space and to concentrate on the surface of the paper through the soft rounded quality of the form, the richness of the brushwork. Unexpectedly, the painting is on gold-flecked paper, which enhances the interest on the surface of the paper itself. Even in a sparse composition such as the 1512 handscroll "Summer Retreat in the Eastern Grove" in the John M. Crawford collection (P1. 237), Wen Cheng-ming was more interested in the creation of an exact compositional order and a cool remoteness through a series of discrete islets and trees spread across the picture plane.[43] As Richard Edwards has pointed out:

> We admire it····not because of a ready pictorialism or an open emotional content but because of a kind of careful skilled reserve--the order, the sense of propriety····He moves from rock, from separate tree to separate tree, along a carefully planned course and in the end our summer's reward is the vertical drop and the sure arcs of a waterfall, the vibrant surface of a cliff and the remote cool of the empty paper from which to enjoy it.[44]

Like Shen Chou, the significance of Wen Cheng-ming's reinterpretation of the mode of Ni Tsan, as we have seen, was to show that in addition to the stereotyped Ni Tsan formula of simple composition with a deserted hut, a few foreground trees, and distant mountains, the mode of Ni Tsan was more complex and could be adapted to a fuller and more crowded composition. Wen Cheng-ming's emphasis on a rhythmical disposition of motifs on the surface of the picture plane, as we shall see, was to be important to Hung-jen's transformation of the mode of Ni Tsan.

Lu Chih's large hanging scroll entitled "Taoist Retreat in Mountain and Stream: An Autumn Landscape after Ni Tsan," dated to 1567 and now in the Cleveland Museum of Art, is perhaps the best example of his reinterpretation of the mode of Ni Tsan (Pl. 8).[45] Like Shen Chou, Lu Chih was bold enough to take the more difficult but fruitful path: emulating Ni Tsan, but not painting like him. It was not by imitating overtly the formal structure of Ni Tsan, but by concentrating on expanding and developing it according to his own expressive purpose, that Lu Chih was able to transform Ni Tsan's inimitable style. In the Cleveland painting, he used dry angular brushstrokes, usually with many small, sharp, vertical accents, to construct a painting of clarity and refinement.

Lu Chih's painting, like Ni Tsan's 1363 hanging scroll "Mountains Seen from a River Bank" in the National Palace Museum in Taipei, can be divided into two discrete halves. The lower half of the hanging scroll is rather flat, with forms isolated against the blank paper and removed from the lower edge of the scroll, whereas the upper half of the scroll has more spatial recession. However, unlike most traditional hanging scroll compositions, the upper half of Lu Chih's painting carries more weight than the lower half. The lower half is merely a flat display of discrete forms on the surface of the paper, whereas the upper half, with its major mountain mass on the left and the water-village on the right, becomes the focus of interest in the whole composition. In addition to the diagonal raised causeway, there is strong juxtaposition of verticals and horizontals in the painting. In its clarity of forms and measured rhythm, Lu Chih's transformation of Ni Tsan was to inspire Hung-jen in the seventeenth century. While Hung-jen's transformation of the mode of Ni Tsan must have drawn on Suchou interpretations, it is even more intimately related to the stylistic context of his own time and place. Thus we should turn to Anhui in the late sixteenth and early seventeenth centuries.

To begin with, it is important to note here that Chan Ching-feng, the impeccable and discriminating connoisseur from Hsiu-ning, always saw the mode of Ni Tsan in a critical, sometimes even negative, perspective. Unlike many others who would follow and worship Ni Tsan without enough art-historical and critical understanding, for instance, Chan Ching-feng commented in his *Tung-t'u hsüan-lan-pien* 東圖玄覽編 on two Ni Tsan paintings he saw in this manner: " [I see] two paintings of old trees, bamboo, and rocks on paper by Ni Tsan. The small one is like new and good. The larger one has wonderful bamboo and rocks, but the branches of the two old trees are rather bad. One long drawn-out brushstroke on the flat river bank is just like a dead eel. Now I see that Ni Tsan's paintings sometimes can be inadequate."[46] It is not possible to determine which paintings were the subject of Chan Ching-feng's critical comments. Nevertheless, his catholic taste and fair judgement seen in his writings suggest that he was probably referring to genuine paintings. In any case, the comments above indicate that he was not afraid to criticize Ni Tsan when, to him at least, the apotheosized Yüan master failed to live up to his reputation. Elsewhere Chan Ching-feng criticized Yüan painting in general as "descending into a pure delicacy and especially lacking in an air of vigor and expansiveness." The commonly held opinion of Yüan as a whole as restrained, elegant, and lofty was contradicted by Chan Ching-feng's opinion that it was listless and over-refined. He preferred, it would seem, the expansiveness of the Sung to the often abused "easygoing feeling" and "lofty desolation" of the Yüan.[48]

It is perhaps this critical spirit that led Chan Ching-feng and other Anhui artists, among them Hung-jen, to a new interpretation of the mode of Ni Tsan, a mode that had so often been slavishly imitated since the late fourteenth century. For instance, Chan Ching-feng saw correctly that the style of Ni Tsan included dense forests and the piling-up of mountain peaks which he would associate with artists such as Fan K'uan.[49] At any rate, it seems that Chan Ching-feng and others in Anhui preferred to go back to the monumental style of the Five Dynasties and the Northern Sung Dynasty in their attempt to transform the mode of Ni Tsan.

Hung-jen's style can be traced to specific currents in Anhui in the late sixteenth and early seventeenth centuries, particularly in relation to a modification of the mode of Ni Tsan. One of these stylistic trends is exemplified in a hanging scroll entitled "Parting at End of the Year" (*Sui-mu fu-pieh* 歲暮賦別) dated 1585 by Ting Yün-p'eng (Pl. 9).[50] The foreground trees are aligned on the same picture plane and with almost the same interval between, thus giving an impression of sparseness. But this remote stylistic reference to Ni Tsan is immediately confronted by the major mountain masses in the middle ground. The highest peaks are built up with layers of angular shapes and sway to the right. In front of these peaks are a series of conical mounds with minimal interior texturing. The spatial recession on the left, though enhanced by the connecting verticality of the central tree in the foreground, seems to be negated by the swaying high peaks. This painting prefigures a kind of conscious manipulation of conventional type-forms that achieves a dramatic mountain landscape while at the same time maintaining the flatness and linearity so characteristic of the Anhui school in the late sixteenth and early seventeenth centuries.

It should be mentioned in this connection that Ting Yün-p'eng and other Anhui artists at the turn of the seventeenth century contributed designs for woodcut prints. These widely published prints, in the form of catalogs of designs on ink-cakes, were to exert a considerable influence on the austere and severe style of Hung-jen. For example, one ink-cake design composed by Ting Yün-p'eng and published by Fang Yü-lu in *Fang-shih mo-p'u* in 1589 is characterized by precarious peaks built up with interlocking angular and geometric rocks (Pl. 10).[51] The cubic rocks are placed in a rhythmic order on a flat picture plane. Granted that these wood-cut designs are by nature linear and flat, the style is strikingly similar to that seen in Ting Yün-p'eng's 1585 hanging scroll and must have been popular at that time among the literati who were the buyers of these catalogs and the ink-cakes advertised therein.

Only a few years after Ting Yün-p'eng's hanging scroll "Parting at End of the Year" was painted, Chan Ching-feng, who was a mentor to Ting Yün-p'eng, painted in 1597 a hanging scroll which provides us with another opportunity to closely examine the genesis of the Anhui School.[52] This hanging scroll (Pl. 16), like Ting Yün-p'eng's, uses the compositional scheme of Ni Tsan as its point of departure. Again, one can see the attempt to resolve the artistic problem of achieving structural clarity and a sense of monumentality at the same time through the building up of a few simple type-forms. The dry and spidery brush-strokes tend to give an insubstantiality to the forms. The foreground tree and solitary figure futher impart a sense of sparseness to the painting. The mountain masses in the far distance contribute to a high degree of linear rhythm and angularity. Chan Ching-feng seemed to be less interested in an outright duplication of Ni Tsan's manner than in creating his own interpretation. He was conscious of his inventiveness when he wrote in the inscription that, though Ni Tsan did not include figures in his painting because he felt there was no true man in this world, he now included a figure because the recipient of the hanging scroll was a true man. Beyond this ostensible reason, it seems that the artist's inclusion of the figure, more than the same modification of the Ni Tsan mode in Shen Chou's "Walking with a Staff," is meant to emphasize the powerful scale of nature.

In his youth, Hung-jen might have come into contact with Ch'eng Chia-sui 程嘉燧 (1565–1643), the eminent poet and painter from She-hsien. Ch'eng

Chia-sui once praised Hung-jen for having captured the spirit of Ni Tsan.[53] Though he lived in Chia-ting and did not return to She-hsien until toward the end of his life, Ch'eng Chia-sui's eminence as a poet and as a painter, who commanded respect for his moral character, his poetic sensibility, and his transformation of the mode of Ni Tsan, must have exerted considerable influence on artists from Anhui. His mature style is characterized by a clarity achieved through clean and spare forms and a tranquil and restrained expression. The lucidity and simplicity in his style can be seen, for instance, in his hanging scroll entitled "Horse-riders on the Willow Bank" (*Liu-t'i k'uan-ma*柳堤款馬) dated to 1639 and now in the T'ien-tsin Museum (P1. 20).[54] The elongation of the composition is typical of the early seventeenth century and especially of the Wen Cheng-ming School. Ch'eng Chia-sui seems to be more interested in the building up of an abstract and relatively flat overhanging cliff that dominates the whole painting. At the same time, however, he maintains a structural clarity. The same style could be seen later in Hung-jen's 1663 landscape now in the Liaoning Provincial Museum (P1. 21).

Ch'eng Chia-sui's 1639 album, now in the National Palace Museum in Taipei, also provides us with several examples for a study of the early stages of the Anhui School.[55] The album is, in general, characterized by dry brushwork, angular forms, and simple composition. Like the 1639 hanging scroll just mentioned, these small paintings are often concerned with some kind of human activity. In particular, in leaf no. 9 from the album, the angular overhanging cliff rises through layers of charcoal-like crumbly brushwork. The rocks are minimally modelled, thus giving the impression of being flat. The overlapping of chute-like rocks seems to suggest a recession to the far distance on the left where a crevice opens between the rock walls. On the other hand, the cliff seems to hang over the two horse-riders in the foreground. As we shall see, the interplay between two-dimensional and three-dimensional forms was to become an important feature of Hung-jen's style.

It was in the same year, in 1639, when Hung-jen and four other artists from Anhui painted the joint handscroll now in the Shanghai Museum (P1. 23). The four additional artists were Li Yung-ch'ang 李永昌, Wang Tu 汪度, Liu Shang-yen 劉上延, and Sun I 孫逸.[56] It was painted for a certain Li Sheng-pai 李生白 on his fortieth birthday, and Hung-jen's section was signed "Chiang T'ao," Hung-jen's secular name before he became a monk (P1. 49). Thus the joint handscroll provides us with an excellent opportunity to study Hung-jen's immediate stylistic context, particularly during his early years when he had perhaps more chance than he did after becoming a monk to meet other literati. Liu Shang-yen, who participated in the 1639 joint handscoll, was from She-hsien but later lived in Wu-hu.[57] Apart from his section in the joint handscroll, his paintings are almost nonexistent. His style is represented by a leaf from a joint album painted for Ch'en Ying-ch'i 陳應頎 from Wu-hu to which Hung-jen also contributed one leaf.[58] Liu Shang-yen's leaf is not dated, but as other leaves are from 1647 to 1653, it is likely that it can be placed within this time. It is a simple picture in the manner of Ni Tsan and depicts a sparse scene. The brushwork is dry and crumbly, and there is little interior texture on the rocks and mountains. The untextured tree trunks in the foreground enhance the linearity and sparseness of the painting.

Like Liu Shang-yen, the paintings by Li Yung-ch'ang whom Kung Hsien praised for his importance in the "T'ien-tu School," are extremely rare.[59] Apart from the section in the 1639 joint handscroll, a hanging scroll entitled "In the Manner of Yüan Masters" (*Fang Yüan-jen pi-i* 仿元人筆意) formerly in the Huang Pin-hung collection[60] provides some insight into the stylistic ambience of Hung-jen's early years. Judging on the basis of the poor reproduction, the 1634 hanging scoll (P1. 19) suggests some similarity to the late Ming followers of Wen Cheng-ming. Compare, for instance, the Li Yung-ch'ang hanging scroll with Hou Mou-kung's 侯懋功 (active late sixteenth century) undated hanging scroll also entitled "In the Manner of Yüan Masters" in Taiwan (P1. 18).[61] Both paintings emphasize surface design; minute details crowd the picture. There is a strong and swaying movement from the foreground to the background.

The Liu Shang-yen album leaf and the Li Yung-ch'ang hanging scroll provide us with examples of two of the main stylistic currents in the Hui-chou areas. On the one hand, there is, in the Liu Shang-yen album leaf, reduction of details and the emphasis on contour lines. On the other hand, in the Li Yung-ch'ang hanging scroll, the tension and restlessness on the picture plane are reminiscent of Wen Cheng-ming and his followers from the Su-chou area.

For a little more than ten years after 1369, there is a hiatus in Hung-jen's artistic production. Nothing seems to have survived. Part of the reason may be the turmoil before and following the fall of the Ming Dynasty. In the meantime, other artists in Anhui further developed the style of the older generation, e.g. Ch'eng Chia-sui. A comparison of Ch'eng Chia-sui's 1639 hanging scroll "Horse-riders on the Willow Bank" (P1. 20) and Hsiao Yün-ts'ung's 1648 hanging scroll "Walking with a Staff in a Sparse Forest," both in T'ien-tsin (P1. 26),[62] suggests that a characteristically sparse and linear style was current in Anhui during this time. The overhanging cliff in Hsiao Yün-ts'ung's painting is done with a minimum of interior texturing and therefore appears linear and austere. The cliff's sheer size and its towering position create a skeletal mountain shape that is unable to sustain any mass or weight; it seems ready to crumble at any time. The foreground trees and the figure underneath them suggest a sparseness often found in this type of composition ultimately derived from Ni Tsan.

Perhaps because of its stylistic affinity to the Hsiao Yün-ts'ung hanging scroll, the undated hanging scroll of landscape by Wang Chih-jui (P1. 22), a pupil of Li Yung-ch'ang, can be placed with some confidence to about his time, i.e., the late 1640s. The details of Wang Chih-jui's life are little known, but he seemed to be active from about the late 1640's to the early 1650's.[63] His paintings are hard to find, and so far only three are available for study: the undated hanging scroll, a leaf dated 1649 from the joint album for Ch'en Ying-ch'i already mentioned, and an undated leaf.[64] The hanging scroll (P1. 22) immediately recalls the tradition of Ni Tsan. The prominent position of the line of trees in the foreground greatly contributes to the sense of simplicity and spareness; most of the trees are leafless and their verticality is emphasized. As in Liu Shang-yen's album leaf, the untextured tree trunks in the foreground enhance linearity and spareness. The distant mountains are built up with dry and apparently angular brushwork; the plateaux and the cliffs form a pattern of verticals and horizontals. The dual spatial organization of Ni Tsan, with the lower half of the hanging scroll flat and the upper half having spatial recession, a feature we also noted in Lu Chih, was successfully captured. Like Chan Ching-feng's 1597 hanging scroll (P1. 16), this painting is a good example of the formal construction of landscape by Anhui artists by means of expanding upon the full as well as the sparse aspects of the mode of Ni Tsan; it must ultimately owe a great deal to Suchou traditions.

Wang Chih-jui was quoted by the eighteenth-century critic Chang Keng as saying: "A richness that is not the result of plethora and a spareness that does not originate in paucity--these are the desiderata in painting."[65] The 1649 album leaf, small as it is, suggests that Wang Chih-jui practiced what he preached (P1. 24). The compact composition is divided into two halves by a diagonal sheet of blank paper that we read as water extending from the lower left corner to the middle of the right-hand border of the painting. In the foreground on the right, three tress are sketchily drawn with broad but crumbly brushstrokes; a simple bridge links the foreground to the masses of rocks on the opposite bank. The major rock formations consist of three slanted and plateau-topped promontories. The plateaux are mainly left untextured, although their edges are quite simply accented with a few spiky vertical strokes to indicate vegetations. To the left, three cubic rocks overlap one another. There is little spatial recession and the sense of the paper surface is very strong. The forms are mostly angular and abstract. The configuration is compressed and, like his undated hanging scroll, there is a strong sense of sparseness.

These works by Liu Shang-jen, Hsiao Yün-ts'ung, and Wang Chih-jui are but a few examples of a predominant tendency in early seventeenth-century painting in Anhui, that is, the reductionist approach to forms which are usually angular and geometric and drawn with dry brushwork and little interior texture. This aesthetic ideal is well described by Cahill: "Paintings of this kind implied a degree of rejection of tangible reality, and offered imaginative escape into worlds that made no strong demand on the senses or emotions. They appealed to an eye in which non-involvement became, for many people, a pervasive ideal."[66] In a statement that spells out the aesthetic basis of his own pristine and bare forms, the contemporary artist Yün Hsiang 惲向 (1586-1655) wrote:

> Painters hold simplicity and cleanness to be superior qualities. Simplicity means being simple in your images, not in your ideas. The utmost simplicity is really the utmost in rich beauty. Cleanness means wiping away all the clouds and mists, leaving only loneliness and nobility.... Some take an extreme sparsity of brushstrokes for simplicity, but that is not it.[67]

These remarks echo Wang Chih-jui's two desiderata in painting: "a richness that is not the result of plethora and a sparseness that does not originate in paucity." This aesthetic canon, when pursued to its logical conclusion, however, can result in an extreme and minimal painting such as Wang Chih-jui's undated album leaf (P1. 25), especially in its planar arrangement of insubstantial forms sketched out with crumbly brushstrokes.[68] As we shall see, Hung-jen was to develop from this sketch-like planar arrangement of forms to a more constructive and lucid articulation of forms.

In 1651, Hung-jen painted a section with other artists from Anhui in the joint handscroll entitled "Joint Handscroll of Landscape for Ko-an" (*Ko-an shan-shui-ho-chin chüan* 格庵山水合錦卷). Its whereabouts is not known (Pls. 31-39, 49-51).[69] This joint handscroll was painted at the request of Fang Shih-yü 方式玉 (died 1654) to present to a certain Mr. Ko-an 格庵. The artists, in the order that they appeared, include Wang Chia-chen 汪家珍, Wang Jan 汪然, Chu

Ch'ang 祝昌, Chang An-pao 張安苞, Wu Hsin 吳星, Hung-jen, Ho Yü 何顒, and Wang Hsüan-tu 王玄度.[70]

The large number of artists that participated in this joint handscroll offers an excellent opportunity to study the stylistic context of Hung-jen. At least five out of the eight artists came from Anhui. As a whole, the joint handscroll exemplifies stylistic currents that were known to Hung-jen. The section by Wang Chia-chen (P1. 31) is, according to the artist's inscription, painted in the manner of Wang Wei. Stylistically, however, it is closer to Ch'eng Chia-sui in its dry and refined brushwork, slender and attenuated but vigorous trees, and the angularity of its rock formations (cf. P1. 20). The section by Wang Jan (Pls. 32-33), a cousin of Wang Chia-chen's, is apparently derived from Huang Kung-wang in the dry, long drawn-out brushstrokes in depicting the flat-topped eroded banks, the building-up of the mountain masses, and the sense of movement. Chu Ch'ang's section (P1. 34), in its tendency to overlap the rather flat, almost rectangular and geometric rock formations which have very little interior texturing, is similar to the woodcut print series entitled "Landscapes of T'ai-p'ing" (*T'ai-p'ing shan-shui t'u* 太平山水圖) designed by Hsiao Yün-ts'ung and published in 1648 (Pls. 13-15).[71] Exhibiting the same interest in the interplay of shapes, Chang An-pao's painting (Pls. 35-36), however, is static and schematic. Painted with wet ink, the section reveals a conscious formal construction (e.g. strong contrast of light and dark, of horizontal and vertical).

Among Hung-jen's predecessors in Anhui are artists whose style is clearly influenced by the sixteenth-century master Wen Cheng-ming and his followers in the early seventeeth century. Indeed, the style of Wen Cheng-ming and his followers was so pervasive that, despite the efforts of such an assertive iconoclast as Tung Ch'i-ch'ang (whose style was nevertheless conditioned by the Wen Cheng-ming school),[72] a great number of artists, including even an "individualist" like Hung-jen, were bound to fall under its influence. Chan Ching-feng, whose 1597 hanging scroll (Pl. 16) has been briefly discussed as a precedent of the Anhui School style, is said to have followed the style of Wen Cheng-ming.[73] Li Yung-ch'ang's stylistic affinity to the Wen Cheng-ming school is demonstrated in his 1634 hanging scroll (Pl. 19). Ch'eng Chia-sui undoubtedly derived his elongated hanging scroll format and piling-up of forms from the late works of Wen Cheng-ming and some of Wen Cheng-ming's followers. The wood-cut prints of design for ink-cake composed by the little known Hui-chou artist Yü Chia-shen 俞嘉諶 and published in 1589 (Pl. 12) also suggest the legacy, though on a more popular level, of the late Wen Cheng-ming style and of Wen Chia.[74] This is evident when one compares Yü Chia-shen's designs with, for example, Wen Cheng-ming's 1548 handscroll entitled "The Red Cliff after Chao Po-chü" and the 1548–50 hanging scroll entitled "A Thousand Cliffs Contend in Splendor," both in the National Palace Museum in Taiwan,[75] and Wen Chia's 1555 hanging scroll entitled "Temples on Immortals' Mountains" in the Ching Yüan Chai collection.[76]

In addition, Sun I has been called the "reincarnation of Wen Cheng-ming."[77] Like Li Yüng-ch'ang, Sun I also participated in painting the 1639 joint handscroll in Shanghai and his works are similarly hard to find today. Sun I was in all probability older than Hung-jen and was often mentioned together with Hsiao Yün-ts'ung as "Sun-Hsiao" by his contemporaries. His earliest dated painting, a landscape in colors painted in 1643 and now in the Palace Museum, Peking, is indeed derived from the Wen Cheng-ming School.[78] Moreover, as pointed out in Chapter I, Sun I is said to have copied T'ang Yin's paintings. His indebtedness to Wen Cheng-ming can be readily seen in a comparison between Sun I's hanging scroll entitled "Flowers at End of Spring" (*Mu-ch'un hua-shih* 暮春花事) dated 1654 and formerly in the Huang Pin-hung collection (Pl. 40),[79] and Wen Cheng-ming's 1507 hanging scroll "Spring Trees after Rain" (*Yu-yu ch'un-shu* 雨餘春樹) now in the National Palace Museum (Pl. 46).[80] Both paintings are characterized by a strong sense of surface, a measured rhythm in the arrangement of forms, and a fastidiousness in the execution of details.

In the 1654 hanging scroll by Sun I, a dramatic quality is achieved through the juxtaposition of the prominent dike-path and the over-hanging cliff. The top of the dike-path and the plateaux is left untextured. However, the overhanging cliff itself receives only a minimum of interior texture. The sharp juxtaposition of the dike-path and the cliff is echoed by the intertwining pine trees at the lower right corner. The sparseness of the picture is relieved only by the measured rhythm created by three groups of deciduous trees accentuating the dike-path in the middle ground. Under these trees and on top of the tilted-up plateaux are stick-like trees.

As we shall see in Chapter III, it is in Sun I's 1657 hanging scroll "Cinnabar Peak" in the Chien-lu collection (Pls. 41–45), however, that we find the closest example of the stylistic affinity between Sun I and Hung-jen. The hanging scroll begins at the lower right-hand corner with a figure on an elevated dike-path. Two deciduous trees grow on the steep wall of the dike-path; in more accurate terms, the trees are only attached precariously to the dike-path; A few houses drawn with straight lines can be seen beyond the foreground expanse of water and the prominent flat repoussoir rock. Except for the space created by the group of trees around and above these houses, the picture appears rather transparent. Its basic structure is created by overlapping flat areas outlind with fluent and dry, sometimes split ("flying white"), brush strokes. These areas have a minimum of interior texure and are accentuated by a few dots. The spiky pine trees are drawn with a similarly fragile brushwork that reinfores the linear but tightly woven structure of the mountain masses.

An examination of Hung-jen's sixty-leaf album "Mt. Huang" (Pls. 65-124), to be discussed in detail in Chapter III, indicates Hung-jen's sources. By illustrating specific scenes at Mt. Huang and naming each site, Hung-jen was simply following a tradition which can be traced back to the T'ang Dynasty. The famous "Wang-ch'uan Villa" (*Wang-ch'uan t'u* 輞川圖) handscroll, attributed to Wang Wei 王維 (699-759) and later copied by hand or preserved in engraving, is but one early example of that tradition.[81] Another example is the series of paintings entitled "Ten Views of a Grass Hut" (*Ts'ao-t'ang shih-chih* 草堂十志) attributed to Lu Hung 盧鴻 (active ca. 713-741) and now in the National Palace Museum in Taiwan.[82]

In Wang Wei's painting, there is an attempt to isolate certain objects. Each site is identified by a label beside it. In Lu Hung's album, each section depicts an individual spot around the thatched studio of Lu Hung; again a descriptive label is provided for each site. Lu Hung's figures often are either placed in centralized diagrammatic enclosures that are uptilted, or framed by trees that form pairs, for emphasis and clarity. The brushwork in the Lu Hung, which is very likely a Northern Sung copy, is dense, heavy, and very different from Hung-jen's drier and more linear brushwork. As we shall see in Chapter III, however, Hung-jen's isolation of objects and lucid graphic design seem to be related to the T'ang style.

The closest precedent to the Hung-jen album is perhaps the album depicting Mt. Hua 華 in Shensi painted in the late fourteenth century by Wang Li 王履 (born 1322).[83] The album originally consisted of about forty leaves and is dated to 1383-84. Only fourteen leaves have survived and are now in the Palace Museum in Peking and the Shanghai Museum. Judging from the surviving leaves, it can be said that the album as a whole shows the artist's synthesis of Northern Sung monumental style, Southern Sung naturalism, and the more subjective and linear mode of the Yüan Dynasty masters. In other words, the images of Mt. Hua as depicted by Wang Li were circumscribed by Wang Li's experience of paintings from previous times. Although Wang Li wrote in his inscription on the album that an artist should learn his art not so much from the ancients or from nature as from his own mind, it seems that mind did not completely account for his style; rather, it was through a combination of experience of nature and experience of art that Wang Li was able to create his style. As we shall see, the same artistic process can be discerned in Hung-jen's album of Mt. Huang.

The interest in depicting actual scenery was revived by Shen Chou and followed by many Suchou artists of the sixteenth and early seventeenth centuries, including Lu Chih, Wen Chia, Wen Po-jen 文伯仁 (1502-1575), Ch'ien Ku 錢穀 (1508-1572), Chu Lang 朱朗 (active mid-sixteenth century), and Shen Shih 沈仕 (active mid-sixteenth century). The predilection for a close-up view and the rendering of the external world through a combination of personal observation and stylistic convention in Hung-jen's album of Mt. Huang can be compared with the same traits in several different paintings by Suchou artists. Among the latter are Shen Chou's albums "Famous View of Two Rivers" now in the Shanghai Museum[84] and "Twelve Views of Tiger Hill" in the Cleveland Museum of Art,[85] Lu Chih's 1554 album entitled "Journey to Mt. Pai-yüeh" now in the Fujii Yurinkan Museum, Kyoto,[86] and Ch'ien Ku's album "Eight Views of Stone Lake" now in the National Palace Museum in Taiwan.[87] Hung-jen's combination of descriptive directness with elements of styles derived from earier artists is particularly similar to Lu Chih's approach in his album of Mt. Pai-yüeh which, incidentally, is part of the Mt. Huang complex.[88] On the other hand, the dramatic thrust of cliffs and the dynamic rhythms of Shao Mi 邵彌 (active ca. 1620-1640), a more innovative artist from Suchou, also had certain repercussions in several leaves in Hung-jen's album.[89]

An illustrated guidebook to famous mountains entitled *Ming-shan-t'u* 名山圖 published in 1633 is nearly contemporary with Hung-jen. The guidebook contains wood-cut prints to accompany the text entitled "Grand Views of Famous Mountains under Heaven" (*T'ien-hsia ming-shan sheng-kai chi* 天下名山勝概記). Among the fifty-five pictures it contains, there is one of Mt. Huang designed by Cheng Chung and one of Mt. Pai-yüeh by Wu T'ing-yü.[90] Cheng Chung, who was from She-hsien, worked as a professional painter and was very much influenced by the Wen Cheng-ming School. This influence can be seen in his 1632 handscroll, now in the Ch'eng Ch'i collection in Tokyo, which is characterized by the use of boulder-strewn mountains rising on overlapping diagonals and dramatic and abstract patterns on a tightly wound, compressed, and often tense pictorial surface.[91] His design depicting Mt. Huang published in the *Ming-shan-t'u* (P1. 240) is so similar to several leaves from Hung-jen's album that Hung-jen might have seen Cheng Chung's work.[92] Compare, for instance, Cheng Chung's rendition with Hung-jen's leaves nos. 1, 2, 18, 35, and 40 (Pls. 65, 66, 82, 99, 104).

It is also possible to relate Hung-jen's album to the work of Hsiao Yün-ts'ung. Significantly, Hsiao Yün-ts'ung wrote a colophon for the album in 1665 praising Hung-jen and saying: "I am an older painter. In the art of painting, I have not yielded my claims to any predecessors. But when I see this album, I cannot help but withdraw my hand."[93] A pupil-teacher relationship between Hung-jen and Hsiao Yün-ts'ung was suggested by Ts'ao Yin 曹寅 (1659-1712) in a colophon attached to Hung-jens's 1657 hanging scroll "Ten Bamboo Studio" (*Shih-chu-chai t'u* 十竹齋圖) (P1. 167).[94] This view, however, has not been confirmed by any other sources. An undated colophon by T'ang Yen-sheng to Hsiao Yün-ts'ung's 1656 handscroll "Cloud Terrace and Sparse Trees," now in the Nanking Museum, merely suggests that Hsiao Yün-ts'ung, Hung-jen, and Sun I were friends.[95] Furthermore, the fact that Hsiao Yün-ts'ung addressed Hung-jen as "Master Chien" [Chien-kung 漸公] in his 1665 colophon to the album of Mt. Huang suggests that Hsiao Yün-ts'ung regarded Hung-jen as his equal rather than as his pupil. In any case, there seems to be no evidence to indicate that Hung-jen studied with Hsiao Yün-ts'ung, although there is no doubt that they knew of each other's work.

Hsiao Yün-ts'ung seemed to begin his schooling with the archaistic style, for his earliest dated painting is a blue-and-green hanging scroll now in the Palace Museum in Peking.[96] His archaistic leaning can also be seen in his long colophon on a blue-and-green handscroll by the early Ming artist Shih Jui 石鋭 (active ca. 1425), now in the Cleveland Museum of Art.[97] In his colophon, however, Hsiao Yün-ts'ung misattributed the handscroll to the T'ang artist Li Chao-tao 李昭道 (active ca. 670-730); perhaps one should regard this misattribution not as a flaw in Hsiao Yün-ts'ung's connoisseurship but as the artist's homage to the ancient master of the blue-and-green mode. His interest in the archaistic and angular style is further exemplified in two leaves from an album dated 1645 and now in the Chih-lo-lou 至樂樓 collection in Hong Kong.[98]

The attempt by Hsiao Yün-ts'ung to achieve a linear and sparse landscape has already been noted in our discussion of his 1648 hanging scroll in the T'ien-tsin Museum (P1. 26). It is, however, his 1648 album of landscape woodcut prints entitled "Landscape of T'ai-p'ing" (discussed above in connection with the 1651 joint handscroll for Ko-an) that bears fruitful comparison with Hung-jen's album of Mt. Huang (Pls. 13-15). Like Hung-jen, Hsiao Yün-ts'ung depicted specific scenes in the area of T'ai-p'ing near Hui-chou. As might be expected from woodblock illustrations, the pictures are strongly episodic. The episodic content is often enhanced by the dramatic undercut escarpments. The mountain masses are generally marked by angular rock formations repeated over and over again. This patten of repetition is also a means of portraying the vegetation motifs. Diminutive figures, alone or in groups, wander through the landscape, or are engaged in their respective activities such as travelling, farming, or fishing, seemingly oblivious of their surroundings. Each scene is rendered in an idealized and idyllic manner, an intention made explicit in the inscribed poems, for these poems were written in various dynastic periods and often described unrelated sites or occasions.

More than the album "Landscapes of T'ai-p'ing," Hsiao Yün-ts'ung's handscroll entitled "Travellers in Autumn Mountains" dated to about 1647 (according to the artist's colophon added in 1657) now in the Tokyo National Museum (Pls. 27-28)[99] suggests his indebtedness to the Wen Cheng-ming School. One finds frequent references to the stylistic features of the Suchou artists such as the attenuated pine trees, the soft and light color schemes (mostly in delicate mauve or pale gray-purple), the strong episodic elements, and the atmospheric effects. Hsiao Yün-ts'ung's re-interpretation of the Wen Cheng-ming mode seems to have provided Hung-jen with a readily available vocabulary.

Hung-jen's style can ultimately be traced to Wen Chen-ming in his similar emphasis on surface design and attention to descriptive details. However, he seems to have derived his abstraction and pictorial tension primarily from Wen Cheng-ming's followers, particularly those who were active during the late sixteenth and early seventeenth centuries. For instance, Chang Fu 張復 (1546-after 1631), one of Wen Cheng-ming's more conservative followers, was interested in the reductionist approach to composition as well as in geometric forms and anticipated Hung-jen' style. For instance, in his 1596 hanging scroll entitled "Landscape with Waterfall" in the Ching Yüan Chai collection (P1. 238), Chang Fu used diluted washes and a minimum of interior texture on his flat and angular mountain forms that are unrelieved by any pronounced tonal content. The painting as a whole is marked by a flattened pictorial space.[100] As we shall see, Hung-jen was to expand on this kind of style and infuse it with his own more lucid and austere vision.

Judging from available literary and visual evidences, there seems little influence from Tung Ch'i-ch'ang on Hung-jen. Conspicuously absent from Kung Hsien's statement on the origins of the Anhui School with which we opened this chapter is the name of Tung Ch'i-ch'ang. Hung-jen himself never mentioned Tung Ch'i-ch'ang in his writings. It is true that Tung Ch'i-ch'ang exerted tremendous influence on later Chinese painting. Both Wang Shih-min 王時敏 (1592–1680) and Chu Ta 朱耷 (1625–after 1705) not only acknowledged his influence in their writings but also demonstrated it in their paintings. Tung Ch'i-ch'ang's influence has in our time led Sherman E. Lee to maintain that "there is no doubt that we must speak of later Chinese painting as before Tung Ch'i-ch'ang and after Tung Ch'i-ch'ang."[101] It has been customary among scholars to view much of later Chinese painting as growing out of Tung Ch'i-ch'ang. For example, Cahill has credited him with "stylistic innovations···· so radical····[and]manipulation of natural form so extreme, that he opened an age of individualism at the same time that he established a new orthodoxy." With that generalization he then goes on to suggest that "the Anhui masters carried to its logical conclusion one aspect of Tung Ch'i-ch'ang's new landscape, its sparse, 'stripped down' quality"and that Tung Ch'i-ch'ang "is responsible for the most significant among the derived elements of style" in Hung-jen.[102]

However, whether or not Hung-jen derived his style from Tung Ch'i-ch'ang needs to be clarified. Arguing for the stylistic affinity between Hung-jen and Tung Ch'i-ch'ang, Max Loehr has written of one leaf from Hung-jen's sixty-leaf album depicting Mt. Huang (Pl. 103):

> It is composed of units of slopes and boulders which are repeated all over and lacks all individuality. Their ordered relationship is all that counts. A stream is recognizable as such only through the boulders strewn in its path. As an image of a ravine, the picture is inexpressive. As a graphic design, on the other hand, it holds well, regardless of what it represents. Its basic orientation is that of a structural study, well compatible with the tendency observable in the work of Tung Ch'i-ch'ang. It does not appear to be either a transposition of an older picture or study from nature, but rather an attempt to come to terms with the functional effect of commonplace and conventional elements of landscape design.[103]

Despite the formal similarity noted by Loehr between Hung-jen and Tung Ch'i-ch'ang, Hung-jen's style is generally derived from a different source and intended for a different artistic purpose.

As we have suggested, the graphic design on the pictorial surface in Hung-jen grew rather out of the mode of Wen Cheng-ming . A comparison of

Hung-jen's handscroll "Bamboo Banks and Reed Shores" (*Chu-an lu-p'u* 竹岸蘆蒲), dated 1652 and now in the Sumitomo collection (Pls. 60-63), with Wen Cheng-ming's 1558 hanging scroll "Red Cliff" in Detroit, reveals Hung-jen's stylistic indebtedness to Wen Cheng-ming.[104] In both we can find the simultaneous presence of surface pattern and spatial recession, the conscious sense of design, the cool open serenity, and the reduction to essentials; a detailed analysis will be given in Chapter III. Hung-jen's sixty-leaf "Mt. Huang" album, moreover, contains many leaves that can be related to the mode of Wen Cheng-ming. For instance, leaf no. 33 (Pl. 97) "White Sand Mountain-range" (*Pai-sha-ling* 白砂嶺) is, like Sun I's 1657 hanging scroll "Cinnabar Peak" in the Chien-lu collection (Pl. 41), derived from Wen Cheng-ming in its strong sense of surface with limited recession, its measured rhythm in the arrangement of forms.

Furthermore, in these two examples by Hung-jen, we can see that, unlike Tung Ch'i-ch'ang's more abstract and formal construction of landscape, Hung-jen's art is more intimately related to his experience of landscape at Mt. Huang. For example, Tung Ch'i-ch'ang painted a hanging scroll "Landscape," now in the Drenowatz collection, after his visit to the tomb of a Ming Emperor. He wrote in his inscription: "After my return from a ceremony at the Imperial Mausoleum in the ninth month of 1624, I have painted what I saw." What we see in the painting, however, has nothing to do with the Imperial Mausoleum; rather, it is a display of the richness of his brushwork and the arbitrary manipulation of his abstract forms. The painting thus best exemplifies Tung Ch'i-ch'ang's strong interest in abstract formal exploration and his disregard for representational elements of nature.

In his theoretical writings, Tung Ch'i-ch'ang also advocated the formalistic approach to landscape painting. For instance, he wrote:

> In painting landscapes, it is necessary to understand the principle of 'proportioning' and 'uniting' [*fen* 分 and *ho* 合]. The 'proportioning' brushstrokes mark the general framework of the composition. There is the 'proportioning' of the whole painting and 'proportioning' within each individual section. If one understands these aspects, one has already grasped the larger half of the way of painting.

He also wrote:

> The form of the mountain should first be outlined and then texture strokes applied. People nowadays pile up small bits to make up a large

> mountain. This is one of the worst mistakes. When an ancient master worked on a large scroll, he made only three or four large 'proportionings' and 'unitings,' and in that manner accomplished the whole composition. Although within the composition there are many small parts, the principal aim is to grasp the momentum [or force, *shih* 勢] of the forms.[105]

Tung Ch'i-ch'ang's significance in both theory and practice has led observers to slip into the easy generalization that he influenced almost everybody after him.

Contrary to Loehr's assertion, the album "Mt. Huang" and many other Hung-jen paintings were intended as depictions of real scenery or "studies of nature," however formalized they may be. For instance, in leaf no. 3 (Pl. 67) depicting "Cloud-Gate Peak (*Yün-men feng* 雲門峰)" at Mt. Huang, the two peak form a gate through which the cloud passes. Between the two peaks, built up with angular rock formation defined by sharp contour lines and little interior texture, Hung-jen inscribed the name of the site as if to draw the viewer's attention to the specificity of his subject, the concrete existence of the peaks and the "gate."

Through concentration and isolation, Hung-jen often brought out most clearly the specific site he was depicting. In leaf no. 8 (Pl. 72), the Tz'u-kuang Temple (*Tz'u-kuang-ssu* 慈光寺) is brought more into focus through the screen of pine trees in the foreground and the surrounding mountains. In leaf no. 35 (Pl. 99), the two diagonals of mountains converge at the lower edge of the picture to highlight the Ts'ui-wei Temple (*Ts'ui-wei-ssu* 翠微寺). The isolation of a central motif, the sense of concentration, can also be seen in Hung-jen's 1658 short handscroll "Plum-Blossom Pavilion" (Pl. 169).

Hung-jen's impressive hanging scroll "Pines and Rocks at Mt. Huang," dated 1660 and now in Shanghai (Pls. 183–187), indicated that he was more concerned with the illustrative content of his subject matter than with the momentum (*shih* 勢) of abstract forms as advocated by Tung Ch'i-ch'ang. Hung-jen's compression of his subject onto a two-dimensionally conceived painting results in great clarity, which intensifies what he had to say about the perseverance of the pines precariously perching on the steep cliffs that seem to reach the sky.

Thus, whatever "structural study" one may find in Hung-jen that is, according to Loehr, "compatible with the tendency observable in the work of Tung Ch'i-ch'ang," Hung-jen seemed to have developed it without necessary connection to Tung Ch'i-ch'ang. Hung-jen's invention will become clear as we turn to an examination of his paintings in Chapter III.

CHAPTER III

THE PAINTINGS OF HUNG-JEN

It is difficult to divide Hung-jen's artistic career into well-defined stylistic periods, mainly because most of his reliable dated paintings, except a few earlier paintings, are from a relatively short time-span, that is, a twelve-year period, from 1651 to 1663, and secondly because there is little stylistic change during that period. Although it is possible to place a great number of undated paintings within that period, often on stylistic grounds alone, the dating can only be approximate. Furthermore, in the course of conducting the present study, it has not been possible to examine all of his extant paintings, which means the conclusion reached here can only be incomplete. Within the range of material now available, however, it is hoped that the present study will lay the foundation for a fuller, more comprehensive study in the future.

The earliest dated painting by Hung-jen is a small fan in light colors on gold paper and mounted as an album leaf, now in the Shanghai Museum.[1] Entitled "Quiet Living in an Autumn Mountain" and dated 1634, it exhibits a strong influence of earlier Anhui artists such as Chan Ching-feng and Liu Shang-yen. Another early work is his section in the 1639 joint handscroll presented to a certain Mr. Li Sheng-pai 李生白 on his fortieth birthday, now in the Shanghai Museum (Pl. 49). Hung-jen was then twenty-nine years old. The painting is above all a modest one. The center of the composition is occupied by a steep-sided bluff that almost touches the upper edge of the scroll, similar to leaf no. 9 from Ch'eng Chia-sui's 1639 album in Taipei, mentioned earlier in relation to Hung-jen's backgrounds. A zigzag dike-path leads the viewer's eye from the right of the painting to the bluff and then behind it. After that brief hiatus, the path reappears at the left where a rock blocks its continuation, thereby leading the eye back to the central bluff. The miniature open pavilion on the dike-path is derived from the same toy-like motif of Ni Tsan. In combination with the two pine trees on the bluff, the pavilion provides a rhythmic interval without disturbing the essentially stable composition. In short, the quiet and static composition and its angular shapes and dry brushwork reveal that it is a carefully wrought work. However, certain motifs such as the raised dike-path and the angular shapes do suggest the direction Hung-jen was to take in his later years.

From 1639 to 1651, a twelve-year time span, no reliable dated paintings by Hung-jen have survived. Hung-jen's section, dated 1651, in the joint handscroll *Ko-an shan-shui ho-chin chüan* (Pl. 50–51), in its more tightly structured composition, is more interesting than the Shanghai painting of 1639. On the left, the raised dike-path is steep and it zigzags as it tilts upward. The dike-path is much more elevated than the same path in the Shanghai handscroll. In addition, the open pavilion and two spare trees, perching precariously on the path as it suddenly goes up, impart a sense of drama; the Shanghai handscroll, by comparison, appears more static.

In the middle ground and to the right is the main mountain brought to the front of the picture plane. The mountain masses are constructed out of a series of overlapping areas; Hung-jen's interest in two-dimensional form is already present. On the far right, a pointed mound is placed beside a group of small boulders and is combined with a flat-topped plateau. The linear and angular forms and the dramatic disposition of these forms were to characterize Hung-jen's later paintings.

Stylistically close to this painting is Hung-jen's undated leaf (Pl. 48) from the joint album for Ch'en Ying-ch'i. From the lower right-hand corner, a man walks towards the middle of the foreground where a steep rock formation arises, seemingly blocking his way. Then, immediately, behind the steep conical rock appears a zigzag dike-path which enters from the lower edge of the painting and leads diagonally and precariously to the upper right-hand corner of the painting. The characteristic raised dike-path is minimally textured on its side while its surface is left untextured. On the left half of the painting, a cave with overhanging spiky rocks is topped by another similar cave, thereby imparting a dramatic effect to the composition. Behind the cave arises a steep cubic mountain mass. To its left, an empty and see-through straw-hut is flanked by the same firmly drawn pine trees as in the lower right-hand corner. The straw-hut is situated on a slope which is left almost untextured. The brushwork of the painting is mostly dry.

Within a small format, Hung-jen placed towering forms. Two diagonals are set up by the tall mountains and the elevated dike-path. The space is shallow; no room is available for a recession with so many forms supporting each other. The effect of fantasy—particularly because of the dike-path and the cave—might have been a result of Hung-jen's experimentation with the fantastic style of Wu Pin 吳彬, the late Ming artist from Fukien where, one recalls, Hung-jen stayed from about 1646 to 1651. It is possible that Hung-jen's leaf might have been painted around 1652, presumably when Hung-jen was visiting Wu-hu where Ch'en Ying-ch'i lived.

Closely related to the album leaf for Ch'en Ying-ch'i is an undated leaf (Pl. 52). Compositionally it is no less complex, but here Hung-jen filled the entire picture plane with a different type of rectangular forms. There is again a crossing of diagonals to organize the composition—both diagonals are anchored in the lower corner. The forms are somewhat more textured and cluttered; overlapping angular shapes dominate the mountain mass. The picture, however, is more spacious since the foreground houses and trees are set back from the front; the viewer's eye moves from the pavilion on the left, to the two adjoining houses on the right, and then to the empty pavilion on the other bank. The small size of the pavilion on the zigzag dike-path in the background suggests a spatial recession. Thus, like the album leaf for Ch'en Ying-ch'i, the juxtaposition of the steep dike-path on the right and the flat rocky wall on the left reveals Hung-jen's interest in the interplay of two-dimensional and three-dimensional forms.

Hung-jen's handscroll entitled "Bamboo Banks and Reed Shores" (*Chu-an lu-p'u* 竹岸蘆蒲), dated to the ninth month of 1652, is now in the Sumitomo collection (Pls. 60-63). The artist's inscription tells us that he was depicting the scenery of Pai-lu-chou 白鷺洲 ("White Crane Islet") from memory. The area is near Nanking where he visited in 1651, as suggested in Chapter I. However, the exact location of the Taoist Temple Pi-hsia 碧霞 where he painted the handscroll remains to be identified. Nevertheless, since Hung-jen spent the winter of 1652 in Wu-hu, the temple might have been located in that region.

The handscroll begins with a slope, followed by a dense bamboo grove in which a few rustic houses can be seen. A figure is sitting by the window of a house in the front yard, looking out toward the river and mountains. The bamboo grove is defined by a fine brushwork which forms a delicate and exquisite pattern, combining descriptive details with firm design, both expressed with extraordinary clarity. The atmospheric effects are subtly suggested by a few bands of blank areas placed horizontally across the bamboo grove. On the other hand, across the water from the bamboo grove are a series of overlapping and undercut cliffs which do not actually recede into space but are cut away by the upper edge of the handscroll. They form a pattern of boulders with a minimum of interior texture. The tendency to create pattern can also be seen in the clusters of firmly drawn reeds in front of the cliffs.

These reed clusters with their screen-like design become a major pictorial theme in the second half of the handscroll, in spite of the incursion of slight atmospheric effects. This sense of atmosphere is created by hiding the lower parts of some boats behind the reeds and by making some other boats smaller. The attention to detail, the delicate pattern of the reeds and bamboo, and the careful arrangement of forms on the picture plane, are certainly reminiscent of Wen Cheng-ming's late style.

Hung-jen's indebtedness to his immediate predecessors as well as his invention can be seen clearly in the undated sixty-leaf album depicting Mt. Huang now in the Palace Museum in Peking (Pls. 65–124)[2]. Several leaves can be compared to Sun I's 1657 hanging scroll "Cinnabar Peak" in the Chien-lu collection (Pl. 41). For example, there is one leaf (no. 4; Pl. 68) "Cinnabar Spring" (*Chu-sha-ch'üan* 朱砂泉) which seems to depict the same site as Sun I's painting, although apparently from a different viewpoint. The Hung-jen leaf comes very close to the Sun I not only in its motifs but also in its style. Note for instance the deciduous trees, the pine trees, the dottings, and the dry and split brushstrokes. The prominence of the dike-path with its walking figure in the Sun I hanging scroll can be found in the scene of the "White Sand Mountain-range" (*Pai-sha-ling* 白砂嶺) from Hung-jen's album (leaf no. 33; Pl. 97). Here the dike-path is an even steeper and more dominant part of the whole composition. The zigzag angularity of the dike-path is echoed by the flat rock formation on which the dike-path stands, and by the pine trees that accentuate the path itself. This comparison highlights the effect of Sun I's style on Hung-jen, an effect that should be taken into consideration along with that of the Hsiao Yün-ts'ung style.

Elements that Hung-jen might have derived from Hsiao Yün-ts'ung can be seen in a comparison of Hsiao Yün-ts'ung's ca. 1647 handscroll "Travellers in Autumn Mountains" in the Tokyo National Museum (Pls. 27–28) with the following leaves from the sixty leaf album of Mt. Huang: the construction of rocks through repetition of angular flakes of light and dark areas (leaf no. 20; Pl. 84); the posture of stubby pine trees (leaf no. 46; Pl. 110); the use of outlined archaistic cloud (leaf no. 3 and no. 10; Pls. 67, 74); and a solitary figure in reddish-brown clothes (leaf no. 25 and no. 46; Pls. 89, 110). However, whereas Hung-jen's style is marked by clarity and lucidity, Hsiao Yün-ts'ung's paintings tended to be more detailed and agitated.

Often Hung-jen's own sense of design seems to prevail. This can be seen in leaf no. 34 (Pl. 98) where the repetitive swelling pattern on the left is relieved by the angular rock formations on the right and by the central waterfall. Moreover, Hung-jen's leaves tend to be characterized by the angular and cubic blocks of the distant mountain masses as well as by architecture that is enclosed or flanked by vegetation. The trees surrounding or flanking the architecture serve the function of calling our attention to the human residence, even though the houses are often empty. Apart from this function, trees in Hung-jen's picture often serve as pointers in the organization of the mountain masses or rock formations. Their positioning usually imparts a measured rhythm to the composition. Numerous examples can be cited. In leaf no. 50 (Pl. 114), for instance, the trees on the path leading from the lower right to the left give emphasis to the diagonal of the path itself in balance with the verticality of the cliffs. In leaf 35 (Pl. 99), the pine trees, mostly in groups of two or more, give enlivening accents to the otherwise barren rock formations. The same can be said of leaves nos. 11, 22, and 23 (Pls. 75, 86, 87). More interesting is the way in which the pine trees in leaf no. 33 (Pl. 97) are used to animate the steep zigzag dike-path in the middle ground; furthermore, the rhythmic placement of these trees is reiterated by groups of deciduous trees in the foreground.

The album exemplifies Hung-jen's well-articulated formal arrangement and structural clarity. For instance, in leaf no. 22 (Pl. 86), he put the rising sun in the cleft between two steep walls of rock peaks. Above the sun, he carefully inscribed the name of the site, "Gate to the Western Sea [of Cloud]" (*Hsi-hai-men* 西海門), as if to draw the viewer's eye to the cleft and to the sun. In leaf no. 7, (Pl. 71) the central verticality formed by the waterfall and the figure is delicately balanced by the horizontal rock and bridge. The subject, "Sounding String Water-fall" (*Ming-hsien-ch'üan* 鳴絃泉) is thus forcefully brought into focus. In addition to the above-mentioned compositions that concentrate on the more balanced and central subject, the album also contains compositions with emphasis on diagonals which impart more dramatic effects to the pictures such as leaves nos. 21 and 33. Whatever the compositional types, there is always a focus in the picture for the viewer's contemplation. In leaf no. 56, it is the Chao-chou Temple (*Chao-chou-an* 趙州庵); in leaf no. 29, it is the "Unsettling Dragon Pine" (*Jao-lung-sung* 擾龍松); in leaf no. 6, it is the "Drunken Rock" (*Tsui-shih* 醉石); and in leaf no. 5, it is the "Well of Elixir" (*Tan-ching*丹井).

Thematically related to the album depicting Mt. Huang is an undated hanging scroll "Landscape, " now in the Freer Gallery of Art (Pl. 59). The artist tells us in the inscription that he painted at the Yün-ku Temple 雲谷寺 at Mt. Huang. It can be seen as an attempt to depict the real scenery of Mt.

Huang. But nature seems to be circumscribed by the conventional compositional scheme: trees in the foreground, the steep mountain mass in the background, and a stream connecting the foreground with background and at the same time imparting a limited recession to the picture. The whole painting is cluttered with details: trees, rocks, pavilion, figure. It might be a result of Hung-jen's experiment with the fusion of relatively naturalistic Mt. Huang landscape and essentially non-naturalistic surface configurations. For instance, numerous trees are arranged like a screen across the entire foreground. They serve to directly support and reinforce the background mountain that is the focus of the composition. The stream does not cut a path dividing the painting into two sections, but rather serves to indicate that the foliage is penetrable and somewhat distinct from the background. The waterfall at the left, echoing the shape of the mountain, defines the stream's source and allows the viewer to follow its route from the top out through the center of the painting. There is an alternation of light and dark areas on the major mountain mass. The primary pictorial interest is the surface; essentially two-dimensional, angular shapes are imposed on irregularly shaped larger forms to define their contours as well as their textures.

Hung-jen's fine "Landscape" fan painting dated to the fifth month of 1655 is now in the Liu Tso-ch'ou collection in Hong Kong[3]. According to the artist's inscription, it was based on an album by Wang Fu. It is not possible now to identify the specific Wang Fu painting to which Hung-jen referred as his model. Wang Fu, as discussed above, was an early follower of the mode of Ni Tsan. And it is thus appropriate for Hung-jen to use Wang Fu as his model.

The fan painting (Pl. 115) represents a great step from Hung-jen's 1639 short handscoll in Shanghai (Pl. 49) in which he attempted only a little more than the conventional mode of Ni Tsan. Like the 1639 painting, it is composed of flat areas defined with a minimum of interior texture. Unlike the earlier painting, however, there is much of Hung-jen's own style now in the fan painting. The earlier stability is here replaced by a more rhythmic movement from one area to another. In place of an isolated central bluff removed from the lower edge of the 1639 handscroll, flat areas in the 1655 fan are built up into an interlocking pattern. The left half, including the cliff upon which rests a simple house flanked by two groups of trees, is to be seen in his leaf no. 8 from the album entitled "Landscapes, Flowers, and Plants, " dated 1656, now in the Drenowatz collection in Zürich, to be discussed later on.

The 1655 fan can be seen also as turning away from the more experimental and fantastic forms found in his album leaf for Ch'en Ying-ch'i dated to around 1652. Situated on a flat terrace, the stable house and its flanking trees further present an interesting visual contrast to the pattern of flat areas that fill most of the picture plane. Here, one finds a feature that was to appear again and again in Hung-jen's later paintings—device of flanking or surrounding a house with trees, thus bringing out the architecture through a contrast between the man-made structure and the natural vegetation. This device is an example of Hung-jen's sense of design. It can be seen almost as a picture within a picture. That is to say, Hung-jen often tried to accentuate his composition with a well-wrought, tight, and concentrated motif as the center of interest without disrupting the unity and primacy of the whole picture. Furthermore, unlike the cluttered composition seen in the Freer hanging scroll, the Liu Tsao-ch'ou fan has a quiet clarity; the forms are placed discretely and yet tightly on the picture plane.

The majority of Hung-jen's dated and undated paintings can be placed between 1656 and 1663. In the year 1656 we have a hanging scroll and two albums by Hung-jen, making it possible to establish the range of his style at this time and providing a foundation for a study of his later work. This is so because the two albums contain a large number of leaves—twenty-two altogether. One album entitled "Landscapes, Plants, and Flowers, " of which no. 9 is dated to the summer of 1656, is now in the Drenowatz collection in Zürich (Pls. 125-137). The other, entitled "Feshermen Album, " dated to the autumn of 1656 and is now in the Vannotti collection in Lugano, Switzerland (Pls. 138-149)[4].

The hanging scroll of landscape, now in Shanghai (Pl. 116), exemplifies Hung-jen's mature style; together with an undated album of eight landscapes, now in Peking (Pls. 117-124), which can be compared to the "Mt. Huang" album, it demonstrates Hung-jen's power of synthesis.

The first leaf in the Drenowatz album (Pl. 125) depicts an open, low level landscape with limited space; there is nothing behind the bridge. A river flows forward from the background, takes a turn in the middle ground, and then zigzags its way out of the picture plane at the lower edge of the painting. In the right foreground are three deciduous trees; on the left, another deciduous tree seems to reiterate the serene horizontality of the whole composition. The brushwork is relaxed, but the structure is nevertheless coherent. Just above the bridge in the middle, Hung-jen put his round seal, like a pictorial accent. He wrote his inscription on the riverbank in the foreground: "Amid the mist in the forest of willow trees, a wood bridge crosses the river like a rainbow. [I] feel that although my brushwork is simple, the picture has conveyed profound ideas. " The tone of the inscription suggests Hung-jen's self-confidence despite the seemingly calculated casualness of the painting. More importantly, the discrepancy between the visual image and the verbal intention as stated in the artist's inscription—there is neither mist nor rainbow in the painting—points to the symbolic nature of Hung-jen's style.

In the second leaf (Pl. 126), the artist mentioned in his inscription "the branches with birds and cloudlike rocks—these are the things to enjoy in Ho-yang [Kuo Hsi 郭熙, active ca. 1060-1075] ." In the painting, the artist referred subtly to the style of the Northern Sung master, for the swelling rock formation in the foreground does suggest something like clouds.[5] Upon close examination, however, this leaf also exhibits influence from Huang Kung-wang in its long drawn-out brushstrokes in depicting the mountain masses.[6] The presence of explicit reference to Kuo Hsi as well as unstated reference to Huang Kuang-wang indicates Hung-jen's syncretism. The painting thus exemplifies the art-historical aspect of Hung-jen's style. As Richard Edwards has pointed out: "As a general rule it is safe to affirm that any Ming artist working in the manner of a specific Yüan master more often than not would include suggestions of other great masters of that time. The goal was broad understanding, not narrow antiquarianism."[7] In a similar way, Hung-jen included suggestion of Hung Kung-wang while working in the manner of Kuo Hsi.

Huang Kung-wang was the inspiration for the fourth leaf (Pl. 129), where his name is invoked in the artist's inscription: "The ink and brush here are rough and ragged, scattered and loose. This is something like the style of the Old Fool [Huang Kung-wang]." Again this leaf can be regarded as an art-historical statement, its dry and crumbly brushwork and angular zigzag plateaux recalling Huang Kung-wang's handscroll "Dwelling in the Fu-ch'un Mountains" (*Fu-ch'un shan-chü* 富春山居) dated 1350 and now in the National Palace Museum in Taipei.[8]

The inscription on the third leaf (Pls. 127-128) clearly spells out the artist's intention and is worth quoting here: "Ch'ang-kung 長公 [Su Shih 蘇軾, 1036-1101] said of the prose of Tzu-yu 子由 [Su Ch'e 蘇轍, 1039-1112] that it is 'like the waterfall plunging down a cliff, making one shiver with cold.' When a painter depicts the charging of a waterfall, how can he be careless? This painting, I hope, is close enough." In other words, Hung-jen hoped to achieve the same powerful effect as Su Ch'e's prose. The plunging waterfall as well as the contours of the angular and frontal rock-formation are drawn with rather fluent and firm brushstrokes. The strong contrast of dark and light between the waterfall and the cliff, as well as that between the front rock and the background mountain range, contribute to the dramatic configuration and give concrete visual expression to the quality of being "cold" mentioned in the artist's inscription. This leaf, indeed, embodies the bold simplification that was to characterize most of the artist's works (Pl. 128).

The same boldness of composition can also be seen in the eighth leaf (Pls. 134–135), painted to accord with the couplet composed by his close friend Wu Chieh 吳揭 who was also from She-hsien. The couplet may be translated as: "Bamboos and rocks are like the mind of the man living here; and wind and waterfall are like the heart of the quiet man." The delicated balance between the vertical and angular plateau on the left and the horizontal water on the right is expressive of the serenity and quietude of the man referred to in the poem. The front of the plateau, which supports a two-storied house, is modeled by dry brushwork. Texture strokes, some pale and some dark, are placed in an angular and yet repetitive rhythmical pattern. The rocks in the water around the islet create a level-distance recession leading to the dark gray mountain range in the background (Pl. 135). The composition, though similar to that of the 1655 fan in the Liu Tso-ch'o collection, is more compact.

Hung-jen's interest in an angular and rhythmical two-dimensional pattern can be seen in the sixth leaf (Pl. 131–132). He carefully constructed masses of rocks with layers of dry brushwork, accentuated by a few dark horizontal dots. In the inscription, the artist stated his intention to depict "emanciated trees and beautifully shaped rocks" in the Hsin-an River 新安江. In the painting he succeeded in depicting the rectilinear form of the sharply cut and deeply fissured rocks along the river. The verticality of the trees and the contours of rocks is balanced with the horizontality of the bridge on the left and the fishing boat on the right. The construction of the rocks is reminiscent of the broken-sash brushstrokes (*Che-tai-ts'un* 折帶皴) of Ni Tsan, but Hung-jen was more interested in the close-up view of the linear pattern (Pl. 132).

The fifth leaf (Pl. 130) indicates his stylistic source. He wrote in his inscription:

> The hills and gullies of Yün-lin [Ni Tsan] are based entirely on those of Ching [Hao] and Kuan [T'ung], and Tung [Yüan] and Chü [-jan]. He was able to paint in a most impressive way. In his later years, his style became relaxed and natural, and his brushwork simple and untrammeled. I am incapable of mastering his style, so I have only fol-

> lowed his simplicity. Those who emulate the Eccentric Old Man [Ni Tsan] should not take me as their standard.

The painting itself seems to be an essential visual-symbolic statement about his stylistic source: the low southern view derived from Tung Yüan and Chü-jan and the angular and brisk brush reminiscent of Ching Hao and Kuan T'ung. Compared with his section in the 1639 joint handscroll in Shanghai (Pl. 47), however, the mode of Ni Tsan was here rendered with a new playfulness and sureness. The sensitive brushstrokes, however dry they may appear at first glance, are effectively employed to convey a compelling sense of desolation and simplicity.

The "Fishermen Album" in the Vannotti collection (Pls. 138–149) was painted to accord with a series of poems on the life of fishermen and woodcutters composed by Hung-jen's close friend Hsü Ch'u 許楚 (1605–1676) during the latter's self-exile in the area of Wu 吳 and Yüeh 越 (modern Chekiang and part of Kiangsu and Fukien provinces) after the fall of the Ming Dynasty. On each leaf, Hung-jen inscribed one poem by Hsü Ch'u. The artist's inscription reads:

> After the collapse of his family, Hsü Lü-t'ing [Hsü Ch'u] wandered in the area of Wu and Yüeh. He stayed for a long time near the Lake Ming-shen [West Lake, near Hangchou, Chekiang]. Every day he would carry his brocaded bag and ramble among the lakes. When excited, he would often sing together with the fishermen and the woodcutters. In this way, he finished forty poems about fishermen. I like these poems and would recite them to myself. I now make twelve pictures to illustrate those poems that are more adaptable, so that he may put them in his gourd-ladle. Someday when he gets together with his friends to compose poems, may these pictures serve as his inspiration.

The narrative content of this album can be related to an extremely long handscroll entitled "Going Home and Abroad Are the Same" by Hsiao Yün-ts'ung now in the Drenowatz collection (Pls. 29–30).[9] Dated 1655, Hsiao Yün ts'ung's handscroll was only one year earlier than Hung-jen's album. Both are concerned with the idyllic activities of fishermen, who exemplify the utopian way of life in Chinese literature and art. The theme was quite appropriate in a time of turmoil like the fall of the Ming Dynasty and the conquest of China by the Manchus. Both works are similar in the emphasis on narrative and treatment of certain motifs. Both, for instance, are populated by the same stubby toy-like figures. One also finds a prominent use of long dike-paths and cubic rock formations.

Unlike the Drenowatz album, the "Fisherman Album" is concerned with a central theme. In the tenth leaf, a fisherman sitting in a boat with wine bottle and ink-stone on a table under a tilted shelter is seen through a screen of trees in the foreground. Despite the horizontal shore lines that indicate recession, the steep cliff on the right is connected with the foreground trees to form a space cell in which the hermit-scholar-fisherman is brought clearly into focus. Hung-jen inscribed Hsü Ch'u's poem, to be read from left to right:

> I don't long for luxury of life; I only need a shelter from cold. Everyday I row a boat with peace. I relax myself, with no thoughts in mind. Only the volume of *Li-sao* (*Encountering Sorrow*) to company the fishing rod.

The unstated reference to a past pictorial style is paralleled by the allusion in the inscribed poem to *Li-sao*, a series of poems composed by Ch'ü Yüan 屈原, the nobleman-poet of the kingdom of Ch'u from the late fourth and early third centuries B.C., who died in exile after being dismissed by his prince. Ch'ü Yüan "was a man of undeviating righteousness who devoted all of his loyalty and all his knowledge to the service of his prince; yet he was traduced by false witness."[10] The identification of the artist with the poet-fisherman in exile is clear not only in the inscribed poem but also in the persistent focus on the fisherman through the screen of trees in the foreground of the painting.

In the seventh leaf (P1. 144), Hung-jen used the cloud-like rock in the foreground that is reminiscent of Kuo Hsi and the archaistic outlined cloud. In the inscription he quoted Hsü Ch'u's poem which reads in part:

> From the curved bays, I can see all the clouds and mountains, which far surpasses the "Ten Grass-huts" of the Lu family [Lu Hung].

As mentioned in Chapter II, Lu Hung was the T'ang artist to whom the ten-leaf album "Ten Views of a Grass-hut" has traditionally been attributed. That album, as suggested earlier, could have inspired Hung-jen's sixty-leaf album of Mt. Huang. Hung-jen thus gave a clue to his art-historical understanding. Nevertheless, he believed in the supremacy of the direct experience of nature.

The "Fishermen Album" exhibits several stylistic features that are distinctively Hung-jen's. There is a strong sense of linearity accentuated by the short bare stick-like trees in leaf nos. 1, 2, and 8 (Pls. 138, 139, 145). In leaf nos. 2 and 8 (Pls. 139, 145), one finds flat overlapping plateaux that create an angular linear pattern. Furthermore, as mentioned earlier, trees often serve as important organizing elements in Hung-jen's composition. In leaves nos. 2, 5, 7, and 8 (Pls. 139, 142, 144, 145), the placement of the overhanging tree provides the picture with its enlivening accent. In leaf no. 1 (Pl. 138), trees are placed so as to create a measured rhythm.

One of the most striking and original compositions in the Vannotti album is the ninth leaf (Pl. 146), where the dominant motif of an empty path is placed horizontally across the lower part of the picture plane. As additional elements, there are only a few firmly drawn reeds dotting the path on the right and two small rock masses in front of and behind the path. By representing a deserted path as the focal area, and leaving most of the picture empty, Hung-jen demonstrated his poetic sensiblility in giving a concrete imagery to the desolation and austerity in the inscribed poem which reads in part: "Sing a noble song [as you pass] along the dike-path. Do not leave [your] name in this world!" Above all, this leaf exemplifies Hung-jen's economy and concentration.

Hung-jen's 1656 hanging scroll, now in the Chih-lo-lou collection in Hong Kong, was painted after his return from a trip to Mt. Huang (Pl. 150). In the foreground, two crossed pine trees grow on a rocky land spit. Here, as elsewhere, Hung-jen used contrast in scale to create an overwhelming and powerful landscape. The fishing boat seen hiding behind the foreground repoussoir rocks is so small that the pines appear gigantic in comparison. This contrast in scale is also evident as our eye passes the mountain masses in the middle ground, over toward the right of the picture. Here, juxtaposition of the architecture (i.e., the house and the stone pagoda) with the major mountain makes the latter so much the higher and grander. The mountain masses are carefully built up through an interplay of horizontal plateaux and vertical boulders. The structure also gains in cohesiveness through the accents of deciduous trees. More subtle contrast is achieved between the inscription on the upper right in standard script and the date in seal script on the rock in the left margin of the picture. The composition can be regarded as anticipating the 1661 handscroll "Landscape of Feng River" in the Sumitomo collection (Pls. 202-206) and the undated hanging scroll "The Coming of Autumn" in Honolulu (Pl. 219).

The undated album (Pls. 151-158) reproduced in 1920 under the title *Chien-chiang shang-jen shan-shui ts'e* 漸江上人山水冊 and now in the Palace Museum in Peking can be dated to about 1656 because of its stylistic affinity with both the Drenowatz and the Vannotti albums.[11] The album does not have any signature of Hung-jen but contains the artist's seals that are comparable to those on the Drenowatz and the Vannotti albums. Despite the rather hazy quality due to the more reflective gold paper used by the artist, the album demonstrates Hung-jen's distinct style.

The first leaf (Pl. 151) employs the constructive approach introduced in the 1651 joint handscroll and repeated in the undated album leaf painted for Ch'en Ying-ch'i around 1652 (Pl. 48). The combination of angular rock formations takes up most of the foreground and the middle ground. On the left, the round forms derived apparently from Huang Kung-wang provide a contrast to the angular ones on the right. The two houses on the tilted-up plateau also partake of the linear and angular quality of the rocks. The distant mountains defined by washes help emphasize the linear pattern in the foreground and the middle ground.

The second leaf (Pl. 152) is strikingly similar to the third leaf (Pl. 127) from the Drenowatz album, sharing its boldness (but not the strong contrast between light and dark) and the severe frontality of the huge angular rock in the center of the picture plan. The waterfall and the ripples on the river are drawn with firm brushwork. Small angular rocks jutting out of the two major rock formations further enhance the shape of the central rock.

The fifth leaf (Pl. 155) comes close to the sixth leaf (Pl. 131) from the Drenowatz album, both in composition and in the construction of rocks. In the undated album, the fifth leaf (Pl. 155) is perhaps the most characteristic of Hung-jen's mature style. The elevated dike-path on the right is now a familiar motif. What is more striking is the narrow crevice between the rock walls. Both the dike-path and the crevice no doubt contribute to the strong sense of design. While the crevice seems to indicate an inviting world beyond the rocky walls, one is constantly pulled back to the front of the picture plane. On the rocky wall near the crevice, Hung-jen placed his rectangular seal as if to enhance the angularity of his forms. The execution of the rocky walls is a fine example of Hung-jen's method of formal construction, particularly in the interplay between the large flat areas which are left almost untextured, and the smaller and textured areas. Hung-jen was less interested in the volume itself than in the pattern created by the layered rocks and their fissures running horizontally and perpendicularly.

The undated album also contains several leaves that suggest Hung-jen's syncretism. The third leaf (Pl. 153), for instance, faintly suggests the style of Wang Meng 王蒙 (1308-1485) in its profusion of dots and the furry appearance of the rock formations. Here Hung-jen created a compelling focus by placing the two pine trees in the center of the picture plane. Though connected by a wooden bridge with the land in the back, the islet and its empty house in the foreground, like the two pine trees, have been isolated and brought to our attention by their very position in the picture plane.

The mode of Wu Chen 吳鎮 (1280-1354) is apparent in the fourth leaf (Pl. 154), which is filled with broad, wet brushwork, especially in the depiction of vegetation. The eighth leaf (Pl. 158) recalls the reinterpretation of the mode of Ni Tsan by Wen Cheng-ming. Compare Hung-jen's leaf with, for instance, Wen Cheng-ming's undated handscroll entitled "Autumn Mountains" in the Art Institute of Chicago (Pls. 235-236).

In the summer of 1657, Hung-jen stayed at the Hsiang-shui-an Temple in Nanking where he painted an eight-leaf album for a Mr. Tsung-hsüan 宗玄 (Pls. 159-166); the whereabouts of the album is not known.[12] The first leaf (Pl. 159) immediately strikes the viewer with its economy and compactness. An empty house is diagonally placed on the angular, reverse-L-shaped mesa in the lower lefthand corner. A twisting gnarled old tree juts out on top of the house. Behind the tree, a central bluff rises, partly submerged in mist. The bluff consists of a few overlapping and somewhat transparent areas defined by fluent vertical contour lines and short horizontals. Interior texture is minimal, making the bluff appear insubstantial, particularly in contrast to the dark tree. This leaf exemplifies Hung-jen's approach to mountain masses that was to characterize much of his later works such as the undated hanging scroll "The Coming of Autumn" in Honolulu. The mountain is a coherent structure that recedes from the right to the left in a step-like progression. One is immediately struck by the flatness of the barely modelled rock formation on the right edge. While the artist tried to build a convincing bluff in the center of the picture plane, he also wanted to emphasize the flat and abstract qualities of his rock on the right edge. This contrast is reiterated in the different directions of the branches of the old trees in the foreground.

In the second leaf (Pl. 160), the high, frontal mountain peaks with their mass of dots to indicate vegetation appear to be his shorthand for the style of Fan K'uan 范寬 (active ca. 1030). The tight pictorial structure, however, is totally Hung-jen's own. He coherently integrated much of his vocabulary into the compact composition: the overlapping pine trees and rectangular rocks jutting out of the cliff, the stick-like spare trees on top of the central peak, the sharply drawn pagoda hidden behind trees, and the fluent though occasionaly split brushstrokes for contouring rocks. The two oddly-shaped towers of mountain, we recall, first appear in Hung-jen's album leaf for Ch'en Ying-ch'i (ca. 1652) (Pl. 48). The one on the right is pointed and the other on the left is flat-topped. Here, Hung-jen used these motifs with more clarity.

One of the most inventive pictures in this album is perhaps the third leaf (Pl. 161) in which the foreground is occupied by a pond encircling a nearly parallelepiped-shaped island mesa that is flanked by sensitively drawn reeds. This horizontal openness of foreground is contrasted with the vertical linearity of three motifs in the middle ground: the deciduous trees, the open pavilion, and the cliff on the left edge. The mesa in the foreground pond with its flanking reeds becomes the focal area of the picture, isolated but prominent. The same composition appears in an undated hanging scroll now in Peking (Pl. 193).

Similarly, the treatment of the houses flanked by trees in the sixth leaf (Pl. 164) reveals Hung-jen's predilection for enclosing severe architecture with a dense growth of vegetation, partly to highlight the space around the architecture and partly to reinforce the tension between the linear and the nonlinear forms.

The fifth leaf (Pl. 163) reveals Hung-jen's interest in the style of Tung Yüan, particularly in the level-distance composition and rounded and soft land forms. Despite the presence of the empty pavilion on the right in the foreground that is reminiscent of Ni Tsan, the composition is closely related to the depiction of a spacious landscape of water-village of the region south of the Yangtze River. The horizontal shore lines, the diminishing size of the land masses, and the gentle curves of the water create a river landscape that is different from the compressed mountain landscape such as the first leaf and the second leaf that are probably derived from Ching Hao, Kuan T'ung, and Fan K'uan.

The seventh leaf (Pl. 165) reveals Hung-jen's persistent interest in the contrast between two-dimensional and three-dimensional forms. There is a convincing spatial recession on the left, with the viewer's eye moving from the repoussoir rock in the foreground, passing the bridge and following the zigzag movement of the river, through the cleft formed by the two screens of trees. That sense of spatial recession is immediately blocked by the mountain masses in the background. The limited spatial recession is enhanced by the rock formation on the right of the foreground that overlaps with the background mountains.

Later in 1657, in the ninth month, Hung-jen painted a hanging scroll (Pl. 167) entitled "The Ten-Bamboo Studio" (*Shih-chu-chai t'u* 十竹齋圖) for Hu Chih-kuo 胡致果 (Pls. 167–168), son of Hu Yüeh-ts'ung 胡曰從, a famous publisher noted for his *Ten-Bamboo Studio Painting Manuals*. Presumably Hung-jen painted this work while he was staying in Nanking. Painted in a cold and rainy morning, it was inspired by a couplet by the T'ang poet Ts'en Shen 岑參 (716–770): "On the window are the wavering shadows of all sorts of trees; in the shade of the wall can be seen a mountain peak."[13] It is a fine example of the artist's style in the late 1650's. One notices, first of all, that the foreground, the bamboo grove, and the trees all are arranged along a diagonal and silhouetted against the flat mountain immediately attached, so to speak, to the wall. There is no distance between the more intimate courtyard and the distant mountain. A narrow two-storied house is squeezed between the trees and the flat mountain mass (Pl. 169). There is the now-familiar contrast between the rectangular and the pointed shapes on the mountain mass. This contrast is in turn echoed by the more naturalistic depiction of vegetation and the more abstract rendering of rock formations. Despite these contrasts, the whole composition is tightly knit. While there is an open space within the courtyard, the spatial recession is blocked by the enclosing walls and the flat mountain mass. The transparency and angularity of the mountain mass and the measured rhythm of the vegetation along the courtyard wall poignantly express the imagery in the inscribed poem.

In the second month of 1658, Hung-jen was still in Nanking. An excursion into the southern suburb of the city when plum blossoms were in full bloom occasioned the short handscroll entitled "Plum-Blossom Pavilion" (*Mei-hua-t'ing* 梅花亭) (Pl. 169). Even from the poor (and probably incomplete) reproduction published in 1930, the sense of concentration and firmness of brushwork in the handscroll is unmistakably characteristic of Hung-jen's late style. By contrasting the plum branches that cross one another and the squared man-made stone table in the foreground, the artist forcefully elicited the perseverance and self-renewal symbolic of the plum tree. The stone table, with its predominant horizontality and linearity, imparts a sense of pause and clarity to the picture, while the vertical tree trunk just behind the stone table brings the contrast into a delicate equilibrium.

In the winter of 1658, Hung-jen painted a hanging scroll, "Frosted Trees," depicting three deciduous trees and a cluster of bamboo underneath on a shallow slope (Pl. 170). In this hanging scroll now in the Musée Guimet, the predominant motif is the three spare trees. Two of them intertwine, and their arrangement presents crystalline order. The bamboo underneath the trees, like houses or figures in other paintings, provide a contrast in scale and enhance the verticality and importance of the trees. The viewer's eye is drawn from the slender trees to the artist's inscription in which he tells us that he painted this scroll after having looked all day at the snow-turned ice on the trees.

The hanging scroll entitled "Plum House and Pine Spring" (*Mei-wu sung-ch'üan* 梅屋松泉), dated 1658 and now in the Shanghai Museum (Pl. 171), has the firm brushwork and ordered structure so characteristic of his later work. In the foreground, three pine trees intertwine energetically, underscoring the compactness of the composition. Behind the pine trees, a firmly drawn house with a few simple furnishings is located on a tilted-up platform. The visual clarity of the painting occurs largely through the contrast between the background and the middle ground. The mountain masses in the background are rendered with a tender and soft "lotus-leaf texture stroke" (*Ho-yeh ts'un* 荷葉皴); their subtle suggestion of a light-and-dark alternation contrasts with the light-and-dark angular shapes and sharp-edged contour lines of the mountain masses in the middle ground. Harrie Vanderstappen has written of it as a work that exemplifies Hung-jen's "often capricious but well-articulated and telling forms." He wrote:

> There is a boldness in Hung-jen's painting. Insisting on rhyming shapes, he largely ignores gradation of size and confronts the viewer with a setting which pre-empts the confrontation of large and small, light and dark. The contour and textures allow only tiny variations and reinforce the austerity of the artist's vision of the world.[14]

The same complex, yet ordered composition may be seen in an undated hanging scroll now in the Burke collection in New York (Pl. 172). It was painted for his close friend Ch'eng Shou 程守 (1619–1689) as a birthday present. Hung-jen mentioned in his inscription that he and Ch'eng Shou had been "sworn brothers" for twenty years. For that reason, he painted the picture in the manner of Ni Tsan so as to express a sense of perseverence and his best wishes for Ch'eng Shou.

The composition is compact and tight, and the pace from foreground to back-ground is quick. Despite the richness of individual motifs, the picture is marked by a strong sense of economy and concentration. In the right foreground, one finds a series of repoussoir rocks and the edge of a fenced courtyard. A tree in the middle of the foreground introduces the steep cliff which rises sharply up-ward on the left. In front of the cliff, a small stone pagoda is drawn with firm and crisp brushwork, as is the vegetation on the cliff. The open space of the courtyard makes the houses in the back and the surrounding trees and bamboo appear more outstanding; this is a compositional device we have already seen in Hung-jen's earlier paintings such as in the third leaf from the 1657 album painted in Nanking (Pl. 161). The mountain masses in the background are built up mostly of flat geometric shapes, echoing the rock and cliff in the foreground.

Hung-jen's sense of concentration is most evident in the manner in which the houses are surrounded, yet, paradoxically, revealed by the enclosing trees and bamboo. This is achieved by contrasting the curves of the trees and bamboo to the severe lines of the man-made objects (houses, tables, and pagoda), a feature also in Hung-jen's 1658 short handscroll "Plum-Blossom Pavilion." The twisted form of the old tree to the left of the houses links it to the same motif in the first and fourth leaves from the 1657 Nanking album. In addition, the pagoda and the vegetation on the cliff in the foreground recall the same motifs in the second leaf from the Nanking album. Therefore, the hanging scroll for Ch'eng Shou can be stylistically dated to about 1657 or 1658.

It is but a simple step from the compactness and complexity of the 1658 hanging scroll "Plum House and Pine Spring" to the monumentality of the 1659 hanging scroll entitled "The Unending Songs of Birds" (*Niao-yü mien-man* 鳥語綿蠻) now in the C. C. Wang collection in New York (Pl. 173). The entire composition is a firmly structured configuration, climaxed by a dominant mountain peak with a rectangular massive rock. The formal construction is gradually built up from the lower part of the picture. The pace is casual at first, with trees as indicators of a rhythmic growth of the forms. In the middle part of the picture, the pace seems to gather momentum, with a series of zig-zag plateaux contributing to the speed of the development. Then, as this development becomes an intricate pattern of pull and thrust, the sheer presence of the stele-like rectangular cliff arises just behind a dense area of trees surrounding a few houses. The towering massive cube of rock not only consummates the rhythmic growth of the whole mountain mass, but also imparts a sense of austerity and fantasy to it. Hung-jen himself seemed to be aware of his own inventiveness; his inscription reads in part: "My eyes have never seen the great oceans. I make up this steep mountain of mirage." How does one relate the great oceans to the mountain depicted here? Perhaps Hung-jen wanted the viewer to see the grandeur and power of his mountain landscape as essentially the same as those of the oceans, with the same permutations and self-renewal. In many ways, "The Unending Songs of Birds" testifies to Hung-jen's ever-deepening personal vision and his formal construction expressive of that vision.[16]

Later in the spring of the same year, Hung-jen painted a hanging scroll that fully exemplifies the artist's late style. The painting "Landscape," formerly in the Huang Pin-hung collection, is now in the Shanghai Museum (Pls. 174–177). In the foreground, an angular rock formation and two twisting trees arise from the edge of the water and introduce the viewer to the two mountain masses immediately behind them. The compactness of the composition, like that of the small hanging scroll painted for Ch'eng Shou in the Burke collection, is achieved by eliminating the middle ground. The only connection between the foreground rock in the water and the massive cliffs behind it are the toy-like houses on the right and the fishing boat on the left. The two major mountain masses, one squarish and the other conical, are variations of the same basic geometric forms that appear in Hung-jen's earlier work, e.g. the album leaf of ca. 1652 for Ch'en Ying-ch'i (Pl. 48) and the second leaf from the 1657 album painted in Nanking (Pl. 160). The emphasis on contour lines of the rock formation, the lack of interior texture, and the rhythmic arrangement of vegetation combine to create a lucidity that is so characteristic of Hung-jen's late work.

Another hanging scroll from 1659 is entitled "Quiet House amid Bamboo and Rocks" (*Chu-shih yu-chü* 竹石幽居) and is now in the T'ien-tsin Museum (Pl. 179). The center foreground is dominated by three pine trees which partially screen a house alongside a grove of bamboo. The house is slightly elevated on a square raised platform. In the right foreground are two intertwining old trees. Immediately behind the trees and the house is a screen of mountain masses built up with geometric shapes and topped by flat plateaux. The composition as a whole is not as dramatic and powerful as the hanging scroll "The Unending Songs of Birds" of the same year, but it is no less compact in its overall structure. The strong verticality of the foreground pine trees and the angular shapes of the platform and plateaux, moreover, give the picture a sense of well-ordered austerity.

Stylistically speaking, the undated hanging scroll "Landscape," painted for a certain Shen Shou-kuo 沈壽國[17] (Pl. 180) and now in the Li Ch'u-li collection, is similar to the 1659 hanging scroll "Quiet House amid Bamboo and Rocks" and therefore can be placed to about the same year. One more reason for proposing this date is that the use of lotus leaf texture strokes (*Ho-yeh-ts'un* 荷葉皴) is similar to that in the 1658 hanging scroll "Plum House and Pine Spring" discussed above. In this regard, both paintings reveal Hung-jen's versatility in combining angular and smooth brushwork.

The 1660 hanging scroll "Heavenly Capital Peak" (*T'ien-tu-feng* 天都峯), now in the Nanking Museum (Pl. 181), is similar both in composition and style to the 1659 hanging scroll "The Unending Songs of Birds." It was painted, according to the artist's inscription, for a certain Mr. Hsüan-i 玄疑. Its forms are multiplied and built up in a far more complicated manner than are these of the 1659 hanging scroll. And yet the complexity of individual motifs is a unified complexity. The twisting pine trees in the foreground seem to echo the clusters of rocks that rhythmically progress along the central mountain mass up to the top of the painting. The masses of rocks on both sides of the picture enhance the verticality of the central mountain. There is little spatial recession; the artist, rather, is interested in the surface pattern created by overlapping and juxtaposition.

The hanging scroll "Landscape in the Style of Ni Tsan" painted for a certain Mr. Yü-yeh 于埜 is dated to the eighth month of 1660 (Pl. 182). Here Hung-jen has transformed the Ni Tsan mode into his own graphic design. It is characterized by flatness of forms. Beyond the toy-like house, a few bamboos and a group of trees in the shallow foreground, the mountain masses rise upward on the right. These elements recall the 1659 hanging scroll "The Unending Songs of Birds" (Pl. 173) and the 1660 hanging scroll "Heavenly Capital Peak" (Pl. 181) where, however, the major mountain masses are placed in the center of the picture. The diagonally placed pavilion seems to reiterate the foreground headland which recedes from the lower left-hand corner to the right-hand side near the major mountain mass. Again, like in many of Hung-jen's late works, the tree group in the foreground serves to direct the viewer's eye; the two trees on the right enhance the verticality of the mountain mass, whereas the one stooping to the left-hand side points to the low-level recession on the left. In both the 1660 "Landscape in the Style of Ni Tsan" (Pl. 182) and the 1659 "The Unending Songs of Birds" (Pl. 173), the mountain masses are climaxed by an almost precariously placed rock formation, imparting sharp dramatic tension to the total effect.

The imposing hanging scroll "Pines and Rocks at Mt. Huang" (*Huang-hai sung-shih* 黃海松石), dated to 1660 and now in the Shanghai Museum, is certainly characteristic of Hung-jen's late style (Pls. 183–187). Its dramatic quality is enhanced by taut, continuous lines that define an angular cliff, and gnarled pine trees with their sharp and pointed leaves. The pine trees grow directly out of the cliff-side in a very dynamic manner. The asymmetrical placement of the main mass to the left leaves a large part of the right half of the pictorial plane blank. This positive and constructive use of void is complemented by the light gray wash which highlights the surface of the cliffs. All these elements reveal the boldness in Hung-jen's conception and the sureness of his execution.

A comparison of this hanging scroll with a modern photograph of "Three Islands of Immortals" (Pl. 188), a famous site at Mt. Huang, indicates how Hung-jen transformed the scenery in nature into his own poignant visualization. By bringing the square mountain mass on the left to the top of the picture, and by bringing to the front of the picture plane the two pointed peaks which in reality are far in distance, Hung-jen was able to create a compact and powerful design. Moreover, the dancing branches of the two projecting pines seem to be engaged in a dialogue with the two pointed peaks below. Hung-jen's favorite motifs, e.g. the squared peak and the pointed peak in the second leaf from the 1657 album (Pl. 160), are here forged to express his experience of the fantastic scenery at Mt. Huang. However, it is not so much physical relationship of the mountain peaks and the pines as the majestic impression of them that Hung-jen was trying to convey through his design. As he himself wrote in a poem: "Often I dream of [Mt. Huang's] thirty-six peaks [as beautiful as] hibiscus. My casual traces of ink perhaps are worthy gifts. Only I cannot name for sure which peak I have depicted."[18]

The 1660 "Pines and Rocks at Mt. Huang" is one of the best examples of how Hung-jen manipulated scale to suit the requirements of his composition in an attempt to achieve effects of grandeur and drama. An undated album of landscapes in the T'ien-tsin Museum (Pl. 189-192) is quite similar in style to the hanging scroll, especially in the emphasis on close-up view and dramatic effect of forms. The album contains a number of leaves which display the angular, flat-topped cliffs and simple houses rendered with fluent brushwork. Especially outstanding is the second leaf depicting a pine tree growing downward from the very summit of a peak (Pl. 190).

Other leaves in the T'ien-tsin album reveal Hung-jen's persistent interest in a flat linear pattern. For instance, in the third leaf (Pl. 191), the surface pattern of a twisted old leafless tree is brought to the front of the picture plane. The importance of the tree is enhanced by the contrast in scale with the bamboo growing besides the flat rock formation which provides a papercut-like backdrop against the angular pattern of the tree. The staccato movement of the contour lines of the rock echoes the angularity of the tree branches. In the first leaf (Pl. 189), the flat, untextured rock similarly provides a backdrop to the trees in the foreground. On the left, the fishing boat near the rock wall in the background, like the bamboo in the third leaf, provides a contrast in scale that emphaiszes the foreground trees. Despite the slight indication of spatial recession by the stream on the right, the overlapping pattern on the left seems to dominate the small album leaf.

Hung-jen's interest in the interplay of two-dimensional and three-dimensional forms can be seen in the fourth leaf (Pl. 192). The repoussoir rocks in the foreground and the water beyond them provide a spatial recession. That spatial recession, however, is immediately cut short by the wall of steep cliffs in the back. The cliffs themselves are minimally textured and the outlines of their angular shapes are emphasized. They form a two-dimensional pattern that is somewhat relieved by the dark accents of the rocks and trees jutting out of the cliff in the center and the waterfall on the right edge. The formal arrangement on the picture plane in this album leaf can be related to Hung-jen's powerful hanging scroll "Landacape of Wu-i" (Pl. 194). The similarity in composition and style of this leaf to the 1660 hanging scroll (Pl. 183) makes it possible to place this album to about the same year.

The undated hanging scroll "Old Trees and Short Reeds" (*Ku-ch'a tuan-ti* 古槎短荻), formerly in the Chang Heng 張珩 collection and now in the Palace Museum in Peking (Pl. 193), can be considered one of Hung-jen's major late works. The scenery is the residence of Ch'en Ying-ch'i for whom Hung-jen painted this hanging scroll and the ca. 1652 album leaf in a joint album (Pl. 48). The composition is similar to that of the third leaf from Hung-jen's 1657 album painted in Nanking (Pl. 161), particularly in such motifs as the pond in the foreground and the two trees in the middle ground. The hanging scroll, however, seems to be grander in conception; the diagonal of the pond, starting with the gate in the lower right-hand corner, leads to an empty house flanked by two vertical trees. Although empty, the house is the center of interest. Like an empty pavilion in Ni Tsan's painting, it symbolizes the lofty character of the absent resident Ch'en Ying-ch'i. The importance of the house, as in many of Hung-jen's late paintings, is brought about by the flanking trees. The two trees, the one on the left being thin and conical and the one on the right truncated, seem to be engaged in a dialogue. They also recall the now-familiar motifs in Hung-jen—a pointed peak and a squared one. The delicate equilibrium between the verticality of the two trees and the horizontality of the empty house, between the vegetation and the man-made structure, is similar to that in the 1658 short handscroll "Plum-Blossom Pavilion (Pl. 169). The groups of delicately drawn reeds in the pond provide a sense of measured rhythm to the picture. The fluent brushwork in depicting the reeds and contouring the rocks is very close to that in the 1660 hanging scroll "Pines and Rocks at Mt. Huang" in the Shanghai Museum (Pls. 183-187). It is thus possible to place "Old Trees and Small Reeds" in the years around 1660.

The undated hanging scroll "Landscape at Mt. Wu-i" (*Wu-i shan-shui* 武

夷山水) is a prime example of Hung-jen's lucidity and simplicity in his late work (Pl. 194). A comparison of the painting with a photograph taken in the early twentieth century at Mt. Wu-i (P1. 195) reveals how Hung-jen's vision of poetic solitude transforms the natural scenery into the austere silhouettes of angularity and linearity. Without committing the mistake to suppose that a landscape painting is admirable in proportion to the degree that it conforms to the scenery depicted, one will be able to see the two-dimensional balance in Hung-jen's painting as the result of definite transfiguration of space.

The artist's inscription suggests that the painting was inspired by his experience of Mt. Wu-i. He wrote: "How could Creation have composed the Nine Bends River [at Mt. Wu-i]? The spindly peaks and bulging cliffs are glassy in appearance. The Taoist's brush lies unused on the boat window sill; He can only stare with wide eyes, too bemused to poetize."[19] Despite the reference to a specific site, the painting itself exemplifies Hung-jen's conscious manipulation of forms. First of all, one immediately notices the contrast in scale between the small boat moored near the mesa in the foreground, and the sheer steep cliff that dominates the whole composition. The placement of three groups of vegetation around the triangular mesa is well thought-out; they provide the starting point to explore the spatial relationships of forms in the whole picture. On the right, the artist provides a step-like progression in depth by means of overlapping and receding forms. The spatial penetration, however, is pulled back by the angular steep cliff on the left which immediately brings the viewer's eye back to the foreground. The sheer drop of the left edge of the precipitous cliff negates the spatial recession achieved by the placement of rocks and trees on the surface of the cliff. The emphasis on overlapping of shapes results in intricate variations of straight and slightly curved lines. The sense of clarity and grandeur of this hanging scroll relates it to the 1659 "Landscape" (Pl. 174) and 1660 "Pines and Rocks at Mt. Huang" (Pl. 183) both in Shanghai and enables one to place it in the years around 1660.

A few paintings should be mentioned before we end the discussion of Hung-jen's paintings in 1660. One painting is a hanging scroll of pine, bamboo, and rocks dated 1660 (Pl. 196). The other is a hanging scroll of an old tree, bamboo, and rock which is undated but "inscribed again" in 1660 by Hung-jen and presumably datable to shortly before that date (P1. 197). The former was painted as a birthday present for a friend and seems to be more carefully executed, whereas the latter is painted to express the desolation and remoteness suggested in the couplet by the T'ang poet Wei Cheng 魏徵 (580–643) quoted in the artist's inscription: "Birds sing in the cold among old trees; Deep in mountains, one hears the cries of monkeys at night."[20] Despite their different intent and medium (the 1660 hanging scroll is painted on satin which partly explains its stiffness), both pictures are dominated by a tree which assumes an important part in the whole composition. The tall pine tree in the 1660 hanging scroll, for instance, immediately takes on a symbolic meaning of perserverence, appropriate for the occasion of birthday celebration. The undated hanging scroll is notable for the fluent brushwork on the contour lines of rocks and the old deciduous tree.

Close stylistically to these two pictures is an undated hanging scroll entitled "Red Forest and Burrowed Rocks (*Tan-lin k'o-shih* 丹林窠石) formerly in the Fujii Seido 藤井靜堂 collection (Pl. 198). It is lightly colored and depicts six maple trees in autumn. The rocks on the left side are done in the technique of "boneless" (*mo-ku* 沒骨) brushwork. Like the two paintings mentioned above, the composition emphasizes a close-up view with a tilted pictorial plane, apparently in keeping with the tendency to flat surface design in Hung-jen's late work.

The interest in the interplay of two-dimensional forms can be seen more clearly in Hung-jen's 1661 hanging scroll entitled "The Sound of Spring in a Quiet Valley" (*Yu-ku ch'üan-sheng* 幽谷泉聲) now in the collection of John Crawford, Jr. in New York (Pl. 199). Despite some damage in the lower left-hand corner and the grayish paper, the painting remains a prime example of Hung-jen's late style. Its many overlapping forms are tightly woven together into a mosaic-like construction. The slightly twisted, dry brushwork in the contour outlines of the mountain masses is a Hung-jen characteristic found in many of his paintings. The painting exemplifies Hung-jen's ordered and finely wrought design in which flat or simply-modelled areas are arranged across the picture plane with a strong emphasis on linear construction.

The flatness in the picture is largely due to the squared and vertical mountain mass that occupies almost three-quarters of the picture plane. The flatness is mitigated only by the stream, the houses, and the waterfall in the middle distance. The shifting spatial relationship between two-dimensional and three-dimensional forms can be seen not only in the pull between the small opening on the right edge and the major mountain mass, but also in the mountain itself in which the cleft in the center is pulled back by the flat shapes on two sides. The verticality of the tall foreground trees also contributes to the flatness of the picture. The trees, moreover, serve as pointers regarding the division in the mountains and give scale to other forms such as the two houses.

Looking at this painting more closely, one finds there is a delicate balance between a naturalistic scenery and a more abstract formal construction. On the one hand, the artist divided the mass of rocks into several layers and gave each layer only minimal interior texture. In order to suggest spatial recession, he placed a flat-topped ridge in the center of the picture. Along the edges of this ridge, he put a waterfall to indicate the depth of the cleft in the rock. On the other hand, the mountain mass seems to be wrapped up and enclosed in an impossibly narrow and flat space. The sheer rise of the central bluff brings it closer to the picture plane and its overwhelming size creates a compelling impression to the viewer. A very similar work, also dated 1661, is now in the Palace Museum, Peking (Pl. 200). Entitled "Snow on the Pines at Western Cliffs," it further exemplifies Hung-jen's powerful formal construction of landscape in his late work.

Because of the similarity in style, the date of the Crawford hanging scroll seems to support the placement of the undated hanging scroll "Landscape of Mt. Wu-i" to about 1660. Furthermore, the Crawford painting can be related to the undated hanging scroll in the Chih-lo-lou collection in Hong Kong (Pl. 201). The foreground is compressed and a tall and angular rock formation is immediately introduced and tilted to the left-hand side. The relationship between this rock formation and the mountain mass in the background is essentially planar; there is very little spatial recession. Whereas the trees on top of the angular rock formation serve to enclose the empty pavilion, the pine trees behind the houses squeezed between the mountain masses serve as pointers to the rock forms, echoing their placement on the picture plane.

One of the best examples of Hung-jen's late works is the handscroll "Landscape of Feng River" (*Feng-ch'i-t'u* 豐溪圖) dated to the eleventh month of 1661 and now in the Sumitomo collection (Pls. 202–206). The handscroll begins with groups of deciduous trees on a slope from which point one can see distant mountains across a river. It is a low-level view with the sky dominating. This view, however, is immediately blocked by a foreground rock formation, drawn with sharp brushstrokes and imparting a sense of flatness. As the rock formation recedes, pockets of house and trees are revealed. At the end of the handscroll, a distant mountain is placed very high, contrasting with the low-level view at the beginning of the scroll. The climax at the end is similar to the sheer presence of the stele-like cliff on the summit of the central mountain in the 1659 hanging scroll "The Unending Songs of Birds" (Pl. 173).

As in a number of Hung-jen's late works, there is a shifting relationship between two-dimensional and three-dimensional forms in the "Landscape of Feng River." In this regard, the middle section of the handscroll is particularly revealing (Pl. 204). The volumes of the mounain are constantly denied by the flatness of the contour lines. In many places, groups of trees are carefully made to bring out the enclosed architecture—a treatment found in many of Hung-jen's late works. The spaciousness thus gained, however, is in contrast to the flat rectangular rock formations that dominate the picture plane. The use of groups of trees as indicators of progression of forms can be related to the 1659 hanging scroll "Landscape" in the Shanghai Museum (Pl. 177). Furthermore, the spare use of colors in the "Landscape of Feng River" not only contributes to the austerity of the expression but also helps the viewer concentrate on the formal construction. Cahill has described the outstanding quality of this handscroll by Hung-jen:

> He draws his angular rock towers in overlapping faces with long, straggly lines, lays on dilute washes of ink and occasionally the palest of colors, adds a sparse growth of wispy trees and bushes, a few toy-like houses and a pavilion on the shore. The resulting landscape is like a construction of thin wires and glass, fragile and without solid substance. Still, it imparts a mood of lonely quiet, and seems a cool, very sensitive abstraction of mountain scenery viewed through clear air.[21]

Also dated to the eleventh month of 1661 is the handscroll entitled "Dawn on the River" (*Hsiao-chiang feng-pien* 曉江風便) now in the Anhui Provincial Museum (Pls. 207-218). Hung-jen painted it as a farewell present for his close friend Wu Hsi who was to leave for Yang-chou. Stylisically, it is very close to the 1661 handscroll "Landscape of Feng River." It can also be roughly divided into three sections, with the middle section more outstanding in the whole composition. As in the "Landscape of Feng River," the middle section exhibits a shifting relationship of two-dimensional and three-dimensional forms (Pl. 212). The mosaic-like angular rock formations, climaxed by a dense group of trees and a half-hidden pagoda, are set in contrast to the flowing, overlapping hills to the left. The sparse arrangement of predominantly dry and linear forms evokes sentiments of loneliness.

Two paintings from the last three years of Hung-jen's life provide us two examples of his transformation of the mode of Ni Tsan. In his 1661 hanging scroll entitled "Quiet Pavilion and Elegant Trees" (*Yu-t'ing hsiu-mu* 幽亭秀木), now in the Palace Museum in Peking (Pl. 224), an empty pavilion, ultimately derived from Ni Tsan, is carefully placed in a space cell formed by trees in the foreground. The viewer's eye is drawn persistently to the pavilion, a symbol of the noble hermit. The contrast between the man-made structure and the vegetation in nature, which we have seen in many of Hung-jen's paintings, is here masterfully employed to create a sense of serenity and quietude. The other painting, a small leaf from a joint album dated to 1663, now in the Ching Yüan Chai collection in Berkeley (Pl. 225), is a sketchy and modest painting. Beyond a few sparse trees in the foreground, an empty pavilion is located on a platform which in turn is placed on a raised terrace. The pavilion is flanked by bamboo on two sides. The discrete placement of the rocks and trees in the foreground and the pavilion in the background, separated by open space, creates a sense of remoteness and stability.

As an example of the unique style of Hung-jen's late years, the undated hanging scroll entitled "The Coming of Autumn," now in the Honolulu Academy of Arts , is certainly worth close examination (Pls. 219-223). In this painting, the towering vertical mountains, supported by slender, tall trees, appear to reach higher and higher into the sky. The feeling of height and monumentality is tremendous. The paper is twice as long as its width and its vertical dimensions are almost overpowering.

The tall mountain at upper right occupies the largest single area. The void or negative space is one vital geometric shape, which when viewed upside down, mirrors the tall mountain. In contrast, there is an unusual "S" submotif, as the water cascades in a zigzag manner into and across the foreground. In addition, there is still another reason for the great height, and that is the contrast in scale. It is a device we have often seen in other paintings by Hung-jen. Here, a small object like the house, in relation to the tall mountain, tends to make the latter seem still taller (Pl. 220).

While the vertical lines give validity to the rigid rocks, the lines themselves are delicate, subtle, and sensitive (Pl. 221). In counterpoint to the uncompromising mountains, the trees supply a rhythmic movement to the picture plane. Particularly interesting is the way in which the trees are grouped, repeated, and almost delightfully spaced. Starting with the lower right, it continues on directly above to a smaller group and, finally, across to the left side, where smaller and smaller groups keep appearing in the distance.

Hung-jen, with his unique vision, limits himself to a small variety of brushstrokes and an equally narrow range of tonal values. Yet, within this self-imposed range, he creates a grand composition, and totally transforms the style of the Northern Sung artists such as Fan K'uan. The poem inscribed by the artist himself reads:

In season comes the time for desolation;
A wooden hut is simple peace.
A mountain wind is off the mountain stream,
And in cold consonance are heard the stalks and branches.[22]

The fusion of poetic solitude with the austere landscape makes this painting one of the most memorable works among Hung-jen's *oeuvre* and among all of seventeenth-century Chinese painting. Although it is undated, a stylistic comparison of it with the 1661 hanging scroll "The Sound of Spring in a Quiet Valley" (Pl. 199) and the 1661 handscroll "Landscape of Feng River" (Pls. 202-206) leaves one in no doubt that it can be placed to the years around 1660.

The compelling landscape in "The Coming of Autumn" invites a comparison with the equally impressive hanging scroll "Travellers among Mountains and Streams" (*Ch'i-shan hsing-lü* 溪山行旅) by Fan K'uan in the National Palace Museum in Taiwan (Pl. 234). The similarity in details and total effect is striking: the division of the background masses into three similarly shaped sections, the tilted peak, and the slightly faded base of the central major mountain mass. In addition, the solidity, the continuous and unified surfaces of the large mountain, and the proportions of the forms are certainly reminiscent of Fan K'uan's masterpiece.

However, Hung-jen's painting has some unsettling spatial ambiguities. These ambiguities can be seen as certain deliberate tensions arising between the spatially coherent arrangement of the elements and a simultaneous play with essentially two-dimensional outlines. We have already seen many examples of this Hung-jen characteristic before. Here, within the precise and restrained landscape, Hung-jen seemed to feel the need to have an opening somewhere in the picture—the need, in other words, for suspense in this rather static arrangement of forms. He accomplished this by building some of the forms out of receding elements and then flattening them by outlining the whole unit with relatively straight lines. This unusual abstract device can be seen clearly in the right-hand mountain. The tension, in this spatially clear and structured landscape, between flat and three-dimensional forms distinguishes Hung-jen from his Northern Sung prototype. In addition, in the economy of brushstrokes, in the simplification of forms, and in texture and tonal values, Hung-jen is closer to Ni Tsan than to Fan K'uan.

Whereas "The Coming of Autumn" represents Hung-jen's transformation of the Northern Sung style, the undated hanging scroll "Trees by the Pool," depicting a figure walking on a raised zigzag path, now in Köln, exemplifies his transformation of the mode of Ni Tsan (Pls. 227–229). The trees were taller than in Ni Tsan's painting, and Ni Tsan's open pavilion was replaced by a solitary figure walking with a staff. In "The Coming of Autumn," the tall trees in the foreground are in sharp contrast to the empty house behind them (Pl. 220). In the Köln hanging scroll, similarly, the trees in the foreground are set in contrast to the figure. The raised dike-path, winding from the lower foreground toward the right and then turning to the left, seems to bring the viewer's eye back to the foreground; graphic design on the picture plane seems more important than the indication of spatial recession. The reeds on the path, like those in the tenth leaf from the Vannotti "Fishermen Album" dated 1656 (Pl. 146), provide a measured rhythm to the composition. The austerity of Hung-jen's visualization is reflected in the poem inscribed in the upper right-hand corner:

> I thrust my toes into straw sandals
> And take a few provisions with me.
> Thus I go out through the gate
> And leave my comfortable bed.
> How empty everything is, and how frosty and cold!
> Wither shall I go?
> Only the precarious nest on the jagged heights.[23]

Even in painting a subject like the plum blossom, Hung-jen was primarily interested in the angular and linear pattern and in the interplay between two-dimensional forms. For instance, in an undated hanging scroll "Plum Blossom," the middle of the main trunk is a vertical placed somewhat to the right of the picture plane, the lower part and upper part of the trunk are connected to the middle part at an almost forty-five degree angle (Pl. 230). The branches thus can be seen as variations on the main trunk. There is some overlapping to indicate the space around the trunk, but most of the branches are brought to the front of the picture plane. The trunk is modelled with a minimum of crumbly interior texture and horizontal dots on the contour lines. The delicate pattern formed by the slender branches and the few petals come close to Hung-jen's intention in the inscription:

> The courtyard is empty, the moon is without shadow;
> But my dream is warm, for snow has brought forth fragrance.

It also exemplifies the highly symbolic and abstract nature of Hung-jen's art, for there is no courtyard, no snow—only the fragrance in the cold.

CHAPTER IV

HUNG-JEN AND HIS FOLLOWERS

Hung-jen's artistic achievement was so great that he was recognized by his older contemporaries and his style attracted many followers from Anhui. We have already mentioned the 1664 colophon to Hung-jen's sixty-leaf album of Mt. Huang written by Hsiao Yün-ts'ung, who was fifteen years senior to Hung-jen. Hsiao Yün-ts'ung called Hung-jen "Master Chien-chiang [Chien-kung 漸公]" and went on to say "I am an older painter. In the art of painting, I have not yielded my claims to any predecessors. But when I see this album [by Hung-jen], I cannot help but withdraw my hand." [1] When Hsiao Yün-ts'ung wrote this, he was already seventy. Another older contemporary Ch'eng Sui, a great painter and collector himself, wrote in his colophon to the same album:

> The orthodox tradition of painting in our province [Anhui] was founded through cultivation of mind. Only Chien-chiang [Hung-jen] was without a rival····In recent years, people have been promoting Hung-jen's nephew Chiang Chu who gratefully inherited his painting abilities. Now it appears that in every corner of the earth, everyone is painting in the Chiang family style and that within the empire each member has established himself firmly as a patriarch. Chien-chiang is still alive—it is as if he had never departed.[2]

Hung-jen's contemporary Chou Liang-kung, a great art patron who knew a large number of important artists of his time, wrote of Hung-jen's critical reputation in his biographical sketches of contemporary painters: "People in the Chiang-nan [south of the Yang-tze River] region regard the possession of Hung-jen's work as the proof of good taste, just as they would a Ni Tsan work. All agree that to own a painting by Hung-jen is the same as a Ni Tsan." [3] Inevitably, great demand for Hung-jen's work led to a profusion of forgeries. As Chou Liang-kung put it, "The forgeries made by his [Hung-jen's] followers are abundant. But they all look skeletal." [4]

In art history, as Max J. Friedlander has demonstrated, the lesser masters often are indispensable in our attempt to understand the great or major masters.[5] It is with the purpose of both further defining the unique style of Hung-jen as well as assessing his influence that we turn to his recognized followers. Selected paintings by the following artists will be discussed: Chu Ch'ang 祝昌, Cha Shih-piao 查士標, Chiang Chu 江注, and Yao Sung 姚宋.

Among the followers of Hung-jen, Chu Ch'ang was perhaps the oldest, for by 1651 he was already associated with the older master. In 1651, he participated in painting the joint handscroll for Mr. Ko-an with Hung-jen and some other artists; the handscroll has been briefly discussed in Chapter II (Pls. 31-39). His exact dates are not known: we only know that he became a *chin-shih* in 1649, and was active well into the 1670's.[6] He lived for a considerable period of time in She-hsien where he naturally came into contact with Hung-jen. According to Chang Keng who wrote his biography, Chu Ch'ang later expanded his stylistic range into "various masters" of the Yüan dynasty.[7] His extant works are extremely rare. In addition to the section in the 1651 joint handscroll just mentioned, few works have survived. There are an undated hanging scroll of landscape published in *Shen-chou ta-kuan* 神州大觀 (Pl. 248),[8] a handscroll dated 1666 now in the Suchou Museum (Pls. 249-266),[9] an undated hanging scroll in the Hashimoto Sueyoshi 橋本末吉 collection,[10] and finally a handscroll dated 1662 (Pls. 241-246) and an album leaf datable to 1663 (Pl. 247, in the same album for Mr. Chung-fu to which Hung-jen also contributed one leaf) now in the Ching Yüan Chai collection in Berkeley.

Chu Ch'ang's section (Pls. 34-35) in the 1651 joint handscroll is largely constructed with flat and angular shapes accentuated only by vegetation strokes and dots. While it seems, in style, to anticipate the geometric forms in his later works, the composition is rather static. The influence from Hung-jen was hardly present. In 1663, however, it is the bold division of the pictorial surface into a few large units that distinguishes Chu Ch'ang's album leaf from his 1651 landscape. In this process of stylistic development of Chu Ch'ang, Hung-jen must have played a formative role. The following examples shall offer a convincing illustration.

According to Chu Ch'ang's inscription, the 1662 handscroll (Pls. 241-246) in the Ching Yüan Chai collection (formerly in the Frank Caro collection) was inspired by a handscroll entitled "A Panoramic View of Rivers and Mountains" (*Chiang-shan sheng-lan* 江山勝覽) by Huang Kung-wang which he had earlier seen in the collection of a certain Wu Ts'an-ju 吳燦如.

The ostensible reference to the Yüan master, however, should in no way distract our immediate attention from Chu Ch'ang's stylistic indebtedness to Hung-jen. The handscroll begins with an overlapping, or more exactly a superimposition, of cliffs one on another, clearly setting forth the dynamic orchestration of angular shapes in the rest of the whole handscroll; this tight formal construction certainly was derived from Hung-jen. Even the overlapping trees on the cliffs suggest the older master's preference for the eccentric and compressed pictorial statement. In fact, Chu Ch'ang seemed to have been so carried away by the sheer force of Hung-jen's style that he became overcomplicated. Again, the very beginning section of the handscroll offers some hints. The path, immediately coming behind the opening cliff and connected to the slightly arched bridge, is at once obscured by a series of rocks and then is broadened into the familiar dike-path which in turn is hidden behind a wedge-shaped rock formation in the foreground and close to the lower edge of the handscroll.

The whole handscroll is built up with a series of thrusts and counterthrusts which gives it a sense of movement onward, until we reach the cliff at the end where our eye is forced to stop abruptly and descend, as if too early, from our enjoyment of the panoramic view of the landscape; the overhanging pines on the cliffs at the end of the handscroll echo those same trees at the beginning.

Chu Ch'ang was evidently interested in dividing the pictorial surface into a few large units and in the cohesiveness of the total structure. He borrowed a number of motifs from Hung-jen, such as the dike-path (section 5), the open pavilion on the edge of the land spit or a plateau (section 4), the squarish cliff (section 4). The structural clarity of the older master, however, has been disquietingly overwhelmed by a profusion of details in the Chu Ch'ang handscroll. It is when Chu Ch'ang was not so much concerned with details themselves but with the structural interplay of shapes on the pictorial surface, for instance in the 1663 album leaf and the undated hanging scroll published in the *Shen-chou ta-kuan* (Pl. 248), that he was most successful in using Hung-jen's style for his own purpose. Stylistically, the *Shen-chou ta-kuan* hanging scroll is so close to the 1662 handscroll, particularly the beginning section, that it can be dated also to the 1660s.

Chu Ch'ang's 1666 handscroll entitled "Streams and Mountains without End" (*Ch'i-shan wu-chin* 溪山無盡), now in the Suchou Museum (Pls. 249–255), is among the more successful adaptations of Hung-jen by the younger artist. In particular, similar to the *Shen-chou ta-kuan* hanging scroll, the strong diagonal movements of shapes create an exuberant rhythm throughout the first half of the handscroll. The second half was apparently derived again from

Huang Kuang-wang, as the artist has told us in his inscription at the end of the handscroll. He succeeded in integrating several motifs from Hung-jen's repertoire, such as the twisting trees behind the open pavilion (section 6), the overhanging pine trees (section 6 and section 7), in a Huang Kuang-wang framework.

Although Hung-jen's influence can be decidedly felt in a great number of works by Cha Shih-piao, the latter was more independent-minded and even more eclectic than other followers of Hung-jen such as Chiang Chu 江注 and Yao Sung 姚宋 to be discussed later on. Cha Shih-piao often painted in a broader style which is closer to the orthodox style derived from Tung Ch'i-ch'ang whom he copied extensively, for example, in an undated album in the Wang Chung-wei 王壯爲 collection in Taipei.[11] He nevertheless was interested in following Hung-jen's more austere and angular style. And when he chose to do so, he could come very close to the older master.

The personal contacts between Hung-jen and Cha Shih-piao might have dated from the early 1650's. Together with Hung-jen and other Hsin-an artists such as Wang Chih-jui, Cha Shih-piao contributed a leaf to a joint album for Ch'en Ying-ch'i briefly discussed in Chapter II; Wang Chih-jui's leaf is dated 1649, Hung-jen's has been stylistically dated to about 1652, and Cha Shih-piao's leaf is dated 1653. Hung-jen's undated hanging scroll for Cha Shih-piao, entitled "Tall *Wu-t'ung* and Quiet Bamboo" (*Kao-t'ung yu-hsiao* 高桐幽篠) with a colophon written by Sun I, formerly in the Hsü Ch'eng-yao 許承堯 (twentieth century) collection is now in the Anhui Provincial Museum.[12] Hung-jen's paintings of plum and bamboo with inscribed poems for Cha Shih-piao has been recorded.[13] An album of paintings to which Cha Shih-piao added calligraphy has also been recorded.[14] Cha Shih-piao's colophons on Hung-jen's other paintings suggest the extent of their artistic contacts.[15] Cha Shih-piao, for example, offered this keen observation on the stylistic development of Hung-jen in a colophon to the latter's sixty-leaf album of Mt. Huang: "The painting style of Master Chien [-chiag] changed when he went to Mr. Wu-i [in Fukien]; after his return to Mt. Huang, his style became even more wonderful. The ancients take the ten thousand things under heaven and on earth as their teachers. Isn't this approach the more appropriate when one considers that mountains and streams can transform one's emotions and can unify oneself with painting? I have heard that, in order to become a master, one should first read ten thousand books and travel ten thousand miles. Today I am convinced when I look at the paintings of Mt. Huang by Master Chien [-chiang]."[16] Cha Shih-piao must have been quite familiar with Hung-jen's work in order to make such a detailed observation on Hung-jen's development.

Both Cha Shih-piao and Hung-jen shared a common interest in the style of Ni Tsan. Moreover, Cha Shih-piao's conception of the "correct" style is very much in tune with that of Hung-jen. He pointed out to his younger contemporary Wang Hung-tu 汪洪度 (1646–died after 1721) that there were two essentials in the art of painting, namely simplicity (*chien* 簡) and austerity (*leng* 冷), adding that only Ni Tsan and Hung Kung-wang, among the "Four Masters of the Yüan," understood these principles.[17]

Several Cha Shih-piao paintings close in style to Hung-jen can now be considered. The landscape fan dated 1659 by Cha Shih-piao (Pl. 256), now in the Steve Addiss collection in Lawrence, Kansas, reveals strong influence from the older master. Though loosely executed, it retains the structure of Hung-jen as seen, for instance, in the first leaf of the 1657 Nanking album for Mr. Tsung-hsüan (Pl. 159). In addition, Cha Shih-piao showed his concern for a tightly knit surface design and his adaptation of such motifs characteristic of Hung-jen as the zigzag plateaux and the dike-path. The contrast of dark washes against empty paper and lightly textured peaks imparts a strong sense of surface pattern to the painting.

Similarly, an undated landscape fan (Pl. 257), now in the K. C. Ho collection in New York, shows Cha Shih-piao's strong stylistic affinity with Hung-jen. The angular forms on a flat surface, carefully but broadly divided into a few large sections, reveals Cha Shih-piao's fascination with the older master's simple yet sophisticated sense of surface design. Moreover, Hung-jen's economy of artistic means is taken over successfully by Cha Shih-piao, whose artistic temperament, though, tended to a freer and looser handling of brush. An album of landscape by Cha Shih-piao, dated 1662, now in the T'ien-yin-t'ang collection in Tokyo, contains several leaves (Pls. 258–260) that demonstrate even more eloquently the influence of Hung-jen's seminal visualization on Cha Shih-piao's style.[18]

Cha Shih-piao's 1678 fan entitled "Pavilion in Autumn Morning" (*Lin t'ing ch'iu-hsiao* 林亭秋曉) now in the Shanghai Museum (Pl. 261),[19] for example, further reveals Hung-jen's influence. It is, above all, characterized by a tendency toward geometric and angular forms to build up a surface design. The solitary pavilion under trees is ultimately derived from the Ni Tsan mode. Cha Shih-piao's interest in the creation of structure through spare linear means, however, is evidently derived from Hung-jen. Note, for instance, the zigzag path and the emphasis on contour outlines of certain rocks.

Marilyn and Shen Fu have compared Cha Shih-piao's 1678 fan to an undated fan painting by Hung-jen, now in the Vannotti collection (Pl. 226). They wrote of the similarities and the differences between the master and the follower:

> The composition and the angular geometric rocks and pathways are almost identical and are definite Hun-jen elements. Differences in the two hands are nonetheless discernible in the brushwork: Cha's overall use of ink is moister, his ink gradations darker, and his brushline much less tense and attentive to the articulation of the joints and angles of forms. He tends to use an upright brush (as opposed to Hung-jen's oblique brushhold) moving in a relatively quick and jagged rhythm (as opposed to Hung-jen's slower, more consistent and penetrating pace).[20]

A younger follower of Hung-jen, Chiang Chu 江注 was born in She-hsien and was a nephew of the older master.[21] His dates are not known. Like his uncle, Chiang Chu was fascinated by the scenery of Mt. Huang and would go there very often to stay for as long as a whole month. His poems on Mt. Huang were highly regarded by the famous contemporary poet and critic Shih Jun-chang 施潤章 (1618-1683).[22] Although primarily known as a poet, Chiang Chu was spoken of by the biographer of Hung-jen, Wang T'ai-cheng 王泰徵 (1600-1675) as having succeeded in transmitting the style of Hung-jen.[23] According to his friend Yin Shu 殷曙 (1623-1692), Chiang Chu lived together with Hung-jen at Mt. Huang and experienced the same scenery as the older master. We also know that they even stayed together and maintained the same residence in the city of She-hsien.[25] Chiang Chu wrote a colophon on Hung-jen's hanging scroll entitled "Quiet Pavilion and Elegant Trees," dated 1661 and now in the Palace Museum in Peking, attesting to their teacher-pupil as well as family relationship.[26] In addition, he also wrote a colophon on Hung-jen's 1656 "Album of Fishermen" in the Vannotti collection.

It is thus not surprising to find that the few surviving works by Chiang Chu all reveal a strong stylistic affinity with Hung-jen. Perhaps the closest is the undated hanging scroll "Pines and Rocks at Mt. Huang" (Pl. 262).[27] The subject matter is a favorite one for both Hung-jen and Chiang Chu. The younger artist, like his master Hung-jen, was interested in a close-up view, showing the rugged rocks and the twisting pines which cling to the rocks precariously and yet grow out of them so triumphantly against a seemingly limitless sky. The rocks, basically wedge-shaped, are outlined with brisk and modulated lines; there is little interior texturing besides a few vegetation dots and spikes. The central rock, in particular, reveals the dynamic patterning by its twisting movement and its slight sway to the right; the verticality is echoed by the rocks on the left, while the twisting movement is paralleled by the rocks on the lower right corner. Compared with Hung-jen's works, for instance, the magnificent hanging scroll "Pines and Rocks at Mt. Huang" dated 1660 and now in the Shanghai Museum (Pls. 183-187), however, Chiang Chu's painting, while taking over Hung-jen's style with a measure of success, seems more cluttered with details and lacks the lucid order as well as the forceful interlocking of the older master. The trunks of some pine trees, for instance, are contorted. Hung-jen's hanging scroll, by taking a much higher point of view and by juxtaposing the near with the very far peaks, creates a more dramatic and powerful visualization.

Chiang Chu's undated hanging scroll entitled "Arranging Rocks and Planting Flowers" (*Hsüan-shih tsai-hua* 選石栽花), formerly in the Richard Bryant Hobart collection in Cambridge , Massachusetts,[28] while still retaining stylistic influence from Hung-jen, indicates the change and deviation that took place even in the first generation of Hung-jen's followers. First of all, the austere linearity of Hung-jen began to be combined with broader brushwork to build up a pattern of geometric shapes on the pictorial surface. The composition, built up of repetitive units, became more static, because the masses often were in a state of heavy and almost sticky plasticity. Compared with Hung-jen's hanging scroll "The Coming of Autumn" in Honolulu (Pls. 219-223), for example, Chiang Chu's painting evidently lacks the surface animation and coherence of the older master. Instead of Hung-jen's artistic economy, moreover, Chiang Chu crowded the composition with details which were repetitively piled up; the forms were much weaker in their articulation.

Yao Sung 姚宋 (*tzu* Yü-chin 雨金, or Yü-ching 雨京, *hao* Yeh-mei 野梅) another close follower of Hung-jen, was born in 1648 in Hsin-an. His date of death is unknown, but must be sometime after 1717. The dates are obtained from his inscription dated 1717 on his hanging scroll entitled "Leisurely View in the Autumn Mountains" (*Ch'iu-shan hsien-t'iao* 秋山閒眺) now in the Nanking Museum (Pl. 266).[29] He lived, however, in Wu-hu. Yao Sung seems to have been too young to have had personal contacts with Hung-jen. According to Chang Keng, he was versed in a variety of subject matters as well as in miniature painting on tiny fruit seeds. Like Chu Ch'ang and Chiang Chu, Yao Sung left very few paintings.[30] One of the most impressive among his extant paintings is, perhaps, the hanging scroll "Landscape" dated 1707 now in the Charles Drenowatz collection in Zürich (Pl. 264). Chu-tsing Li has described it very well:

> A landscape of angular interlocking forms unfolds before the viewer. In the foreground three giant trees rise from a hummock planted near the lower left edge. Their trunks and branches cross each other in sharp turns and angles. Across the river two cliffs rise sharply, echoing the formation and movement of the three pines in verticality and angularity. Beyond, through the cleft between the cliffs, is another group of mountains seen only in fragments. The most outstanding details of this distant scene is a waterfall done in white against especially dark ink tones.[31]

Indeed, as Chu-tsing Li has pointed out, it is characterized by arbitrariness and ambiguity in spatial relationship. Apparently the artist was more interested in the "formal arrangement of an almost surreal quality" and "a world of fantasy." Li seems to see little, if any, influence from Hung-jen in the Yao Sung. A comparison with Hung-jen's grand hanging scroll "The Coming of Autumn" in Honolulu (Pl. 219), however, reveals that Yao Sung was greatly indebted to Hung-jen not only in the composition, but also in the treatment of the major motifs such as the rock formations, the trees, the waterfall, and the pavilion on the edge of the plateau.

Two tiny figures have been added to the sparse and austere world of Hung-jen. They provide a scale in contrast to the overwhelming mountain in the background. The upper part of the central mountain is decorated with a profusion of dots, spikes, and comb-shaped miniature pines, thus creating a patternistic effect. The formula of the building up of geometric shapes, borrowed from Hung-jen, is particularly evident in the steep terrace in the lower left corner where the two figures are conversing, and in the overlapping cliffs on the right.

Yao Sung's undated hanging scroll entitled "High Dwelling in a Quiet Valley" (*Yu-ho kao-ch'i* 幽壑高棲), now in the Hashimoto collection, belongs to the same formula (Pl. 265).[32] He has captured the tight structure characteristic of the older master. Though painted on silk which tends to make the brushwork appear less fluent, Yao Sung's geometric shapes are integrated into the mountain formation. The vegetation accent, however, appears only on the surface of earth or rocks, making the forms less coherent.

The composition and brushwork of Yao Sung's 1717 hanging scroll "Leisurely View in the Autumn Mountain" in Nanking (Pl. 266), mentioned above, are once again derived from Hung-jen. Moreover, we have a number of motifs which are unmistakably based on those of the older master, such as the zigzag plateaux, the leafless twisting old trees, and sharply-drawn reeds and bamboo. In general, however, the Yao Sung painting does not succeed in capturing the older master's lucid visualization. The hasty handling of the brush, for instance, in the repoussoir rocks in the lower part of the foreground as well as in the highest peaks, shows that the artist's intention lies more in the movement of the individual brushstroke than in the integration of forms on the pictorial surface. In the same way, the short horizontal dots for vegetation on the central rock formation only serve to decorate rather than to give form to the masses. The handling of architectural elements, in addition, further points to Yao Sung's jerkier brushwork and less sure hand.

More importantly, perhaps, the tight structure of Hung-jen has been replaced by a pattern of disparate units. In other words, in Yao Sung's painting, the forms are less strongly defined and less firmly joined together. We can compare Yao Sung's painting with Hung-jen's hanging scroll "The Coming of Autumn." One immediately notices Hung-jen's masterful use of the foreground trees through which the house in the middle distance is seen (a device which lucidly brings about the powerful mountainscape in the composition). On the other hand, Yao Sung's two houses are so enlarged that the intended contrast in scale is completely lost. Moreover, Yao Sung's trees, compared with Hung-jen's, are hardened, thickened, and made more self-consciously angular. In the Hung-jen, we find an equilibrium and clarity. Yao Sung's painting, in comparison, is rather bland and crowded with details. Compared with most of his followers, Hung-jen is often more concentrated and absorbed, and is able to render more powerful the grandeur of the landscape scenery he is depicting. Despite the complex composition, moreover, Hung-jen's hanging scroll is clear and is animated by a continuous rhythm, with larger and smaller accents well related and with a distinct indication of recession in space. In contrast, Yao Sung's accents on the major central mountain are less incisive and serve no formal function. The penetration into distance at the left and center is overpowered by the flat mountain, and the overall articulation of form is less clear.

Stylistically related to the 1717 hanging scroll in Nanking is Yao Sung's undated album of eight leaves of landscape now in the Vannotti collection in Lugano, Switzerland, formerly in the J. P. Dubosc collection, Paris (Pls. 267 -274).[33] Hung-jen's tendency to compress his composition, particularly with the format of album leaf, is evident in the Yao Sung album: for instance, the crossing of strong diagonals (leaf no. 2, leaf no. 4, and leaf no. 5; Pls. 268, 270, 271); the opening through narrow cleft in the middle of the pictorial surface (leaf no. 3, leaf no. 8, and to a lesser extent, leaf no. 6; Pls. 269, 274, 272). Even the eccentric placement of artist's seal is evidently derived from the older master (leaf no. 3, leaf no. 4, and leaf no. 6; Pls. 269, 270, 272). The vegetation accents such as dots, spikes, and hooks, which Hung-jen often used to accentuate his rock formations, are overdone here and have become so repetitive as to have a disruptive effect on the whole picture, particularly in leaf no. 3, leaf no. 4, leaf no. 6, leaf no. 7, and leaf no. 8 (Pls. 269, 270, 272, 273, 274). The brushwork also tends to become less firm than Hung-jen's. Compare, for instance, the leafless old tree in Hung-jen's 1657 Nanking album (leaf no. 1; Pl. 159) with those in Yao Sung's leaf no. 1, leaf no. 7, and leaf no. 8 (Pls. 267, 273, 274). Yao Sung's treatment in general is much harsher, and we therefore miss Hung-jen's coherence of design in most of Yao Sung's album leaves.

The style of some of Yao Sung's painting seems to suggest the possible intrusion of the style of the orthodox school led by the Four Wangs (Wang Shih-min 王時敏 [1592-1680], Wang Chien 王鑑 [1598-1677], Wang Hui 王翬 [1632-1717], and Wang Yüan-ch'i 王原祁 [1642-1715]) into the style of artists who followed Hung-jen in the late seventeenth and early eighteenth century.[34] However unique his vision might be, Hung-jen's influence became rather diminished even among the first generation of his followers who were less talented and did not have the same vision as the older master. Chiang Chu's hanging scroll "Arranging Rocks Planting Flowers," formerly in the Hobart collection and discussed briefly above (Pl. 263) showed such a change. The same deviation from Hung-jen's style is also evident in several leaves from an undated album of landscapes by Yao Sung formerly in the Hobart collection (Pls. 275-282).[35] For example, leaf no. 6 (Pl. 280), while preserving some Hung-jen motifs (such as the overlapping rocks in the left, the house under pines in the lower right, the dike-path in the middle ground, and the zigzag plateau), is a far cry from the tightly woven structure of the older master. The repetition of type-forms, the uniformly applied texture strokes, and the rather static composition, which dominates so many of the leaves in the album, are clearly related to the style of the orthodox school.

The greatness of Hung-jen's art, and his difference from his followers, lies in the fact that he never submitted to a multiplication of details that led to mere patternistic form. Instead, he always showed a consistent inclination toward linear expression that emphasized articulated form as well as coherent design. His angular and very often geometric shapes were selected with concentration and artistic economy. As we have seen, he achieved his strength through abstraction and simplification, as well as through a tight compositional order. Unfortunately, in the hands of his followers, his sense of order and structural clarity tended to become weakened, and his brushwork of acuity and resolution became repetitious and meaningless, unrelated to the total rhythm of the composition. As we have briefly pointed out, many accent dots or strokes in the works of Chu Ch'ang and Yao Sung, for instance, became too scattered and rhythmically unrelated to the contour lines of rocks.

CONCLUDING REMARKS

In the first edition of the *Chieh-tzu-yüan hua-chuan* 芥子園畫傳 (*Mustard Seed Garden Painting Manual*) compiled by Wang Kai 王槪 (ca. 1650–ca. 1710) and published in 1679, sixteen years after Hung-jen's death, there is a painting with a signature which reads "Hung-jen" and a double-square seal which reads "Chu-ch'uang" 竹窗.[1] The painting (or, strictly speaking, the woodcut) has touches of light blue and brownish-red colors on rocks and mountains. It is, of course, impossible to determine whether it was indeed based on an original of Hung-jen or whether it was merely a fabrication on the part of the compiler Wang Kai. What is interesting, however, is that Hung-jen's fame in painting had spread to Nanking, where the manual was published and where the Nanking School had been developing under the influence of Kung Hsien. Evidently, Hung-jen's unique style had come to be regarded highly as worth emulating by artists in the Nanking region only sixteen years after the master's death. Significantly, the rest of the "Four Great Painter-Monks," K'un-ts'an, Chu Ta, and Tao-chi, did not appear in this manual. Moreover, for the well-known art collector and patron Chou Liang-kung, Hung-jen seems to have enjoyed a higher position than K'un-ts'an. In his *Tu-hua-lu* 讀畫錄 (1670), he mentioned only two of the four artists, Hung-jen and K'un-ts'an, suggesting Hung-jen and K'un-ts'an were better known than Chu Ta and Tao-chi. The only other painter-monk mentioned by Chou Liang-kung was Fang I-chih. [2]

It was precisely thirty years after Hung-jen's death when, in 1694, Tao-chi wrote, in his inscription on a leaf from his own "Album of Landscape" now in the Los Angeles County Museum of Art, a lengthy evaluation of his older contemporaries, thereby putting Hung-jen into the artistic context of the seventeenth century (Pl. 283). Tao-chi wrote :

> Those who enter through the ordinary gate to reach the *Tao* of painting are nothing special···· But to achieve resounding fame in a given age—isn't that difficult to accomplish? For example, the lofty antiquity of K'un-ts'an, Ch'eng Cheng-k'uei and Ch'en Shu; the pure elusiveness of Cha Shih-piao and Hung-jen; the parched leanness of Ch'eng Sui; the drenched moistness and rare antiquity of Pa-ta-shan-jen [Chu Ta] from Nan-ch'ang; and the untrammeled expressiveness of Mei Ch'ing and

Mei Keng. Of a generation these were all the ones who understood.[3]

Although he rightly recognized the stylistic affinity between Cha Shih-piao and Hung-jen, Tao-chi's characterization of their style as "pure elusiveness" (*Ch'ing-i* 淸逸), apart from paying his homage to the older masters, was vague. Tao-chi offered a more specific view of the tyle of Hung-jen, however, in another comment he made in 1696 in a colphon to Hung-jen's 1661 handscroll "Dawn over the River" now in the Anhui Provincial Museum. Tao-chi wrote:

> Since Ni Tsan, the style of loftiness and gracefulness has rarely found its followers; only Hung-jen was able to transform it. Therefore, those who [claimed] to have followed Ni Tsan actually derived their style from Hung-jen. He stayed at Mt. Huang longer [than other artists]. He was able to capture the true nature and feeling of Mt. Huang. His trees and rocks are indeed true expression of Mt. Huang: angular, cool, and lively····[4]

These two observations by Tao-chi are but some examples of the extent to which the artistic achievement of Hung-jen earned the respect of many contemporaries and younger contemporaries.

The eighteenth-century critic Chang Keng, in his *Kuo-ch'ao hua-cheng-lu* 國朝畫徵錄, offered his evaluation of Hung-jen's art. He wrote:

> Hung-jen was versed in poetry, prose, and landscape painting. He followed Ni Tsan in particular and led the way for many of the Hsin-an painters who then clung to the Ch'ing-pi [Ni Tsan] trandition···· I have seen his authentic painting which, representing layers of cliffs and steep gullies, majestically lofty and imposing, are not like those pictures of scattered bamboo and leafless trees which have been made by the ordinary painters who, nevertheless, call themselves lofty hermits.[5]

It was Chang Keng's critical acumen which enabled him to see the monumentality in Hung-jen's style—"layers of cliffs and steep gullies, maiestically lofty and imposing." For Chang Keng, Hung-jen and K'un-ts'an still represented higher position than Tao-chi, while Chu Ta, though mentioned, was not yet considered as one of the "Four Great Painter-Monks."

By mid-eighteenth century, the school of Hsin-an as established by Hung-jen was regarded as one of the major regional schools by Chang Keng who nevertheless pointed out the weakness in each of them, presumably caused by the less-talented followers of the older master. He wrote:

> The division of painting into a Southern and a Northern School started in the T'ang period, but that division in schools was not in accordance with localities. At the end of the Ming Dynasty the name of the Che school was introduced. This school started with Tai Chin [born 1388] and ended with Lan Ying [1585 - after 1660]. It had four faults; it was hard, stiff, bare, and clumsy. The school of Sung-chiang started in the present [Ch'ing] Dynasty; it followed Tung Ch'i-ch'ang and Chao Wen-tu [Chao Tso, active ca. 1600 - 1630]. The paintings were done in a vaporous and diffused manner and became gradually delicate, soft, sweet, and weak. In the Chin-ling [Nanking] School there were two divisions, one resembling the Che and another resembling the Sung-chiang School. The Hsin-an School attracted many people, since Master Chien [-chiang] excelled in the manner of Yün-lin [Ni Tsan]. If it has not the fault of knotting [tying], it has the fault of scattering brush-work. That is the third school···· These various schools were all famous at their start, but in the course of their transmission they deteriorated through successive imitation.[6]

The decline of Hung-jen's reputation in the eighteenth century was probably due more to the lack of inventiveness among his followers and the lack of his genuine works seen by critics than to change of taste. In fact, as early as the seventeenth century, Chou Liang-kung pointed out: "The forgeries made by his followers are abundant. But they all look skeletal."[7] The late eighteenth and early nineteenth-century artist and critic Hsieh Lan-sheng 謝蘭生 (1760-1831) criticized the followers of Hung-jen and even Hung-jen himself. He wrote:

> One kind of painting by Ni [Tsan] possesses the quality of luxuriance, depth, smoothness, and substance. Recent artists have only imitated the dryness and leanness of his style. They also have praised highly Chien-chiang [Hung-jen], saying that among those who followed Ni Tsan, this monk cannot be surpassed. They do not know that the style of Chien-chiang is among the driest and baldest. Once addicted, this bad habit cannot be cured."[8]

In all probability, however, Hsieh Lan-sheng's adverse criticism of Hung-jen was based on run-of-the-mill imitations of Hung-jen or even downright forgeries. For among the prominent collectors in Kwangtung at that time, such as

P'an Cheng-wei 潘正瑋 (1791-1850), Yeh Meng-lung 葉孟龍 (1778-1832), Liang T'ing-nan 梁廷枏 and K'ung Kuang-t'ao 孔廣陶, only two Hung-jen paintings were known to be collected.[9] Although Hsieh Lan-sheng's own painting reveals some influence from Cha Shih-piao, it is unlikely that he had the opportunity to study any genuine Hung-jen work to arrive at a judicious assessment.[10]

Despite the common association of Hung-jen, K'un-ts'an, Chu Ta, and Tao-chi as the "Four Great [Painter]-Monks," it should be pointed out, their life patterns were not all the same. Tao-chi and Chu Ta were related to the Ming royal family, and therefore their decision to become monks was perhaps more compelling, even necessary, in order to avoid persecution by the Manchus. K'un-ts'an and Hung-jen, on the other hand, became monks mostly out of a conscious choice. K'un-ts'an, in particular, was known for his fervid religiosity and understanding in Buddhism.[11] The later part of their life was also different. Chu Ta, for instance, changed from a Buddhist monk to a Taoist, and might well have returned to secular life, married and raised a family.[12] Tao-chi became involved with the Manchu government and patrons. He twice, in 1684 and 1989, had audience with the K'ang-hsi emperor.[13] In contrast, Hung-jen and Kun-ts'an never compromised.

Chang Keng in the eighteenth century spoke of only three painter-monks, i.e., Hung-jen, K'un-ts'an, Tao-chi, while Chu Ta was not considered a painter-monk with all the connotations of the term, but only regarded as an eccentric painter.[14] The grouping of the "Three Monks" (K'un-ts'an, Hung-jen, and Tao-chi) continued into the twentieth century in critical literature. For instance, Huang Pin-hung still followed the same grouping in his *Ku-hua-wei* 古畫微 published in 1925.[15] Only in 1936 did critics begin to group the four artists together.[16]

But more importantly, what distinguishes Hung-jen from other monk-painters, and from most of the individualist artists in the seventeenth century, is perhaps that he seemed to be the only master who had a number of followers working closely in his unique style, and who eventually became the virtual leader of the Hsin-an School. One explanation would be that his style exhibited a highly personal transformation of the Ni Tsan mode which by this time had become inimitable, something that later artists could only aspire to. Moreover, Hung-jen's unique style, derived from the Wen Cheng-ming School, emerged as a viable solution to the contemporary artistic problem. Still another explanation, which has been previously suggested in Chaper II, would be that the Hsin-an artists seemed to have a stronger sense of regional pride which tended to encourage a regionalism and cohesiveness in their style.

In the study of the so-called "post-Tung Ch'i-ch'ang" era (if indeed there can be such a division) in the history of seventeenth-century Chinese painting,[17] it is easy to see those who follow him as revolutionaries and those who followed the Wen Cheng-ming School as reactionaries. This view may be an oversimplification of the state of scholarship on later Chinese painting, but the fact remains that recent art-historical studies, with few exceptions, tend to overlook the continuity of the Wen Cheng-ming School and its stylistic influence. It is easy to forget that in art, as in politics, it is possible for a reactionary to have virtues and for a revolutionary to have faults. The degeneration of the orthodox school of the teaching of Tung Ch'i-ch'ang would be an example of how the iconoclasm of the master can become the pretext for careless literati dabblings.

Hung-jen's greatness as an artist lies in his integration of a variety of past styles into his own. He expanded the formal as well as expressive possibilities suggested in the work of Ni Tsan. He was also able to develop further the arrangement of shapes on the picture plane of the Wen Cheng-ming School. In Hung-jen's masterful composition, forms are often reduced into geometric shapes. In many cases, the conventions are transformed by Hung-jen through a clear and firm conception of forms and the sensitive subordination of details in the interest of an over-all design. The result is a striking and inventive reinterpretation of tradition and an art of austerity and concentration.

FOOTNOTES TO INTRODUCTION

[1]This appellation is used, for example, as the title of reproductions of their paintings, *Ming-mo ssu-seng hsüan-chi* 明末四僧選輯 (Hong Kong: Pan-tao Chung-kuo shu-hua ch'u-pan-she, n.d.). Cf. Kuo Wei 郭威, comp., *Pa-ta, Shih-t'ao, Chien-chiang, Shih-ch'i ssu-seng hua-chi* 八大石濤漸江石谿四僧畫集 (Taipei: Ya-yün-t'ang, 1972).

[2]As our knowledge of the history of seventeenth-century Chinese painting increases, it now appears more appropriate to describe the style of Hung-jen and his followers simply as the "Hung-jen School" or broadly as the "Hsin-an School." There is a comparable phenomenon in the history of sixteenth-century painting. Recent studies of Wen Cheng-ming have revealed the fact that the "Wu School" of Shen Chou and Wen Cheng-ming simply did not exist and have prompted the art-historian Ellen Laing to ascribe the term "Wen Cheng-ming School"to Wen Cheng-ming and his followers, discarding the unfortunately ambiguous term "Wu School." (Ellen Johnston Laing, "The State of Ming Painting Studies," *Ming Studies,* no. 3 [Spring, 1977], pp. 9-25, at p. 14.) In the same way, it is proposed that the term "Anhui School," too broad to be useful, should be used with caution, for it refers to the artists' geographical origin rather than their specific stylistic affinity.

The term "Hsin-an School" has often been used loosely. For example, in the *Hsin-an-p'ai ming-hua-chi* 新安派名畫集 ("A Collection of Famous Paintings of the Hsin-an School") published by the Shen-chou kuo-kuang-she in Shanghai in 1924, the following artists are included: Hsiao Yün-ts'ung, Hung-jen, Wang Chih-jui, Cha Shih-piao, Cheng Min 鄭旼, Ch'eng Hsün 程壎 (*tzu* Tu-shan 讀山, *chin-shih* 1754), Mei Ch'ing (1623–1697), Mei Keng 梅庚 (*chü-jen* 1681, brother of Mei Ch'ing, both are from Hsüan-ch'eng, Anhui), Huang Lü 黃呂 (*tzu* Tz'u-huang 次黃, *hao* Feng-liu shan-jen 鳳六山人). When used so loosely, the term becomes merely a twentieth-century reference to almost every artist from the Anhui province regardless of historical positon or style. The classifcation is based on regional origin rather than style.

In most cases, however, the term "Hsin-an School" has been correctly used by the authors of leading general texts on the history of Chinese painting published in the early twentieth century, to refer to the "Four Great Masters of Hsin-an" or the "Four Great Masters of Hai-yang" including Wang Chih-jui, Sun I, Hung-jen, and Cha Shih-piao, and to followers of Hung-jen. For example, see Ch'en Shih-tseng 陳師曾 (Ch'en Heng-ch'üeh 陳衡恪), *Chung-kuo hui-hua-shih* 中國繪畫史, (Tsi-nan: Han-mo-yüan mei-shu-yüan 翰墨緣美術院, 1925), p. 46b; P'an T'ien-shou 潘天壽, *Chung-kuo hui-hua-shih* (Shanghai: Commercial Press, 1926), p. 208; Yü Chien-hua 俞劍華, *Chung-kuo hui-hua-shih* 中國繪畫史 (Shanghai: Commercial Press, 1937), pp. 171-172; Cheng Ch'ang 鄭昶, *Chung-kuo hua-hsüeh ch'üan-shih* 中國畫學全史 (Shangai: Chung-hua shu-chü, 1928), p. 440.

In 1958, Osvald Sirén wrote: "The group commonly known as 'The Masters of Hsin-an' (in Anhui) includes four painters active at the end of the seventeenth century who, as far as may be concluded from their still preserved works, had not very much in common from a stylistic point of view, yet formed the nucleus of an important group which included some of the most attractive painters of this epoch." See Osvald Sirén, *Chinese Painting: Leading Masters and Principles,* 7 vols. (New York: The Ronald Press Company, 1956-58), vol. 5, p. 114.

Sirén was right in using the term "The Masters of Hsin-an." There are, however, a few points to be clarfied here. First of all, the four artists were not active at the end of the century, as Sirén suggested, but rather in the middle of the century, with the exception of Cha Shih-piao (1615-1698), whose artistic career extended into the last decade of the seventeenth century. Secondly, the stylistic affinity among them, as will be described in Chapter II and Chapter III, is unmistakable. Sirén grouped Hung-jen and Cha Shih-piao with Hsiao Yün-ts'ung and Mei Ch'ing, saying that the latter two artists "from a stylistic point might serve as their [Hung-jen and Cha Shih-piao]partners better than Sun I and Wang Chih-jui," Sirén's grouping apparently is unfounded. Furthermore, Sirén's decision to dismiss Sun I and Wang Chih-jui as rather unimportant is equally distorting, because, even though very few of their works are extant today, it was they who first established the major stylistic characteristics of the Hsin-an School which later found their fullest expression in Hung-jen and his followers.

[3]Yü-hsiang 予向 [Huang Pin-hung 黃賓虹], "Chien-chiang Ta-shih shih-chi i-wen" 漸江大師事蹟佚聞*Chung-ho yüeh-k'an* 中和月刊, 1.5 (May 1940), pp. 26-33, 1.6 (July 1940), pp. 46-65.

[4]Cheng Hsi-chen 鄭錫珍, *Hung-jen,K'un-ts'an* 弘仁髡殘 (Shanghai: Jen-min mei-shu ch'u-pan-she, 1963). A slightly revised, second edition, with different illustrations, was published in 1979 by the same publisher.

[5]Wang Shih-ch'ing 汪世清 and Wang Ts'ung 汪聰 comps., *Chien-chiang tzu-liao-chi* 漸江資料集 (Ho-fei, Anhui: An-hui jen-min ch'u-pan-she, 1964). A slightly revised, second edition was published in 1984 by the same publisher.

[6] James Cahill, *Fantastics and Eccentrics in Chinese Painting* (New York: Asia House Gallery, 1967), pp. 42-53.

[7]*Dictionary of Ming Biography: 1368-1644* (1976), s.v."Hung-jen," by James Cahill. See also James Cahill, ed., *Shadows of Mt. Huang: Chinese Painting and Printing of the Anhui School* (Berkeley: University Art Museum, 1981) and James Cahill, *The Compelling Image: Nature and Style in Seventeeth-Century Chinese Painting* (Cambridge, Mass,: Harvard University Press, 1982). In his forthcoming book on early Ch'ing painting (fourth in a projected series of five volumes under the title "A History of Later Chinese Painting, 1279-1950", published by John Weatherhill, Inc., of New York and Tokyo), Cahill devotes a chapter to the "Anhui School"; I am grateful to Professor Cahill for making his manuscript available to me.

[8]Marilyn and Shen Fu, *Studies in Connoisseurship: Chinese Paintings from the Arthur M. Sackler Collection in New York and Princeton* (Princeton: The Art Museum, Princeton University Press, 1973), pp. 140-51.

[9]Chu-tsing Li, *A Thousand Peaks and Myriad Ravines: Chinese Paintings in the Charles A. Drenowatz Collection,* 2 vols. (Ascona: Artibus Asiae, 1974), vol. 1 pp. 180-84.

FOOTNOTES TO CHAPTER I

[1]In the Western calendar, his date of death would be January 19, 1664. The basic information about his biography has been gathered in Wang Shih-ch'ing 汪世清 and Wang Ts'ung 汪聰 comps., *Chien-chiang tzu-liao chi* 漸江資料集 (Ho-fei, Anhui: An-hui jen-min ch'u-pan-she, 1964).

[2]Chang Keng 張庚, *Kuo-ch'ao hua-cheng-lu* 國朝畫徵錄 (Shanghai: Yu-cheng shu-chü, 1919), 3.12a.

[3]The epitaph is dated 1691 and published in Wang Shih-hung 汪士鋐, comp., *Huang-shan-chih hsü-chi* 黃山志續集, An-hui ts'ung-shu, nos. 10-15 (Shanghai: An-hui ts'ung-shu pien-yin-ch'u, 1935), 3.1-6. For a biography of Wang Mu-jih, see Hsü Ch'eng-yao 許承堯, comp., *She-hsien-chih* 歙縣志, (Shanghai: Lü-hu t'ung-hsiang-hui, 1937; reprinted., Taipei: Ch'eng-wen ch'u-pan-she, 1975), pp. 1653-54.

[4]Wang T'ai-cheng 王泰徵, "Chien-chiang ho-shang chuan" 漸江和尚傳, in *Chien-chiang tzu-liao-chi*, p. 3.

[5] In *Chien-chiang tzu-liao-chi*, p. 11.

[6] Hsieh Kuo-chen 謝國楨, *Nan-ming shih-lüeh* 南明史略 (Shanghai: Shanghai jen-min ch'u-pan-she, 1957), pp. 131-42.

[7] See Ch'eng Sui's colophon to Hung-jen's sixty-leaf album on Mt. Huang, in *Chien-chiang tzu-liao-chi*, p. 74.

[8] Huang Sheng 黃生 (1622-1696), *I-mu-t'ang shih-kao* 一木堂詩稿, in *Chien-chiang tzu-liao-chi*, pp. 215-16.

[9] Miyazaki Ichisada 宮崎市定, "Chō Fu to sono jidai" 張溥とその時代, *Tōyōshi kenkyū* 東洋史研究, 33:3 (December, 1974), 1-47.

[10] Teng Chih-ch'eng 鄧之誠, ed., *Ch'ing-shih chi-shih ch'u-pien* 清詩紀事初編 (1959; reprint ed., Taipei: Ting-wen shu-chü, 1971), 1.145.

[11] For a recent study on the Fu-she, see William S. Atwell, "From Education to Politics: The Fu She," in Wm. Theodore de Bary, ed., *The Unfolding of Neo-Confucianism* (New York: Columbia University Press, 1975), pp. 333-67.

[12] *She-hsien-chih*, p. 2800: *Chien-chiang tzu-liao-chi*, p. 7.

[13] *Chien-chiang tzu-liao-chi*, p. 59.

[14] Ibid., pp. 93, 99.

[15] Ibid., pp.8, 25.

[16] Wu Chia-chi 吳嘉紀, *Lo-hsüan-shih* 陋軒詩, in *Chien-chiang tzu-liao-chi*, pp. 107-8.

[17] This is evident from Hung-jen's own poems in *Chien-chiang tzu-liao-chi*, pp. 14, 39.

[18] Ch'en Yüan 陳垣, *Ming-chi Tien Ch'ien fo-chiao k'ao* 明季滇黔佛教考 (Peking: Ko-hsüeh ch'u-pan-she, 1962), esp. pp. 108, 238.

[19] *Chien-chiang tzu-liao-chi*, p. 4.

[20] Ibid., p. 120.

[21] Ibid., p. 132.

[22] Ibid., p. 119.

[23] Shih-pai-chai chu-jen 十百齋主人 *Shih-pai-chai shu-hua-lu* 十百齋書畫錄, 11 vols. (n. p., n.d.), hsia han, yin chi 下函寅集, p. 24.

[24] For the historical geography of this area, see *She-hsien-chih*, p. 67-105

[25] See Louise R. Yuhas, "The Landscape Art of Lu Chih (1496-1576)," (Ph. D.

dissertation, University of Michigan, 1979), Pls. 21 a-m.

[26]Ibid., p. 101.

[27]Recorded in Ch'en Chuan 陳撰, *Yü-chi shan-fang hua-wai-lu* 玉几山房畫外錄, Mei-shu ts'ung-shu, vol. 8, no. I/8 (Taipei: I-wen yin-shu-kuan, n.d.), 1.90.

[28]*Chien-chiang tzu-liao-chi*, p. 75.

[29]Ibid., p. 172.

[30]Ibid., p. 80

[31]Ibid., p. 77; tr. courtesy of James Cahill.

[32]Hsü Piao-jan 徐表然, comp., *Wu-i-chih* 武夷志 (Chin-chiang, Fukien: n.p., 1619), vol. 1, preface to the map of Mt. Wu-i. Also see, in particular, illustrations of the Fifth Bend *(Wu-ch'ü* 五曲*)*, the Sixth Bend *(Liu-ch'ü* 六曲*)*, and the Eighth Bend *(Pa-ch'ü* 八曲) in ibid.

[33]The handscroll is published under the title *Ko-an shan-shui ho-chin chüan* 格庵山水合錦卷 (Shanghai: Shen-chou kuo-kuang-she, 1936).

[34]For Wu Pin and other painters from Fukien, see Elizabeth Fuller, "Professional Painters in Fukien and Nanking," in James Cahill, ed., *The Restless Landscape: Chinese Painting of the Late Ming Period* (Berkeley: University Art Museum, 1971), pp. 103-14; also see James Cahill, "Wu Pin and His Landscape Painting," in *Proceedings of the International Symposium on Chinese Painting* (Taipei: National Palace Museum, 1970), pp. 637-98, and Harrie A. Vanderstappen, "The Style of Some Seventeenth-century Chinese Paintings," in Christian F. Murck, ed., *Artists and Traditions: Uses of the Past in Chinese Culture* (Princeton: The Art Museum, Princeton University, 1976), pp. 149-68. The possibility of contact between Anhui and Fukien was made possible by the active commercial links established by Anhui merchants in Fukien; see Fujii Hiroshi, "Shin-an shōnin no kenkyū" 新安商人の研究, *Tōyō gakuhō* 東洋學報, 36.1-4 (1953-54): 1-44, 32-60, 65-118, 115-45, esp. 39.

[35]The revival of Northern Sung style in the seventeenth century has been discussed by James Cahill, "The Early Styles of Kung Hsien," *Oriental Art*, 16.2 (Summer 1970): 51-71. The Hua Shu-ho hanging scroll is reproduced in ibid., fig. 14.

[36]The album has been published under the title *Chien-chiang shang-jen shan-shui-ts'e* 漸江上人山水冊 (n.p., n.d.); also in *Shina nanga taisei* 支那南畫大成, 22 vols. (Tokyo: Kobunsha, 1935-37), vol. 11.

[37]*Chien-chiang tzu-liao-chi*, p. 217. Also see Ch'en Yüan 陳垣, *Ch'ing-ch'u seng-cheng chi* 清初僧諍記 (Peking: Chung-hua shu-chü, 1962), pp. 4-5. The dates of Wu-

ming Hui-ching is provided by Ch'en Yüan 陳垣, *Shih-shih i-nien-lu* 釋氏疑年錄 (Peking: Chung-hua shu-chü, 1964), p. 377. For a recent study on Fang I-chih, see Jao Tsung-i 饒宗頤, "Fang I-chih chih hua-lun" 方以智之畫論, *The Journal of The Institute of Chinese Studies at The Chinese University of Hong Kong*, 7.1 (1974): 113-31.

[38]*Chien-chiang tzu-liao-chi*, pp. 204-5.

[39]Ibid., p. 10.

[40]Hsü Hsia-k'o 徐霞客, *Hsü Hsia-k'o yu-chi* 徐霞客遊記, ed., Ting Wen-chiang 丁文江, 20 vols. (Reprint ed., Taipei: Ting-wen shu-chü, 1972), 1.26. Translation adapted from Li Chi, *The Travel Diaries of Hsü Hsia-k'o* (Hong Kong: The Chinese University of Hong Kong, 1974), p. 81.

[41]Hsü Hsia-k'o, ibid., p. 27; Li Chi, ibid., p. 82.

[42]Hsü Hsia-k'o, ibid., p. 12; Li Chi, ibid., p. 75.

[43]Li Chi, ibid., pp. 244-45.

[44]*She-hsien-chih*, pp. 2701-4.

[45]Ibid., pp. 2659-60.

[46]Reproduced in Kojiro Tomita and Hsien-ch'i Tseng, eds., *Portfolio of Chinese Paintings in the Museum: Yüan to Ch'ing Periods* (Boston: Museum of Fine Arts, 1961), Pls. 27-28.

[47]Ch'en Yüan, *Shih-shih i-nien-lu*, p. 126.

[48]James Cahill, *Chinese Painting* (Geneva: Skira, 1960), p. 171.

[49]Cf. Wang Hung-tu 汪洪度, *Huang-shan ling-yao-lu* 黃山領要錄, Chih-pu-tsu-chai-ts'ung-shu, vol. 150 (Reprint ed., 1921), pp. 1-2, 24, and passim.

[50]Yü Ta-fu 郁達夫, *Yü Ta-fu yu-chi* 郁達夫遊記 (Shanghai: Shanghai tsa-chih kung-ssu, 1948), pp. 261-62.

[51]*Chien-chiang tzu-liao-chi*, p. 179.

[52]Ho T'ien-chien 賀天健, *Hsüeh-hua shan-shui kuo-ch'eng tzu-shu* 學畫山水過程自述 (Peking: Jen-min mei-shu ch'u-pan-she, 1962), pp. 135-39.

[53]Quoted in Chu Hsing-chai 朱省齋, *Hsing-chai tu-hua-chi* 省齋讀畫記 (Hong Kong: Ta-kung shu-chü, n.d.), p. 42.

[54]See the classic study on the merchants from Hui-chou by Fujii, "Shin-an shōnin no kenkyū."

[55]Ku Yen-wu 顧炎武, *T'ien-hsia chün-kuo li-ping shu* 天下郡國利病書, Ssu-pu ts'ung-

k'an, no. 238 (Shanghai: Commercial Press , 1935-36), 9.76ab. Translation amended from Willard Peterson, *The Bitter Gourd* (New Haven: Yale University Press, 1979), p. 70.

[56]See Fu I-ling 傅衣凌, "Ming-tai Hui-chou shang-jen" 明代徽州商人 in his *Ming-Ch'ing shih-tai shang-jen chi shang-yeh tzu-pen* 明清時代商人及商業資本 (Peking: Jen-min ch'u-pan-she, 1956), , pp. 49-91. Cf. Shang Yüeh 尙鉞, *Chung-kuo tzu-pen chu-i kuan-hsi fa-sheng chi yen-pien te ch'u-pu yen-chiu* 中國資本主義關係發生及演變的初步研究 (Peking: San-lien shu-tien, 1956), pp. 145-271.

[57]Hsieh Chao-chih 謝肇淛, *Wu-tsa-tsu* 五雜俎 (Shanghai: Chung-yang shu-tien, 1935), 4.154.

[58]Ibid., p. 154; translation amended from Ping-ti Ho, "The Salt Merchants of Yangchou: A Study of Commercial Capitalism in Eighteenth Century China,"*Harvard Journal of Asiatic Studies*, 17 (1954): 130-68.

[59]Ibid., p. 155.

[60]See Chang Chung-i 張仲一 et al., *Hui-chou Ming-tai chu-chai* 徽州明代住宅 (Peking: Chien-chu kung-ch'eng ch'u-pan-she, 1957), and Mu Hsiao-t'ien 穆孝天, *An-hui wen-fang ssu-pao shih* 安徽文房四寶史 (Shanghai: Jen-min mei-shu ch'u-pan-she, 1962).

[61]Ch'en Yeh 陳野, "Lun Hui-chou shang-yeh tzu-pen ti hsing-ch'eng chi-ch'i t'e-se" 論徽州商業資本的形成及其特色, *An-hui shih-hsüeh t'ung-hsün* 安徽史學通訊, no. 7 (September 1958), pp. 32-42, esp. p. 41; cf. Frederic Wakeman, Jr., *The Fall of Imperial China* (New York: The Free Press, 1975), p. 44.

[62]Chang Chung-i, *Ming-tai Hui-chou chu-chai*, pp. 8-9, 17.

[63]Ibid., pp.33-36.

[64]Wang Shun-min 汪舜民 and P'eng Tse 彭澤, comps., *Hui-chou fu-chih* 徽州府志, Chung-kuo shih-hsüeh ts'ung-shu, no. 15 (Reprint ed., Taipei: Hsüeh-sheng shu-chü, 1965), and Ho Tung-hsü 何東序 and Wang Shang-ning 汪尙寧, comps., *Hui-chou fu-chih* 徽州府志, Chung-kuo shih-hsüeh ts'ung-shu, no. 15 (Reprint ed., Taipei: Hsüeh-sheng shu-chü, 1965), passim.

[65]For example, see Cheng Chung's handscroll "Landscape," dated 1632, in the Ch'eng Ch'i 程琦 collection, Tokyo; published in Cahill, *Restless Landscape*, Pl. 50.

[66]For instance, compare Cheng Chung's undated fan published in *Hsin-an shan-yeh chi-ts'e* 新安扇頁集冊 (Shanghai: Shen-chou kuo-kuang-she, 1931), Pl. 1 with Ch'eng Chia-sui's hanging scroll "Horse-riding on the Willow Bank" dated 1639 in the T'ien-tsin Museum and published in *T'ien-tsin shih i-shu po-wu-kuan ts'ang-hua-chi* 天津市藝術博物館藏畫集, 2 vols. (Peking: Wen-wu ch'u-pan-she, 1963), 2.64.

[67]Wang Shih-chen 王世貞, *Ku-pu-ku lu* 觚不觚錄, Ts'ung-shu chi-ch'eng ch'u-pien, no. 281 (Shanghai: Commercial Press, 1937), p. 17.

[68]Chan Ching-feng, *Tung-t'u hsüan-lan-pien* 東圖玄覽編, Mei-shu ts'ung-shu, vol. 21, no. V/1 (Taipei: I-wen yin-shu-kuan , n.d.) pp. 167-68.

[69]Wu Ch'i-chen 吳其貞, *Shu-hua-chi* 書畫記 (Shanghai: Jen-mim mei-shu ch'u-pan-she, 1963), p. 168.

[70]Ibid., p. 79.

[71]See Wen Fong, "Rivers and Mountains after Snow (*Chiang-shan hsüeh-chi*): Attributed to Wang Wei (A.D. 699-759),"*Archives of Asian Art*, 30 (1976-77): 12-13. The Wang Wei handscroll remained in various Anhui private collections from 1618 to about 1642.

[72]The handscroll is published in *Chung-kuo-hua* 中國畫, no. 9 (1959). For an extensive study of the handscroll, especially its art-historical importance, see Chu-tsing Li, *The Autumn Colors on the Ch'iao and Hua Mountains: A Landscape by Chao Meng-fu*, Artibus Asiae Supplement no. 21 (Ascona: Artibus Asiae, 1965), pp. 53-59.

[73]Wu Ch'i-chen, *Shu-hua-chi*, pp. 74-79, 86-87.

[74]Ibid., pp. 160-61.

[75]For a short biography of Wu T'ing, see *She-hsien-chih*, pp. 1566-67.

[76]See Shen C.Y. Fu, *Traces of the Brush* (New Haven: Yale University Art Gallery, 1977), pp. 241-43.

[77]Kuo-li ku-kung po-wu-yüan pien-chuan wei-yüan-hui 國立故宮博物院編纂委員會, comp., *Ku-kung shu-hua-lu* 故宮書畫錄, rev. ed. (Taipei: National Palace Museum, 1965), 5.210

[78]It was published in two series in 1598 and 1609-14, 24 volumes in total. See Hirayama Kangetsu 平山觀月, *Shin Chūgoku shodoshi* 新中國書道史 (Tokyo: Yuhodo, 1962). pp. 316-17; K'uan-yü 寬予, "Yü-ch'ing-chai fa-t'ieh k'ao-lüeh" 餘清齋法帖考略 in *I-lin ts'ung-lu* 藝林叢錄, 10 vols. (Hong Kong: Commercial Press, 1961-66), 6.204-8.

[79]A partial list of extant paintings and calligraphy from the collection of Wu T'ing is as follows:
(1) Anonymous, "The Emperor Hsüan-tsung Entertaining Li Po and Another Poet" *(Chao-yin-t'u 招飲圖)*, handscroll, stylistically dating from about the fourteenth or fifteenth century, in the Museum of Fine Arts, Boston (reproduced in Tomita and Tseng, *Portfolio,* P1. 93).
(2) T'ang copy of Wang Hsi-chih's calligraphy handscroll *Tung-fang So hsiang-*

chan 東方朔像讚 in the National Palace Museum (recorded in *Ku-kung shu-hua-lu*, 1. 36-37).

(3) Chu Yün-ming 祝允明 (1461-1527), "Poems and Letter" (*Shih-han* 詩翰) for Shen Pien-chih 沈辨之, calligraphy handscroll, in the National Palace Museum (recorded in *Ku-kung shu-hua-lu*, 1.101-4; partially reproduced in Chiang Chao-shen, *Ninty Years of Wu School Paintings* [Taipei: National Palace Museum, 1975], no. 96, p. 106).

(4) *Hsing-jang-t'ieh* 行穰帖, an early T'ang tracing copy of a letter written by Wang Hsi-chih 王羲之, mounted on a handscroll, in the Art Museum, Princeton University (reproduced in Fu, *Traces of the Brush*, no. la).

(5) Chang Chi-chih 張即之, "Diamond Sutra" (*Chin-kang ching* 金剛經), a Buddhist treatise originally written in Sanskrit about 350, in two albums, now in the Art Museum, Princeton University (reproduced in Fu, *Traces of the Brush*, no. lla); also see Fu Shen 傅申, "Chang Chi-chih te chung-k'ai" 張即之的中楷, *National Palace Museum Quarterly* 10.4 (Summer 1976): 43-65, and Pls. 5-6.

(6) Mi Yu-jen 米友仁 (1086-1165), attributed to, "Mountains in Clouds and Mist" (*Yün-shan te-i* 雲山得意), handscroll, in the National Palace Museum (recorded in *Ku-kung shu-hua-lu*, 4.32; reproduced in *Chinese Art Treasures* [Geneva: Skira, 1961-62], P1. 35).

(7) Hui-ch'ung 惠崇 (active ca. early eleventh century), "Mountain Ducks in Autumn" (*Ch'iu-p'u shuang-yüan* 秋浦雙鴛), album leaf in *Li-ch'ao hua-fu chi-ts'e* 歷朝畫幅集冊, in the National Palace Museum (reproduced in *Chinese Art Treasures*, Pl. 22).

(8) Sung (?) anonymous, "Snowscape" (*Hsüeh-ching* 雪景), album leaf in *Sung-jen chi-hui* 宋人集繪, in the National Palace Museum (VA20n).

(9) Wu Ping 吳炳 (active ca. late 12th century), "Mountain Bird and Autumn Leaves" (*Ch'iu-yeh shan-ch'in* 秋野山禽), album leaf dated 1199, in the National Palace Museum (VA23a).

(10) Ma Lin 馬麟 (active ca. mid-13th century), "Winter Birds at Evening" (*Mu-hsüeh han-ch'in* 暮雪寒禽), album leaf in *Ming-kua chi-chen* 名畫集眞, in the National Palace Museum (VA4h).

(11) Sung anonymous, "Market Village by the River" (*Chiang-fan shan-shih* 江帆山市), handscroll, in the National Palace Museum (reproduced in *Chinese Art Treasures*, Pl. 27).

(12) Chao Yung 趙雍 (b. 1289-after 1352), attributed to, "Gathering Water Chestnuts" (*Ts'ai-ling-t'u* 採菱圖), dated 1342, in the National Palace Museum (YV8).

(13) Sung (?) anonymous, "Gathering of Old Scholars" (*Ch'i-ying-hui* 耆英會), in the National Palace Museum (SH35).

(14) Wang Hsien-chih 王獻之, *Ya-t'ou-wan-t'ieh* 鴨頭丸帖, in the Shanghai Museum (described in *I-lin-ts'ung-lu*, 5.151-53).

(15) Wang Meng 王蒙, "The Forest Chamber Grotto at Chü-ch'ü (*Chü-ch'ü lin-wu* 具區林屋), undated hanging scroll, in the National Palace Museum (reproduced in Chang Kuang-pin 張光賓, *Yüan ssu-ta-chia: Huang Kung-wang, Wu Chen, Ni Tsan, wang Meng* 元四大家：黃公望、吳鎮、倪瓚, 王蒙 [Taipei: National Palace Museum, 1975], Pl. 411).
(16) Ni Tsan 倪瓚, "Ink Bamboo," album leaf, undated, in the Freer Gallery of Art.
(17) Anonymous, "River of Szechwan," handscroll (ca. 13th or 14th century), in the Freer Gallery of Art.
(18) Anonymous, "Forest Landscape and Figures," handscroll (ca. 16th century), in the Freer Gallery of Art.
(19) Li T'ang 李唐 (active ca. 1120-40), "Marquis Wen-kung of Chin Recovering His State," undated handcroll, in the Metropolitan Museum of Art, New York (reproduced in Wen Fong and Marilyn Fu, *Sung and Yüan Paintings* [New York: The Metropolitan Museum of Art, 1973], no. 2.)
(20) *K'uai-hsüeh shih-ch'ing-t'ieh* 快雪時晴帖, attributed to Wang Hsi-chih, in the National Palace Museum (recorded in *Ku-kung shu-hua-lu*, 3.1-4).
(21) *Chi-chih wen-kao* 祭姪文稿, one of the "Three Masterpiece Manuscripts" by the T'ang calligrapher Yen Chen-ch'ing 顏眞卿 (709-785), in the National Palace Museum (recorded in *Ku-kung shu-hua-lu*, 1.18-24).
(22) *Shu-ssu-t'ieh* 蜀素帖, calligraphy handscroll, dated 1088 by Mi Fu 米芾 (1051-1107), in the National Palace Museum (recorded in *Ku-kung shu-hua-lu*, 1.55-60; partially reproduced in *Chinese Art Treasures*, Pl. 117).
(23) *Ch'ih-tu-chüan* 尺牘卷, a short handscroll of calligraphy consisting of letters by Mi Fu, in the National Palace Museum (recorded in *Ku-kung shu-hua-lu*, 1.62-63).
(24) Chuang Lin 莊麟 (late Yüan Dynasty), *Ts'ui-yü hsüan* 翠雨軒, a short handscroll of landscape in the *Yüan-jen chi-chin* 元人集錦 joint handscroll in the National Palace Museum (recorded in *Ku-kung shu-hua-lu*, 4.302-5).
(25) Ni Tsan, "Ink Bamboo, " album leaf, undated, in the Freer Gallery of Art (reproduced in Wang Chi-ch'ien, 王季遷, "Ni Yün-lin chih-hua" 倪雲林之畫, *National Palace Museum Quarterly*, 1.3[January 1967]: 29, 42 and Pl. 25).

[80]For a short biography of Wu Chen, see *She-hsien-chih*, pp. 1569-70.

[81]The following calligraphy and paintings were once in Wu Chen's collection: the *Chi-chih wen-kao* 祭姪文稿, one of the "Three Masterpiece Manuscripts" by the T'ang calligrapher Yen Chen-ch'ing 顏眞卿 (709-785), now in the National Palace Museum (*Ku-kung shu-hua-lu*, 1.18-24); two important calligraphy handscroll, the *Shu-ssu-t'ieh* 蜀素帖 dated 1088 and the undated *Ch'ih-tu-chüan* 尺牘卷 by Mi Fu 米芾 (1051-11070), now in the National Palace Museum (*Ku-kung shu-hua-lu*, 1.55-60, 62-62); the anonymous handscroll "Clearing Autumn Skies over Mountains and Valleys" (ca. 11th century), formerly attributed to Kuo Hsi 郭熙 (ca. 1060-1075), in the

Freer Gallery of Art (acc. no. 16.538); the famous short handscroll "Watching the Mid-autumn Moon," undated, by Shen Chou 沈周 (1427-1509), now in the Museum of Fine Arts, Boston (published in Richard Edwards, *The Field of Stones: A Study of the Art of Shen Chou [1427-1509]*, Smithsonian Institution, Freer Gallary of Art, Oriental Studies, no. 5 [Washington: The Freer Gallery of Art, 1962], Pl. 14A,); and the handscroll "Peony" attributed to Ch'ien Hsüan (ca. 1235 to after 1301), now in the National Palace Museum (recorded in *Ku-kung shu-hua-lu*, 4.73-76); it had been in the collection of Wen Cheng-ming before Wu Chen acquired it.

82See Chu Chih-ch'ih 朱之赤, *Chu Wo-an ts'ang shu-hua-mu* 朱臥菴藏書畫目, I-shu ts'ung-pien, vol. 17, no. 157 (Taipei: Shih-chieh shu-chü, 1962), pp. 85, 87; Ko Chin-lang 葛金烺, *Ai-jih-yin-lou shu-hua-lu* 愛日吟樓書畫錄, preface dated 1881 (n.p., 1910), 2. 15-16; Chang Ta-yung 張大鏞, *Tzu-i-yüeh-chai shu-hua-lu* 自怡悅齋書畫錄, preface dated 1832 (n.p., n.d), 15. 15-17. Li Yung-ch'ang's leaf in the joint album for Ch'en Ying-ch'i 陳應頎 by other artists, including Hung-jen, is not dated (*Hsin-an ming-hua chi-chin ts'e* 新安名畫集錦冊 [Shanghai: Shen-chou kuo-kuang-she, 1920], Pl. 8). However some other leaves in the same album are dated to 1647, 1648, 1653 (Ibid., Pls. 1, 9, 4, respectively). Presumably Li Yung-ch'ang's leaf could be dated to about 1650.

83Wu Ch'i-chen, *Shu-hua-chi*, p. 94.

84Ibid., pp. 93-95.

85Ibid., pp. 120-21.

86Ibid., pp. 184-85, In addition, Li Yung-ch'ang owned a calligraphy handscroll "One Thousand Characters" (*Ch'ien-tzu-wen* 千字文) by the Emperor Kao-tsung (reigned 1127-1163) of the Southern Sung Dynasty, now in the Shanghai Museum; it is published as *Sung Chao Kou chen-ts'ao chien-tzu-wen* 宋趙構眞草千字文 (Shanghai: Shanghai Museum, n.d.). It was previously in the collection of Hsiang Yüan-pien 項元汴 (1525-1590).

87Shih-pai-chai chu-jen, *Shih-pai-chai shu-hua-lu*, shang han, i chi 上函乙集, p. 27.

88For a recent study of the hanging scroll, see Alan John Hay, "Huang Kung-wang's *Dwelling in the Fu-ch'un Mountain*: Dimension of a Landscape" (Ph.D. dissertation, Princeton University, 1978), pp. 220-23.

89Wu Ch'i-chen, *Shu-hua-chi*, pp. 350-51.

90Published in Hsieh Chih-liu 謝稚柳, *T'ang Wu-tai Sung Yüan ming-chi* 宋五代宋元名跡 (Shanghai: Ku-tien wen-hsüeh ch'u-pan-she, 1957), Pl. 99.

91Published in *Chien-chiang tzu-liao-chi*, Pl. 3 and *Shina nanga taisei*, vol. 16.

92Chang Ch'ou 張丑, *Ch'ing-ho shu-hua-fang* 清和書畫舫, preface dated 1616 (n.p.,

1875), 11.44, 47; Chang Ch'ou, *Chen-chi jih-lu* 眞蹟日錄 (n.p., 1918), 2.22a.

[93]Published in Cheng Cho-lu 鄭拙廬, *Ni Tsan* 倪瓚 (Shanghai: Jen-min mei-shu ch'u-pan-she, 1961).

[94]Wu Ch'i-chen, *Shu-hua-chi*, pp. 377, 532.

[95]Published in Chang Kuang-pin, *Yüan ssu-ta-chia*, Pl. 328.

[96]Published in Cheng Cho-lu, *Ni Tsan*, Pl. 3; also see Wu Ch'i-chen, *Shu-hua-chi*, pp. 117-18.

[97]Published in Chang Kuang-pin, *Yüan ssu-ta-chia*, Pl. 308; also see Wu Ch'i-chen, *Shu-hua-chi*, pp. 168-89.

[98]Published in Hsieh Chih-liu 謝稚柳, *T'ang Sung Yüan Ming Ch'ing hua-hsüan* 唐宋元明淸畫選 (Kuang-chou: I-shu hua-pao-she, 1963), Pl. 27; also see Wu Ch'i-chen, *Shu-hua-chi*, p. 459.

[99]*Chien-chiang tzu-liao-chi*, pp. 147-48.

[100]For example, see Terese Tse Bartholomew, *I-hsing Ware* (New York: China House Gallery, 1977), nos. 9, 42, 43.

[101]*Chien-chiang tzu-liao-chi*, pp. 150-51.

[102]Ibid., p. 3.

[103]Chan Ching-feng, *Tung-t'u hsüan-lan-pien*, pp. 4, 14.

[104]Wang Chieh 王杰 et al., *Shih-ch'ü pao-chi hsü-pien* 石渠寶笈續編 (1793; reprint ed., Taipei: National Palace Museum, 1971), pp. 166, 264.

[105]Published in Cahill, *Restless Landscape*, Pl. 79. Also see *Style Transformed* (Taipei: National Palace Museum, 1977), Pl. 002-1, and p. 54.

[106] Shih-pai-chai-chu-jen, *Shih-pai-chai shu-hua-lu*, shang han, hsin chi 上函辛集, pp. 3-6.

[107]*Ku-kung shu-hua-lu*, 4.219-29.

[108]Published in Cahill, *Restless Landscape*, Pl. 78.

[109]Yuhas, "Lu Chih, " pp. 383-84, and Pl. 150.

[110]Ibid., pp. 326-29. The handscroll is published in *Kuo-t'ai mei-shu-kuan hsüan-chi* (Taipei: Kuo-t'ai mei-shu-kuan, 1978) , Pl. 6.

[111]Chang Keng, *Kuo-ch'ao hua-cheng-lu*, 1.12-12; Lu Hsin-yüan 陸心源, *Jang-li-kuan kuo-yen-lu* 穰梨館過眼錄 (n.p., 1829), Supplement, 13.1-5.

[112]*She-hsien-chih*, p. 2780.

[113]Ibid., p. 2792.

[114]*Chien-chiang tzu-liao-chi*, p. 59.

[115]Mu Hsiao-t'ien, *An-hui wen-fan ssu-pao shih*; also see Pao Yu-wen 鮑幼文, "Hui-chou ti wen-feng ssu-pao" 徽州的文房四寶, *Anhui shih-hsüeh t'ung-hsün* 安徽史學通訊, nos. 12 and 13 (October 1959), pp. 160-66.

[116]For a biography of Ch'eng Ta-yüeh and that of Fang Yü-lu, see L. Carrington Goodrich and Chaoying Fang, eds., *Dictionary of Ming Biography* (New York: Columbia University Press, 1976), pp. 212-15, and 438-39, respectively.

[117]See Chiao Hung's biography in Arthur W. Hummel. ed., *Eminent Chinese of the Ch'ing Period (1644-1912)* (Reprint ed., New York: Paragon Book Reprint Corp., 1964), pp. 145-46.

[118]*Dictionary of Ming Biography*, pp. 1324-27.

[119]Ibid., pp 1427-30.

[120]Acc. no. 30.76.203.

[121]Wang Chi-chen, "Notes on Chinese Ink," *Metropolitan Museum Studies*, 3 (1930-31): 114-33, and fig. 1; also see Herbert Franke, *Kulturgeschichtliches über die chinesische Tusche* (Munich: Verlag der bayerischen Akademie der Wissenschaften , 1962), pp. 29, 126, and figs. 11-18.

122.Chou Liang-kung 周亮工, *Yin-shu-wu shu-ying* 因樹屋書影, preface dated 1667 (Shanghai: Ch'un-hsüeh-she, n.d.), 1.2a.

[123]See Hsieh Chao-chih, *Wu-tsa-tsu*, 13.208-9; Chang Chung-i et al., *Ming-tai Hui-chou chu-chai*, pp. 27, 33, and Pls. 7 and 51; K. T. Wu, "Color Printing in the Ming Dynasty,"*T'ien-hsia Monthly* 11(1940): 30-44; K. T. Wu, "Ming Printing and Printer," *Harvard Journal of Asiatic Studies*, 7 (1943): 204-60, esp. p. 203-7; *Exposition d'ouvrages illustés de la dynastie Ming* (Peking: Centre franco-chinois d'études sinologiques, 1944), Pls. 8, 11, 14-19, 30, 36.

[124]For a discussion of the technical advancement made by Hu Yüeh-ts'ung, see Hiroshi Higuchi 樋口弘, *Chūgoku hanga shūsei* 中國版畫集成 (Tokyo: Mito sho-oku 味燈書屋, 1967), pp. 22-23. For a biography of Hu Yüeh-ts'ung, see Chih-k'an 智龕, "Shih-chu-chai-chu Hu Cheng-yen" 十竹齋主胡正言, in *I-lin ts'ung-lu* 藝林叢錄, 9.296-99.

[125]For the development of wood-cut print during the sixteenth centuries, see *Chūgoku no Min Shin jidai no hanga* 中國の明清時代の版畫 (Nara: Yamato bunkakan, 1972); Aoki Masaru 青木正兒, Iriya Yoshitaka 入矢義高, and Kōhara Hironobu 古原宏

伸, *Kaishien gaden* 芥子園畫傳 (Tokyo: Chikuma shobō, 1975), pp. 195-264; Robert T. Paine, Jr., "The Ten Bamboo Studio: Its Early Edition, Pictures, and Artists," *Archives of the Chinese Art Society in America* 5 (1951): 39-54; A. K'ai-ming Ch'iu, "The Chieh Tzu Yuan Hua Chuan (Mustard Seed Garden Painting Manual): Early Editions in American Collections,"*Archives of the Chinese Art Society in America* 5 (1951): 55-69. For a useful compendium, see Chou Wu 周蕪 *Hui-p'ai pan-hua-shih lun-chi* 徽派版畫史論集 (Hofei, Anhui: Jen-min ch'u-pan-she, 1984).

[126]Hsieh Chao-chih, *Wu-tsa-tsu*, 12.155.

[127]Published in *Shen-chou ta-kuan*, no. 3; also see *Chien-chiang tzu-liao-chi*, p. 161.

[128]Published in *Chien-chiang tzu-liao-chi*, Pl. 10 ; also see p. 73.

[129]Chang Keng, *Kuo-ch'ao hua-cheng-lu*, 1.12-13.

[130]Wang Po-min 王伯敏, *Chung-kuo pan-hua-shih* 中國版畫史 (Shanghai: Jen-min mei-shu ch'u-pan-she, 1961), pp. 151-57; Kuo Wei-ch'ü 郭味渠, *Chung-kuo pan-hua shih-lüeh* 中國版畫史略 (Peking: Chao-hua mei-shu ch'u-pan-she, 1962), pp. 157-64. Hsiao Yün-ts'ung 蕭雲從, *T'ai-p'ing shan-shui shih-hua* 太平山水詩畫 (Reprint ed., Osaka: Tombo-sha, 1931).

[131]Published in Su Tsung-jen, comp., *Huang-shan ts'ung-k'an* 黃山叢刊, 8 vols. (Peking: Pai-i-yen-chai, 1937), vol. 2; only fifty leaves are published in *Shina nanga taisei*, vol. 13.

[132]For example, see *Huang-shan-chih*, 1.22a and 1.28a; also see *Chien-chiang tzu-liao-chi*, pp. 154-56.

[133]Published in *Shen-chou ta-kuan*, no. 6.

[134]For example, see *Fang-shih mo-p'u*, 3.30, 4.40 , and 6.3.

FOOTNOTES TO CHAPTER II

[1]For example, Wang Ts'ai-pai 汪采白, *Hsin-an hua p'ai* 新安畫派 (Nanking: Chin-ling ta-hsüeh Chung-kuo wen-hua yen-chiu-so, 1948); Ch'in Chung-wen 秦仲文, "Ch'ing-tai ch'u-ch'i hui-hua te fa-chan" 清代初期繪畫的發展, *Wen-wu ts'an-k'ao tzu-liao* 文物參考資料 no. 8 (1958), p. 53-58. The term "Hsin-an-p'ai" was used in a discussion of woodblock prints and books in Anhui by Fu Hsi-hua 傅惜華 in *Exposition*

d'ouvrages illustrés de la dynastie Ming (Peking: Centre franco-chinois d'études sinologiques, 1944), Pls. 14-19.

[2]For example, see James Cahill, *Fantastics and Eccentrics in Chinese Painting* (New York: Asia House Gallery, 1967), pp. 42-53.

[3]The hanging scroll is unpublished, but is listed in Jerome Leslie Silbergeld, "Political Symbolism in the Landscape Painting and Poetry of Kung Hsien (c. 1620-1589)" (Ph. D. dissertation, Stanford University, 1974), p. 247. Silbergeld, however, did not identify both Ch'eng Chia-sui and Li Yung-ch'ang. See also William Wu, "Kung Hsien" (Ph.D. dissertation. Princeton University, 1979), p. 172 and Pl. 40.

[4]Leaves by both Kung Hsien and Wu Ting are published in *Ming-hua fa-shu chi-miao* 名畫法書集妙 (Shanghai: Shen-chou kuo-kuang-she, 1935), Pls. 6 and 2 respectively.

[5]See *Chien-chiang tzu-liao-chi*, p. 204. Only a few of Kung Hsien's acquaintances from Anhui can be mentioned here.
(1) Wang Hsüan-tu 王玄度 (d. 1656) was from She-hsien and painted the 1951 joint handscroll for Ko-an with Hung-jen.
(2) T'ang Yün-chia 唐允甲 (*tzu* Tsu-ming 祖命), who was from Hsüan-ch'eng, wrote a colophon to Hung-jen's album of Mt.Huang in 1674; see *Chien-chiang tzu-liao-chi*, pp. 75-94. Also see Jao Tsung-i 饒宗頤, "Ming-tai hua-chia yü hua-lun" 明代畫家與畫論, *The Journal of The Institute of Chinese Studies at The Chinese University of Hong Kong*, 9.1 (1978): 46-48.
(3) Huang Sheng 黃生 (1622-1969) was from She-hsien. He dedicated a poem to Hung-jen; see *Chien-chiang tzu-liao-chi*, pp. 124-25.
(4) Wang Cheng-yüan 汪徵遠 (1632-1706, *tzu* Fu-ch'en 扶晨 and Li-t'ing 栗亭, later name Shih-hung 士鋐) was from She-hsien. He edited *Huang-shan-chih hsü-chi* 黃山志續集; see *She-hsien-chih*, pp. 1622-23.
For all of these people in relation to Kung Hsien, see Wu, "Kung Hsien," pp. 16-17, 34, 43, 57, 109, 114, 172, 180-81. Also see Wang Shih-ch'ing 汪世清, "Kung Hsien te Ts'ao-hsiang-t'ang chi" 龔賢的《草香堂集》, *Wen-wu* 文物, no. 5 (1978), pp.45-49.

[6]Sherman E. Lee, *The Colors of Ink* (New York: Asia House Gallery, 1974), Pl. 40.

[7]The inscription is reproduced in *Shina nanga taisei*, Supplement 5, p. 233. The whole album is published under the title *Kung Pan-ch'ien shan-shui-ts'e* 龔半千山水冊 (Shanghai: Wen-ming shu-chü, 1931).

[8]Recounted by Huang Pin-hung, quoted in *Chien-chiang tzu-liao-chi*, pp. 82 and 142.

[9]Ibid., pp. 32, 34, 37.

[10]Ibid., pp.149-50.

[11]Susan Arkush, "Ni Tsan and the Image of Yüan Painting" (Ph. D. dissertation, Harvard University, 1978), pp. 161-62.

[12]Ibid., p. 194.

[13]Ibid.

[14]Ernst H. Gombrich, *Art and Illusion*, 2nd ed. (Princeton: Pinceton University Press, 1961), pp. 291-329.

[15]*Chien-chiang tzu-liao-chi*, p. 40.

[16]Idid., p. 37.

[17]Ibid., p. 29.

[18]Ibid., p. 32.

[19]Chou Liang-kung 周亮工, *Tu-hua-lu* 讀畫錄, I-shu ts'ung-pien, vol. 1, no. 14 (Taipei: Shih-chieh shu-chü, 1968), p. 19.

[20]Roger Goepper, *The Essence of Chinese Painting*, tr. Michael Bullock (London: Percy Lund, Humphries & Co., 1963), pp. 66, 219.

[21]James Cahill, *Chinese Painting* (Lausanne: Skira, 1960), p. 174.

[22]Richard Barnhart, *Marriage of the Lord of the River: A Lost Landscape by Tung Yüan*, Artibus Asiae Supplementum 27 (Ascona: Artibus Asiae, 1970), fig. 13.

[23]Laurence Sickman et al., eds., *Chinese Calligraphy and Painting in the Collection of John M. Crawford, Jr.* (New York: The Pierpont Mogan Library, 1962), Pl. 31 and pp. 111-13.

[24]Sirén, *Chinese Painting*, 6-94.

[25]Chang Kuang-pin, *Yüan ssu-ta-chia*, Pl. 301.

[26]For a concise discussion of the Northern tradition of landscape painting in the Five Dynasties and Northern Sung Dynasty, see Richard Barnhart's essay "Sōkin ni okeru sansuiga no seritsu to tenkai" 宋金における山水畫の成立と展開 in *Bunjinga suihen* 文人畫粹編 (Tokyo: Chūō-kōron-sha, 中央公論社 1979), 2: 112-15.

[27]Ni Tsan 倪瓚, *Ch'ing-pi-ko ch'üan-chi* 清閟閣全集 (Taipei; Hsüeh-sheng shu-chü, 1969), p. 434.

[28]*Ku-kung shu-hua-chi* 故宮書畫集, no. 38.

[29]Cheng Ping-shan 鄭秉珊, *Ni Yün-lin* 倪雲林 (Shanghai: Jen-min mei-shu ch'u-pan-she, 1958), Pls. 5-7, This handscroll depicts a specific scenery, i.e., the Lion Grove (*Shih-tzu-lin*) Garden in Suchou. Ni Tsan's inscription, dated 1373, indicates that he had discussed the conception of the work with Chao Yüan, although the passage does not clearly indicate which of the two or both had actually executed the painting. Despite the controversy regarding its authorship, however, it seems that both artists were responsible for the artistic import of the painting. Cf. Chu-tsing Li, "Chao Yüan," *Dictionary of Ming Biography*, eds. L. Carrington Goodrich and Chao-ying Fang, 2 vols. (New York: Columbia University Press, 1976), 1. 136-37.

[30]*Ku-kung ming-hua san-pai chung* 故宮名畫三百種, 6 vols. (Taichung: National Palace Museum, 1959), 4.186. Also see Wen Fong, "Chinese Painting: A Statement of Method," *Oriental Art*, 9.2 (Summer 1963): 73-78.

[31]Tr. in Richard Edwards, *The Painting of Tao-chi* (Ann Arbor: The University of Michigan Museum of Art, 1967), P1. 22.

[32]Richard Edwards, *The Field of Stones: A Study of the Art of Shen Chou (1427-1509)*, Smithsonian Institution, Freer Gallery of Art, Oriental Studies, 5 (Washington, D. C: Smithsonian Institution , 1962), Pl. 23B.

[33]James Cahill, *Parting at the Shores: Chinese Painting of the Early and Middle Ming Dynasty, 1368-1580* (New York: John Weatherhill, Inc., 1978), Pl. 28.

[34]Quoted in Yü Chien-hua 俞劍華, comp., *Chung-kuo hua-lun lei-pien* 中國畫論類編 (Peking: Jen-ming mei-shu ch'u-pan-she, 1957), p. 707.

[35]Quoted in Wen Chao-t'ung 溫肇桐, *Yüan-chi ssu-ta hua-chia* 元季四大畫家 (Hong Kong: Hsing-fu ch'u-pan-she, 1960), "Ni Yun-lin." p. 22.

[36]Edwards, *The Field of Stones*, Pl. 24.

[37]Edwards, "The Arist and the Landscape: Changing Views of Nature in Chinese Painting," *Renditions* 6 (1976): 30-52, at 49-50.

[38]Cf. Mae Anna Quan Pang's discussion in "Wang Yüan-chi (1642-1715) and Formal Construction in Chinese Landscape Painting," (Ph. D. dissertation, University of California, Berkeley, 1976), pp. 166-78.

[39]Quoted in Cheng Pin-shan 鄭秉珊, *Shen Shih-t'ien* 沈石田 (Shanghai: Jen-min mei-shu ch'u-pan-she, 1958), p. 13.

[40]For an excellent study of the concept of *i-p'in* in Chinese Painting, see Shimada Shūjirō 島田修次郎, "Ippin gafū ni tsuite" 逸品畫風について, *Bijutsu kenkyū*

美術研究, no. 161 (1950), pp. 20-46. The article has been translated by James Cahill as "Concerning the I-p'in Style of Painting ," *Oriental Art*, 7.2 (Summer 1961): 66-74; 8. 3 (Autumn 1962): 130-37; 10.1 (Spring 1964): 19-26. For a recent study of the transmutation of the meaning of *i-p'in*, see Susan Nelson[Arkush], "*I-p'in* in Later Painting Criticism, " in Susan Bush and Christian Murck, eds., *Theories of the Arts in China* (Princeton: pp. 397-424. James Cahill has given a stimulating discussion of how "style" became "idea" in his "Style as Idea in Ming-Ch'ing Painting," in Maurice Meisner and Rhoads Murphey, eds., *The Mozartian Historian: Essays on the Works of Joseph R. Levenson* (Berkeley: University of California Press, 1976), pp. 137-56.

41 Ho Liang-chün 何良俊, *Ssu-yu-chai hua-lun* 四友齋畫論 Mei-shu ts'ung-shu, vol. 12, no. III／3 (Taipei: I-wen yin-shu-kuan, n. d.), p. 44.

42 Edwards, *The Art of Wen Cheng-ming* (Ann Arbor: The University of Michigan Museum of Art, 1976), pp. 72-73.

43 Ibid., p. 71.

44 Edwards, in Sickman et al. eds., *Collection of John M. Crawford*, p. 135.

45 Lee, *The Colors of Ink*, Pl. 31. Cf. Louise Yuhas's comparison of the transformation of the mode of Ni Tsan by Shen Chou, Wen Cheng-ming, and Lu Chih, in her "The Landscape Art of Lu Chih (1496-1576)," (Ph.D. dissertation, University of Michigan, 1979), pp. 175-96.

46 Chan Ching-feng, *Tung-t'u hsüan-lan-pien* 東圖玄覽編, Mei-shu ts'ung-shu, vol. 21, no. V／1 (Taipei: I-wen yin-shu-kuan, n.d.), p. 171. See also ibid., p. 149.

47 Ibid., p. 183.

48 Ibid., p. 47.

49 Ibid., pp. 41, 46.

50 Ibid., p. 59. The Ting Yün-p'eng painting is published in *Shen-chou ta-kuan* 神州大觀, no. 6 (August 1914); it was then in the Feng-yü-lou collection (Teng Shih 鄧實). Teng Shih commented that the painting was "rather like Heng-shan 衡山 [Wen Cheng-ming], but with more strength" (ibid., p. 2).

51 *Fang-shih mo-p'u*, 7.14a. Also see ibid., 3.31b.

52 *Kokka* 國華, no. 329 (October 1917).

53 *Chien-chiang tzu-liao-chi*, p. 151.

54 *T'ien-tsin-shih po-wu-kuan ts'ang-hua-chi hsü-chi*, Pl. 64. Cf. *Tsin-T'ang Wu-Tai Sung-Yüan Ming-Ch'ing ming-chia shu-hua chi* 晉唐五代宋元明清名家書畫集 (Shang-

hai: Commercial Press, 1937), Pl. 194.

[55]The University of Michigan, Palace and Central Museums Archives no. MA. 37.

[56]*Chien-chiang tzu-liao-chi*, p. 59.

[57]Ibid.

[58]*Hsin-an ming-hua chi-chin ts'e* 新安名畫集錦冊 (Shanghai: Shen-chou kuo-kuang-she, 1920), Pl. 5.

[59]Li Yung-ch'ang was teacher of Wang Chih-jui; see Chang Keng 張庚, *Kuo-ch'ao hua-cheng-la* 國朝畫徵錄, preface 1739, 5 vols. (Shanghai: Yu-chen shu-chü, 1919), 2. 6b.

[60]*Shen-chou ta-kuan hsü-pien* 神州大觀續編, no. 11 (April 1931).

[61]*Ku-kung shu-hua chi* 故宮書畫集, no. 8; The University of Michigan, Palace and Central Museums Archives no. MV 259.

[62]*T'ien-tsin-shih po-wu-kuan ts'ang-hua-chi hsü-chi*, Pl. 84 .

[63]*Shen-chou ta-kuan hsü-pien*, no. 5 (May 1929).

[64]The leaf of Ch'en Ying-ch'i has been published in *Hsin-an-ming-hua chi-chin ts'e*, Pl. 3 and in *Hsin-an-p'ai ming-hua-chi* 新安派名畫集 (Shanghai: Shen-chou kuo-kuang-she, 1924), Pl. 6 . The undated leaf is published in Wang Ts'ai-pai, *Hsin-an hua-p'ai*, Pl. 5. Three more painting have been published in *An-hui ming-jen hua-hsüan*, Pls. 4-6.

[65]Chang Keng, *Kuo-ch'ao hua-cheng-lu*, 2.6b.

[66]Quoted with permission from James Cahill, "Late Ming Painting" (MS), Chapter V, p. 18.

[67]Translated in Cahill, ibid., p. 56. The inscription is recorded in Ch'en Chuan 陳撰, *Yü-chi shan-fang hua-wai-lu* 玉几山房畫外錄 Mei-shu ts'ung-shu, vol. 4, no. I／8 (Taipei: I-wen yin-shu-kuan , n.d.), p. 98.

[68]Wang Ts'ai-pai, *Hsin-an hua-p'ai*, Pl. 5.

[69]*Ko-an shan-shui ho-chin chüan* 格庵山水合錦卷 (Shanghai: Shen-chou kuo-kuang-she, 1936).

[70]For a biography of Fang Shih-yü 方式玉 (*tzu* Yü-ju 玉如), see *She-hsien-chih*, p. 1714. both Wang Chia-chen 汪家珍 (*tzu* Shu-hsiang, 叔向, *hao* Pi-jen 璧人) and his cousin Wang Jan 汪然 (*tzu* Yü-jan于然) were from She-hsien; see *She-hsien-chih*, pp. 1712-13. Chu Ch'ang 祝昌 (*tzu* Shan-ch'ao 山嘲) was from Shu-ch'eng 舒城, Anhui, but

lived in Hsin-an. He became a *chin-shih* in 1649; see Kuo Wei-ch'ü 郭味渠, *Sung Yüan Ming Ch'ing shu-hua-chia nien-piao* 宋元明清書畫家年表 (Peking: Chung-kuo ku-tien i-shu ch'u-pan-she, 1958), p. 244. Cf. his biography in Chang Keng, *Kuo-ch'ao hua-cheng-lu*, 3.6a. Chang An-pao 張安苞 (*tzu* Tzu-ku 子固) was from Yün-chien 雲間, Kiangsu. Wu Hsin 吳星 was from Chung-shan 鍾山 (modern Nanking). Ho Yü 何顒 (*tzu* Po-ch'iu 伯求) was from Yün-chien. Wang Hsüan-tu 王玄度 (*tzu* Tsun-ssu 尊素, died 1656) was from She-hsien; see *She-hsien-chih*, p. 1663.

The handscroll was started by Wang Chia-chen towards the end of 1650 , probably in She-hsien. Earlier in the following year, Wang Jan added one section, probably still in She-hsien . Chang An-pao contributed his section in the tenth month of 1651 and Wu Hsin painted his in the early winter, both in Ch'u-chou 滁州, Anhui. The last section was painted by Wang Hsüan-tu in the late winter of 1651 in Yang-chou 揚州, Kiangsu. The section by Ho Yü, also painted in Yang-chou, was not dated, but was probably finished in the late winter. Though Hung-jen's section was not dated and the place of execution not indicated, it was probably done in the eleventh month of 1651 in Nanking where he was visiting on his way back to Anhui from Fukien. It comes after Wu Hsin's but before Ho Yü's and Wang Hsüan-tu's contributions. Since Fang Shih-yü probably took the still incomplete scroll with him first to Ch'ü-chou, then went to Nanking, before arriving at Yang-chou, Nanking is a likely place for Hung-jen's contribution.

[71]For a recent study, see Wu-jan 无染 "Hsiao Yün-ts'ung 'T'ai-p'ing shan-shui t'u' " 蕭雲從太平山水圖 *I-lin ts'ung-lu* 10.218–23. Note particularly *T'ai-p'ing shan-shui t'u* 1.5, 1.10, and 2.2.

[72]Cf. Pang, "Wang Yüan-ch'i," pp. 166–78.

[73]Chang Keng, *Kuo-ch'ao hua-cheng-lu*, 1.13a.

[74]An example of Yü Chia-shen's design is in *Fang-shih mo-p'u*, 4.4la.

[75]*Ninety Years of Wu School Painting* (Taipei: National Palace Museum, 1975), Pls. 160 and 170.

[76]Cahill, *Parting at the Shores*, Pl. 129.

[77]Chang Keng, *Kuo-ch'ao hua-cheng-lu*, 1.13a. The exact dates of Sun I are unknown. Since Sun I and Hsiao Yün-ts'ung were associated together by their contemporaries as "Sun-Hsiao", it is likely that they might belong to the same generation. Furthermore, a colophon by T'ang Yen-sheng to Hsiao Yün-ts'ung's 1656 handscroll "Cloud Terrace and Sparse Trees" now in Nanking indicates that Sun I died after Hung-jen, i.e. after 1663; see Hsü Yün-ch'iu 徐澐秋, "Hsiao Ch'ih-mu, 'Yün-t'ai shu-shu t'u-chüan,' " 蕭尺木雲台疏樹圖卷, *I-lin ts'ung-lu*, 8.191–93.

[78]Recorded in Kuo Wei-ch'ü, *Nien-piao*, p. 234. Hsü Pang-ta 徐邦達 stated that he saw a landscape hanging scroll on gold paper, also dated 1643, in his *Li-tai liu-ch'uan shu-hua tso-p'in pien-nien-piao* 歷代流傳書畫作品編年表 (Shanghai: Jen-min mei-shu ch'u-pan-she,1963), p. 126. Both Kuo Wei-ch'ü and Hsü Pang-ta were probably referring to the same painting published in *An-hui ming-jen hua-hsüan*, Pl. 3.

[79]Published in *Shen-chou ta-kuan hsü-pien*, no. 5 and Wang Ts'ai-pai, *Hsin-an hua-p'ai*, Pl. 6. It is also published in *Chin-shih shu-hua* 金石書畫, no. 66 (September 30, 1936), p. 3 where it was indicated as belonging to Huang Pin-hung.

[80]*Ninety Years of Wu School Painting*, Pl. 005. A landscape handscroll attributed to Wen Po-jen in the Ching Yüan Chai collection is so similar in style to Sun I's extant paintings that it has been re-attributed to Sun I. This re-attribution, though still open to debate, is certainly based on an affinity between Sun I and the Wen Cheng-ming School artist.

[81]One version is published in Sirén, *Chinese Painting*, 3.91.

[82]Published in *Ku-kung ming-hua san-pai-chung*, 1.5-14.

[83]For a recent discussion of the album, see Shan Kuo-lin 單國霖 "Wang Li ho 'Hua-shan t'u-ts'e' " 王履和《華山圖冊》, *Mei-shu ts'ung-k'an* 美術叢刊, no. 1 (February 1978), pp. 17-25. Cf. Cahill, *Parting at the Shores*, pp. 5-7 and Pls. 1-2.

[84]Edwards, *The Field of Stones*, Pl. 20A.

[85]Cahill, *Parting at the Shores*, Pls. 39-40.

[86]Yuhas, "Lu Chih," Pls. 21a-m.

[87]Ibid., Pl. 54; The University of Michigan, Palace and Central Museums Archives no. MA 20.

[88]Cf. ibid., pp. 100-5, 136-44.

[89]Compare Hung-jen's leaf no. 48 with Shao Mi's "Album of Landscapes" dated 1638 and now in the Seattle Art Museum, especially the third leaf published in Cahill, *Restless Landscape*, Pl. 19.

[90]See Ho Lo-chih 何樂之, "*Ming-shan-t'u* chieh-shao" 名山圖介紹, in *Ming-k'an ming-shan-t'u pan-hua-chi* 明刊名山圖版畫集 (Shanghai: Jen-min mei-shu ch'ui-pan-she, 1958), pp. 1-8 and Pls. 7-8. One recalls that Wu T'ing-yü also contributed designs for ink-cakes published in *Fang-shih mo-p'u*, e.g. 4.9a and 6.16a

[91]For discussion on Cheng Chung, see Elizabeth Fuller, "Professional Painters in Fukien and Nanking," in Cahill, *Restless Landscape*, pp. 103-7 and Pl. 50.

[92]Cheng Chung is said to have painted the scenic beauty of Mt. Huang at the request of the Buddist monk P'u-men 普門. It was P'u-men who was instrumental in the opening up of Mt. Huang to an increasing number of visitors; see Chapter I above and *She-hsien-chih*, pp. 1704-5. According to Wang Shih-hung's 汪士鋐 preface to Tao-wu's designs of Mt. Huang published in *Huang-shan ts'ung-shu*, vol. 1, Cheng Chung's designs were later copied by Hsiao Ch'en 蕭晨 (*tzu* Ling-hsi 靈曦, *hao* Chung-su 中素, active ca. 1680-1710), who was from Yang-chou.

[93]*Chien-chiang tzu-liao-chi*, p. 74; trans. in Fu and Fu, *Studies in Connoisseurship*, p. 141.

[94]Published in *Shen-chou ta-kuan*, no. 3; see also *Chien-chiang tzu-liao-chi*, p. 161.

[95]Hsü Yün-ch'iu, "Hsiao Ch'ih-mu," 8.191-93.

[96]Unpublished, but recorded in Kuo Wei-ch'ü, *Nien-piao*, p. 137 and Hsü Pang-ta, *Pien-nien-piao*, p. 125.

[97]For a similar handscroll by Shih Jui now in the Hashimoto collection, see *Hashimoto shūzō Min Shin ga mokuroku* 橋本所藏明清畫目錄 (Tokyo: Kadokawa shoten, 1972), Pl. 2.

[98]Published in *Chih-lo-lou ts'ang Ming i-min shu-hua* 至樂樓藏明遺民書畫 (Hong Kong: Art Gallery, Institute of Chinese Studies, Chinese University of Hong Kong, 1975), p. 201.

[99]*Illustrated Catalogues of Tokyo Nation Museum: Chinese Painting*, Pl. 102. For a study of the handscroll, see Akiyama Teruo 秋山光夫, "Sho Shakuboku to shuzan koryo zu maki" 蕭尺木の秋山行旅圖卷, in his *Nihon bijutsu ronkō* 日本美術論考 (Tokyo: Dai-ichi shobō, 1943), pp. 407-34.

[100]Reproduced in Cahill, *Restless Landscape*, p. 47.

[101]Sherman E. Lee, *A History of Far Eastern Art*, rev. ed. (New York: Harry N. Abrams, Inc., 1973), p. 439.

[102]Cahill, *Fantastic and Eccentrics*, pp. 16, 44.

[103]Max Loehr, "Chinese Landscape Painting and Its Real Content," *Rice University Studies* 59.4 (Fall 1973): 92.

[104]Edwards, *Wen Cheng-ming*, pp. 209-11.

[105]Cf. Wai-kam Ho, "Tung Ch'i-ch'ang's New Orthodxy," in *Artists and Traditions*, ed. Murck, p. 128.

FOOTNOTES TO CHAPTER III

[1]Published in *Tou-yun*, no. 15 (October 1987), p. 6.

[2]All sixty leaves are published in Su Tsung-jen 蘇宗仁 comp., *Huang-shan ts'ung k'an* 黃山叢刊), 8 vols. (Peking: Pai-i-yen-chai, 1937), v. 2. Only fifty leaves are published in *Shina nanga taisei*, vol. 13. The numerical order in the *Huang-shan ts'ung-k'an* is followed in the present study. It does not have any signature of Hung-jen, though his seal and label for the site depicted appear on each leaf. The colophons by his contemporaries and later artists including Hsiao Yün-ts'ung (dated 1665), Ch'eng Sui (undated), T'ang Yün-chia 唐允甲 (dated 1674), Cha Shih-piao (undated), Wang Tzu-sui 汪滋穗 (undated), Jao Ching 饒璟 (undated), and Wang Chia-chen (dated 1665), as well as the quality of the paintings themselves, leave no doubt about the authenticity of the album. The album apparently was well-known to other artists and became their model in depicting the scenery of Mt. Huang. Some leaves from Hung-jen's album found their way into the local gazetteer *Huang-shan-chih* 黃山志, compiled by Min Lin-ssu 閔麟嗣 (1628-1704) and first published in 1661. One of the illustrations in the *Huang-shan-chih* is labeled "Unsettling Dragon Pine" (*Jao-lung-sung* 擾龍松), and is referred to as a design by Hung-jen. (Published in *Huang-shan-chih* 黃山志 , An-hui ts'ung-shu 安徽叢書, 1.20a. It can be compared with Hung-jen's leaf no. 29, also labeled "Unsettling Dragon Pine," in Su Tsung-jen, *Huang-shan ts'ung-k'an*, 2.29.) The fact that the *Huang-shan-chih* was published in 1661 dates Hung-jen's album to some time before that year. Furthermore, the modern scholar Hsü Ch'eng-yao 許承堯, who was the editor of the local gazatteer *She-hsien-chih* published in 1937, reported that there was an album on Mt. Huang by Huang Ch'i 黃錤 in the collection of Cheng Wei-chan 鄭渭占 in She-hsien. (Hsü Ch'eng-yao 許承堯, *She-ku* 歙故, quoted in *Chien-chiang tzu-liao-chi*, pp. 154-56. Huang Ch'i [*tzu* Yün-an 筠庵 and Lu-t'ing 麓亭] was from She-hsien; see *She-hsien-chih*, pp. 1716-17.) Huang Ch'i's album bears a date *ting-yu* 丁酉 which most likely corresponds to 1717. An alternate date would be 1657 which would certainly place the Hung-jen album before that date.

[3]Published in Cahill, ed., *Shadows of Mt. Huang*, p. 79, fig. 16.

[4]For the Drenowatz album, see Li, *A Thousand Peaks and Myriad Ravines*, vol. 1, pp. 180-84.

[5]Cf. Kuo Hsi's 1072 hanging scroll "The Early Spring," now in the National

Palace Museum in Taipei; reproduced in Sirén, *Chinese Painting*, 3.175.

[6]For example, Huang Kung-wang's 1350 handscroll "Dwelling in the Fu-ch'un Mountains," now in the National Palace Museum in Taipei; reproduced in Sirén, *Chinese Painting*, 6.67-72.

[7]Edwards, *The Art of Wen Cheng-ming*, p. 46.

[8]Reproduced in Sirén, *Chinese Painting*, 6.57-72.

[9]Reproduced in Li, *A Thousand Peaks and Myriad Ravines*, no. 43.

[10]From Ssu-ma Ch'ien's 司馬遷 biography of Ch'ü Yüan, tr. in David Hawkes, *Ch'u Tz'u: The Songs of the South* (Oxford: Oxford University Press, 1962), p. 12.

[11]Recently republished in *Chung-kuo i-shu* 中國藝術, No. 1 (July 1985), pp. 52-53.

[12]Published (some time before 1950?) as *Chien-chiang seng shan-shui-ts'e* 漸江僧山水冊 (n. p., n.d.).

[13]The whole poem can be found in *Ts'en Chia-chou shih* (1520; reprint ed., Shanghai: Han-feng-lou, n.d.), 3.29b.

[14]Vanderstappen, "Some Seventeenth-Century Chinese Paintings," p. 161.

[15]The close friendship between Ch'eng Shou and Hung-jen can be seen in a poem composed by Ch'eng Shou in 1652. In the preface to the poem, Ch'eng Shou wrote that he and Hung-jen stayed in the same room during a snowy night at Wu-hu. See *Chien-chiang tzu-liao-chi*, pp. 122, 203.

[16]The inscribed poem provides an instructive example of Hung-jen's subtle and selective approach to the past. Although Hung-jen did not mention Ni Tsan in the inscription, the first four characters of the poem ("niao-yü-mien-man 鳥語綿蠻) is apparently taken from the third line of a poem ("ku-niao yü-mien-man 谷鳥語綿蠻) by Ni Tsan; see Chuang Shen 莊申, comp., *Yüan-chi ssu-hua chia shih chiao-chi* 元季四畫家詩校輯 (Hong Kong: The Center of Asian Studies, Hong Kong University, 1973), p. 140.

[17]*Chien-chiang tzu-liao-chi*, pp. 78, 94. Hung-jen painted another hanging scroll for Shen Shou-kuo; see ibid., p. 83.

[18]Ibid., p. 38.

[19]Translation courtesy of James Cahill.

[20]The couplet is taken from a poem composed by Wei Cheng, in *Ch'üan-T'ang-shih* 全唐詩, 12 vols. (Peking: Chung-hua shu-chü, 1960), 1.441.

[21]Cahill, *Chinese Painting*, pp. 171-72.

[22]Translation by Richard Edwards, in Cahill, *Fantastics and Eccentrics in Chinese Painting*, no. 13.

[23]Translation amended from Contag, *Chinese Masters of the 17th Century*, p. 39.

FOOTNOTES TO CHAPTER IV

[1] *Chien-chiang tzu-liao-chi*, p. 74; translated in Marilyn and Shen Fu, *Studies in Connoisseurship*, p. 141.

[2] Ibid., p. 150.

[3] Chou Liang-kung, *Tu-hua-lu*, p. 19.

[4] Ibid.

[5] Max J. Friedlander, *On Art and Connoisseurship*, translated by Tancred Borenius (Boston: Beacon Press, 1960), p. 217.

[6] Kuo Wei-ch'ü, *Sung-Yüan Ming-Ch'ing shu-hua-chia nien-paio*, p. 244. An album and a handscroll by Chu Ch'ang, both dated 1673, have been recorded. (Shih-pai-chai-chu-jen, *Shih-pai-chai shu-hua-lu*, hsia han, ch'eng chi 下函辰集 pp. 24-25 and shang han, ting chi 上函丁集, p. 10.) If these recorded paintings are reliable, Chu Ch'ang probably was active well into the 1670's. According to a letter written by Cha Shih-piao, he died sometime before the eighth month of 1679; see Ma Pin 馬彬, "Ch'ien-t'an Hsin-an hua-p'ai" 淺談新安畫派, *I-yüan to-ying* 藝苑掇英 no. 20 (April 1983): 27 and 35 and *An-hui ming-jen hua-hsüan*, Pls. 19-20.

[7] Chu Ch'ang (*tzu* Shan-ch'ao 山嘲) was not from She-hsien, but the neighboring Shu-ch'eng 舒城 (near modern Ho-fei 合肥); he was also said to have come from Shun-te 順德 (near modern Hsüan-ch'eng 宣城); see Chang Keng, *Kuo-ch'ao hua-cheng-lu*, 3, 5-6.

[8] *Shen-chou ta-kuan hsü-pien*, vol. 10.

[9] *Su-chou po-wu-kuan ts'ang-hua-chi*, Pls. 64-70.

[10]*Hashimoto shūzō Min Shin ga mokuroku* (Tokyo: Kadokawa shoten, 1972), Pl. 94.

[11]Wang Wei-an 王維庵 (Wang Chuang-wei 王壯爲), "Mei-ho ti-i hua-ts'e 梅壑第一畫

冊," *Chung-kuo shu-hua* 中國書畫, no. 9 (December 1969): 9-19.

[12]*Chien-chiang tzu-liao-chi*, pp. 71, 99-100; reproduced in *An-hui ming-jen hua-hsüan*, Pl. 1.

[13]Ibid., pp. 45, 55-56.

[14]Shih-pai-chai-chu-jen, *Shih-pai-chai shu-hua-lu*, hsia han, mo chi, 下函卯集, pp. 20-22.

[15]*Chien-chiang tzu-liao-chi*, p. 102.

[16]Ibid., p. 75.

[17]Ibid., p. 129; *She-hsien-chih*, pp. 1151-52; Huang Pin-hung, *Huang-shan hua-yüan-lüeh*, p. 31.

[18]*T'ien-yin-t'ang ming-hua-hsüan* 天隱堂名畫選 (Tokyo: n.p., 1963), vol. 1, Pls. 72-81.

[19]*Ming-Ch'ing shan-mien-hua hsüan-chi* (Shanghai: Jen-min mei-shu ch'u-pan-she, 1959), Pl. 64.

[20]Marilyn and Shen Fu, *Studies in Connoisseurship*, p. 154.

[21]*Chien-chiang tzu-liao-chi*, pp. 3-5, 8-9; *She-hsien-chih*, pp. 1665-66.

[22]*She-hsien-chih*, ibid.

[23]*Chien-chiang tzu-liao-chi*, p. 3-5; for example, see *An-hui ming-jen hua-hsüan*, Pl. 18.

[24]Ibid., p. 8.

[25]*She-hsien-chih*, p. 162.

[26]*Chien-chiang tzu-liao-chi*, pp. 67-68, 107. The painting was reproduced in Wang Ts'ai-pai 汪采白, *Hsin-an hua-p'ai* (Nanking, 1933), Pl. 26 where it is, however, mislabeled as a painting by Chiang Chu. A close copy is in the Chang Po-chin collection in Tokyo; reproduced in *T'ien-yin-t'ang ming-hua-hsüan*, vol. 1, Pl. 70.

[27]*Shen-chou ta-kuan hsü-pien*, vol. 6.

[28]*The Richard Bryant Hobart Collection of Chinese Ceramics and Paintings* (New York: Parke-Bernet Galleries, Inc., 1969), no. 278.

[29]*Nanking po-wu-yüan ts'ang-hua-chi*, 2. 40.

[30]Chang Keng, *Kuo-ch'ao hua-cheng-lu*, 2. 11-12.

[31]Li, *A Thousand Peaks and Myriad Ravines*, p. 184.

[32] *Hashimoto shuzo Min Shin ga mokuroku*, Pl. 114.

[33] One leaf is reproduced in *Duizend jaar Chinese schilderkunst* (The Hague: Gemeentemuseum, 1960), Pl. 45.

[34] For a most succinct summary of the stylistic denominators of the orthodox school, see James Cahill, "The Orthodox Movement in Early Ch'ing Painting," in Christian F. Murck, ed., *Artists and Traditions: Uses of the Past in Chinese Culture* (Princeton: Princeton University, Art Museum, 1976), pp. 169-181. See also Roderick Whitfield, *In Pursuit of Antiquity: Chinese Paintings of the Ming and Ch'ing Dynasties from the Collection of Mr. and Mrs. Earl Morse* (Princeton: Princeton University, Art Museum, 1967).

[35] *The Richard Bryant Hobart Collection of Chinese Ceramics and Paintings*, no. 286; for additional example, see *An-hui ming-jen hua-hsüan*, Pl. 23.

FOOTNOTES TO CONCLUDING REMARKS

[1] Aoki Masaru 青木正兒 and Iriya Yoshitaka 入矢義高, with commentary by Kōhara Hironobu 古原宏伸, *Kaishi-en gaden shoshū: sansui juseki* 芥子園畫傳初集：山水樹石, 2 vols. (Tokyo: Chikuma shobō, 1975), vol. 1, preceding p.1 (in colors) and p. 338.

[2] Chou Liang-kung, *Tu-hua-lu*, passim.

[3] The leaf is reproduced in Edwards, *The Painting of Tao-chi*, p. 108. The translation is adapted with slight amendation from Marilyn and Shen Fu, *Studies in Connoisseurship*, pp. 52-53.

[4] *Chien-chiang tzu-liao-chi*, p. 69 and Pl. 3.

[5] Chang Keng, *Kuo-ch'ao hua-cheng-lu*, 3.12.

[6] Chang Keng, *P'u-shan lun-hua* 浦山論畫 (about 1750), in Yü Chien-hua, *Chung-kuo hua-lun lei-pien*, p. 223; translation adapted with slight amendation from Sirén, *The Chinese on the Art of Painting* (New York : Schoken Books, 1963), pp. 212-13.

[7] Chou Liang-kung, *Tu-hua-lu*, p. 18.

[8] Hsieh Lan-sheng 謝蘭生, *Ch'ang-hsing-hsing-chai shu-hua t'i-pa* 常惺惺齋書畫題跋 (Macao: Wen-chi t'u-shu kung-ssu, n. d.), 2.58.

[9]Chuang Shen 莊申, "Ch'ing-chi Yüeh-chung ts'ang-hua yü Ch'ing-tai hua-p'ai chih kuan-hsi" 清季粵中藏畫與清代畫派之關係 *Journal of Oriental Studies*, 15.2 (July 1977): 206-231.

[10]For example, Hsieh Lan-sheng's leaf no. 5 in his 1827 landscape album now in the Chih-lo-lou collection in Hong Kong, published in *Landscape Paintings by Kwangtung Masters During the Ming & Ch'ing Periods*, (Hong Kong: Art Gallery, Institute of Chinese Studies, Chinese University of Hong Kong, 1973), no. 53.

[11]See Wu Yin-ming 吳因明, "Wan-Ming chiang-nan fo-hsüeh feng-ch'i yü wen-jen-hua 晚明江南佛學風氣與文人畫," *Hsin-ya shu-yüan hsüeh-shu nien-k'an*, no. 2 (Sept. 1960), pp. 1-22. See also Cheng Hsi-chen, *Hung-jen, K'un-ts'an*, pp. 19-30.

[12]Hsien Chih-liu 謝稚柳, *Chu Ta* 朱耷 (Shanghai: Jen-min mei-shu ch'u-pan-she, 1958), p. 7.

[13]Cheng Cho-lu 鄭拙廬, *Shih-t'ao yen-chiu* 石濤研究 (Peking: Jen-min mei-shu ch'u-pan-she, 1961), pp. 26-26.

[14]Chang Keng, *Kuo-ch'ao hua-cheng-lu*, 3.1.

[15]Huang Pin-hung 黃賓虹, *Ku-hua-wei* 古畫微 (Shanghai: Commercial Press, 1925), pp. 36-39.

[16]For instance, P'an T'ien-shou, *Chung-kuo hui-hua-shih* (1936), p. 206.

[17]Cf. Sherman E. Lee, *A History of Far Eastern Art*, rev. ed. (New York: Harry N. Abrams, Inc., 1973), p. 439.

[18]T'ang Chih-ch'i 唐志契, *Hui-shih wei-yen* 繪事微言, in Yü Chien-hua, *Chung-kuo hua-lun lei-pien*, p. 737.

BIBLIOGRAPHY

Akiyama Teruo 秋山光夫. *Nihon bijutsu ronkō* 日本美術論考. Tokyo: Dai-ichi shobō, 1943.

An-hui ming-jen hua-hsüan 安徽名人畫選. Hofei, Anhui: Anhui Provincial Museum, 1961.

An-hui ts'ung-shu 安徽叢書. Shanghai: Anhui ts'ung-shu pien-yin-ch'u, 1932-36.

Aoki Masaru 青木正兒 and Iriya Yoshitaka 入矢義高. *Kaishi-en gaden shoshū: sansui juseki* 芥子園畫傳初集: 山水樹石. Tokyo: Chikuma shobō, 1975.

Arkush, Susan Nelson. "*I-p'in* in Later Painting Criticism." In *Theories of the Arts in China*, pp. 397-424. Edited by Susan Bush and Christian Murck. Princeton: Princeton University Press, 1983.

————. "Ni Tsan and the Image of Yüan Painting." Ph. D. dissertation, Harvard University, 1977.

Atwell, William S. "From Education to Politics: the Fu She." In *The Unfolding of Neo-Confucianism*, pp. 333-67. Edited by Wm. Theodore de Bary. New York: Columbia University, 1975.

Barnhart, Richard. "Sōkin ni okeru sansuiga no seritsu to tenkai." In *Bunjinga suihen* 文人畫粹編, 2: 112-15. Tokyo: Chūō-kōron-sha, 1979.

————. *Marriage of the Lord of the River: A Lost Landscape by Tung Yüan.* Ascona: Artibus Asiae, 1970.

Bartholomew, Terese Tse. *I-hsing Ware*. New York: China Institute in America, 1977.

Bussagli, Mario. *La pittura cinese*. Milan: Fratelli Fabbri Editori, 1966.

Cahill, James. *Chinese Painting*. Geneva: Skira, 1960.

————. "The Early Styles of Kung Hsien." *Oriental Art* 16.2 (Summer 1970): 51–71.

————. *Fantastics and Eccentrics in Chinese Painting*. New York: Asia House Gallery, 1967.

————. *Hills Beyond a River: Chinese Painting of the Yüan Dynasty, 1279-*

1368. New York: John Weatherhill, 1976.

————. *Parting at the Shores: Chinese Paintings of the Early and Middle Ming Dynasty, 1368-1580*. New York: John Weatherhill, 1978.

————, ed. *The Restless Landscape: Chinese Painting of the Late Ming Period*. Berkeley: University Art Museum, 1971.

————. "Style as Idea in Ming-Ch'ing Painting." In *The Mozartian Historian: Essays on the Works of Joseph R. Levenson*, pp. 137-56. Edited by Maurice Meisner and Rhoads Murphey. Berkeley: University of California Press, 1976.

————. "We Pin and His Landscape Painting." In *Proceedings of the International Symposium on Chinese Painting*, pp. 637-38. Taipei: National Palace Museum, 1970.

————, ed. *Shadows of Mt. Huang: Chinese Painting and Printing of the Anhui School*. Berkeley: University Art Museum, 1981.

————. *The Compelling Image: Nature and Style in Seventeenth-Century Chinese Painting*. Cambridge, Mass.: Harvard University Press, 1982.

Chan Ching-feng 詹景鳳. *Tung-t'u hsüan-lan-pien* 東圖玄覽編. Mei-shu ts'ung-shu, vol. 21, no. V／1. Taipei: I-wen yin-shu-kuan, n. d.

Chang Ch'ou 張丑. *Ch'ing-ho shu-hua-fang* 淸和書畫舫. N.p., 1875.

————. *Chen-chi jih-lu* 眞跡日錄. N.p., 1918.

Chang Chung-i 張仲一 et al. *Hui-chou Ming-tai chu-chai* 徽州明代住宅. Peking: Chien-chu kung-ch'eng ch'u-pan-she, 1957.

Chang Keng 張庚. *Kuo-ch'ao hua-cheng-lu* 國朝畫徵錄. Shanghai: Yu-cheng shu-chü, 1919.

Chang Kuang-pin 張光賓. *Yüan ssu-ta-chia: Huang Kung-wang, Wu Chen, Ni Tsan, Wang Meng* 元四大家: 黃公望、吳鎭、倪瓚、王蒙. Taipei: National Palace Museum, 1975.

Chang Ta-yung 張大鏞. *Tzu-i-yüeh-chai shu-hua-lu* 自怡悅齋書畫錄. N.p., n. d.

Ch'en Chuan 陳撰. *Yü-chi shan-fang hua-wai-lu* 玉几山房畫外錄. Mei-shu ts'ung-shu, vol. 8, no. 1／8. Taipei: I-wen yin-shu-kuan, n.d.

Ch'en Jen-t'ao 陳仁濤. *Chin-k'uei ts'ang-hua-chi* 金匱藏畫集. Hong Kong: T'ung-ying kung-ssu, 1956.

Ch'en Shih-tseng 陳師曾. *Chung-kuo hui-hua-shih* 中國繪畫史. Tsi-nan: Han-mo-yüan mei-shu-yüan, 1925.

Ch'en Yeh 陳野. "Lun Hui-chou shang-yeh tzu-pen te hsing-ch'eng chi-ch'i t'e-she" 論徽州商業資本的形成及其特色. *Anhui shih-hsüeh t'ung-hsün* 安徽史學通訊, no. 7 (September 1958), pp. 32-42.

Ch'en Yüan 陳垣. *Shih-shih i-nien-lu* 釋氏疑年錄. Peking: Chung-hua shu-chü, 1964.

————. *Ch'ing-ch'u seng-cheng-chi* 清初僧諍記. Peking: Chung-hua shu-chü, 1962.

————. *Ming-chi Tien Ch'ien fo-chiao k'ao* 明季滇黔佛教考. Peking: Ko-hsüeh ch'u-pan-she, 1962.

Cheng Ch'ang 鄭昶. *Chung-kuo hua-hsüeh ch'üan-shih* 中國畫學全史. Shanghai: Chung-hua shu-chü, 1928.

Cheng Chen-to 鄭振鐸 comp. *Yün-hui-chai ts'ang T'ang Sung i-lai ming-hua-chi* 韞輝齋藏唐宋以來名畫集. Shanghai: n.p., 1947.

————. comp. *Ming i-min hua hsü-chi* 明遺民畫續集. N.p., n.d.

Cheng Cho-lu 鄭拙廬. *Ni Tsan* 倪瓚 Shanghai: Jen-min mei-shu ch'u-pan-she, 1961.

————. *Shih-t'ao yen-chiu* 石濤研究. Peking: Jen-min mei-shu ch'u-pan-she, 1961.

Ch'eng Hsi 程曦. *Mu-fei ts'ang-hua k'ao-p'ing* 木扉藏畫考評. Hong Kong: Privately published by Ch'eng Hsi and Cheng Te-k'un, 1965.

Cheng Hsi-chen鄭錫珍. *Hung-jen, K'un-ts'an* 弘仁髡殘. Shanghai: Jen-min mei-shu ch'u-pan-she, 1963; rev. ed. 1979.

Cheng Pin-shan 鄭秉珊. *Shen Shih-t'ien* 沈石田. Shanghai: Jen-min mei-shu ch'u-pan-she, 1958.

Ch'eng Ta-yüeh 程大約. *Ch'eng-shih mo-yüan* 程氏墨苑. N.p.: Tzu-lan-t'ang, 1606.

Cheng Yüan-hsün 鄭元勳, ed. *Mei-yu-ko wen-wu* 媚幽閣文娛. Shanghai: Pei-yeh-shan-fang, 1936.

Chi-jang 擊壤. "Huang Pin-hung yü Hsin-an hua-p'ai 黃賓虹與新安畫派." In *I-lin ts'ung-lu* 藝林叢錄, 2.351–54. Hong Kong: Commercial Press, 1962.

Chiang Chao-shen 江兆申. *Ninety Years of Wu School Painting*. Taipei: National Palace Museum, 1975.

Chiang Hsi-chao 蔣希召. *Wu-i-shan* 武夷山. Shanghai: Commercial Press, 1924.

Chien-chiang hua-chi 漸江畫集. Hong Kong: Cafa Publising Co., n.d.

Chien-chiang shang-jen shan-shui-ts'e 漸江上人山水冊. Shanghai: Shen-chou kuo-kuang-she, 1920.

Chien-chiang seng shan-shui-ts'e 漸江僧山水冊. N.p., n.d.

Chien-chiang, Shih-ch'i, Shih-t'ao, Pa-ta-shan-jen shu-hua chi 漸江石谿石濤八大山人書畫集. Taipei: National Museum of Histoy, 1978.

Chih-k'an 智龕. "Shih-chu-chai-chu Hu Cheng-yen" 十竹齋主胡正言. In *I-lin ts'ung-lu* 藝林叢錄, 9.296-99. Hong Kong: Commercial Press, 1962.

Chih-lo-lou ts'ang Ming-i-min shu-hua 至樂樓藏明遺民書畫. Hong Kong: A catalog from an exhibition held at the Art Gallery, Institute of Chinese Studies, Chinese University of Hong Kong in September, 1975.

Ch'in Chung-wen 秦仲文. "Ch'ing-tai chu-ch'i hui-hua te fa-chan" 清代初期繪畫的發展. *Wen-wu ts'an-k'ao tzu-liao*, no. 8 (1958), pp. 53-58.

Chin-shih shu-hua 金石書畫, no. 56 (April 30, 1936).

Chinesische Malerei des 17. Jahrhunderts. Frankfurt am Main: Museum für Kunsthandwerk, 1957.

Chinese Art Treasures. Geneva: Skira, 1961-62.

Ch'iu, A. K'ai-ming. "The Chieh Tzu Yuan Hua Chuan (Mustard See Garden Painting Manual): Early Editions in American Collection." *Archives of the Chinese Art Society of America* 5 (1951): 55-69.

Chou Liang-kung 周亮工. *Tu-hua-lu* 讀畫錄. I-shu ts'ung-pien, vol. 1, no. 14. Taipei: Shih-chieh shu-chü, 1968.

————. *Yin-shu-wu shu-ying* 因樹屋書影. Shanghai: Ch'un-hsüeh-she, n.d.

Chou Wu 周蕪, *Hui-p'ai pan-hua-shih lun-chi* 徽派版畫史論集, Hofei, Anhui: Jen-min mei-shu ch'u-pan-she, 1984.

Chu Chih-ch'ih 朱之赤. *Chu Wo-an ts'ang shu-hua-mu* 朱臥菴藏書畫目. I-shu ts'ung-pien, vol. 17, no. 157. Reprint ed., Taipei: Shih-chieh shu-chü, 1962.

Chu Hsing-chai 朱省齋. *Hsing-chai tu-hua-chi* 省齋讀畫記. Hong Kong: Ta-kung shu-chü, n. d.

Ch'üan-T'ang-shih 全唐詩. 12 vols. Peking: Chung-hua shu-chü, 1960.

Chuang Shen 莊申. *Yüan-chi ssu-hua-chia shih chiao-chi* 元季四畫家詩校輯. Hong Kong: The Center of Asian Studies, Hong Kong University, 1973.

Chuang Shen 莊申. "Ch'ing-chi Yüeh-chung ts'ang-hua yü Ch'ing-tai hua-p'ai-chih kuan-hsi" 清季粵中藏畫與清代畫派之關係 *Journal of Oriental Studies* 15.2 (July 1977): 206-31.

Chūgoku meiga-shū 中國名畫集 Tokyo: Ryūbundō, 1935.

Chūgoku no Min Shin jidai no hanga 中國の明清時代の版畫. Nara: Yamato bunkakan, 1972.

Chung-kuo-hua 中國畫, no. 2 (1958).

Clapp, Anne De Coursey. *Wen Cheng-ming: The Ming Artist and Antiquity*. Ascona: Artibus Asiae, 1975.

Contag, Victoria. *Chinesische Malerei der letzten vier Jahrhunderte*. Hamburg: Museum für Kunst und Gewerbe, 1949.

————. *Chinese Masters of the 17th Century*. Translated by Michael Bullock. Rutland, Vermont: Charles E. Tuttle Co., 1970.

Dubosc, J. P. *Pao Hui Chi: Twelve Chinese Paintings in the Collection of J. P. Dubosc*. Peking: n.p., 1937.

————. *Exposition de peintures chinoises de la collection J. P. Dubosc*. Paris: Bibliothēque Nationale, 1937.

————. *Mostra di pitture cinesi delle dinastie Ming e Ch'ing*. Rome: Is. M. E. O., 1950.

Duizend jaar Chinese schilderkunst. The Hague: Gemeetemuseum, 1960.

Edwards, Richard. *The Art of Wen Cheng-ming*. Ann Arbor: The University of Michigan Museum of Art, 1976.

————. "The Artist and the Landscape: Changing Views of Nature in Chinese Painting." *Renditions* 6 (1976): 30–52.

————. *The Field of Stones: A Study of the Art of Shen Chou*. Smithsonian Institution, Freer Gallery of Art, Oriental Studies no. 5. Washington D. C.: The Freer Gallery of Art, 1962.

————. *The Painting of Tao-chi*. Ann Arbor: The University of Michigan Museum of Art, 1967.

Exposition d'ouvrages illustrēs de la dynastie Ming. Peking: Centre franco-chinois d' d'ētudes sinologiques, 1944.

Fang Yü-lu 方于魯. *Fang-shih mo-p'u* 方氏墨譜. N.p., 1588.

Fong, Wen. "Chinese Painting: A Statement of Method." *Oriental Art* 9.2 (Summer 1963): 73-78.

————. "How to Understand Chinese Painting." *Proceedings of the American*

Philosophical Society 115 (1971): 282-92.

————. "Tung Ch'i-ch'ang and the Orthodox Theory of Painting." *National Palace Museum Quarterly* 2.3 (January 1968): 1-26.

————. "Rivers and Mountains after Snow (*Chiang-shan hsüeh-chi*): Attributed to Wang Wei (A.D. 699-759)." *Archives of Asian Art* 30 (1976-77): 12-13.

————, and Fu, Marilyn. *Sung and Yüan Paintings*. New York: The Metropolitan Museum of Art, 1973.

Fontein, Jan, and Hempel, Rose. *China, Korea, Japan*. Propyläen kunstgeschichte, vol. 17. Berlin: Propyläen Verlag, 1968.

Fu I-ling 傅衣凌. *Ming-Ch'ing shih-tai shang-jen chi shang-yeh tzu-pen* 明清時代商人及商業資本 Peking: Jen-min ch'u-pan-she, 1956.

Fu, Marilyn and Shen. *Studies in Connoisseurship: Chinese Paintings from the Arthur M. Sackler Collection in New York and Princeton*. Princeton: The Art Museum, Princeton University, 1973.

Fu Shen 傅申. "Chang Chi-chih te chung-k'ai" 張即之的中楷. *National Palace Museum Quarterly* 10.4 (Summer 1976): 43-65.

————. "Ming-Ch'ing chih-chi te k'o-pi kou-le feng-shang yü Shih-t'ao te tsao-ch'i tso-p'in" 明淸之際的渴筆勾勒風尙與石濤的早期作品. *The Journal of the Institute of Chinese Studies of the Chinese University of Hong Kong* 8.2 (1976): 579-615.

————. *Traces of the Brush*. New Haven: Yale University Art Gallery, 1977.

Fujii Hiroshi 藤井宏. "Shin-an shōnin no kenkyū 新安商人の研究. *Tōyō gakuhō* 東洋學報 36. 1-4 (1953-54): 1-44, 52-60, 65-110, 115-45.

Franke, Herbert. *Kulturgeschichtliches über die chinesische Tusche*. Munich : Verlag der bayerischen Akademie der Wissenschaften, 1962.

Goepper, Roger. *The Essence of Chinese Painting*. Translated by Michael Bullock. London: Percy Lund, Humphries & Co., 1963.

————. *Chinese Painting: The Later Tradition*. Translated by John L. McHale. Berne: Hallway, 1966.

————, et al. *Kunst der Ch'ing-Zeit, 1644-1911*. Köln: Museum für Ostasiatische Kunst, 1973.

Gombrich, Ernst H. *Art and Illusion*, 2nd ed. Princeton: Princeton University Press, 1961.

Goodrich, L. Carrington, and Fang, Chaoying, eds. *Dictionary of Ming Biography, 1368-1644*. New York: Columbia University Press, 1976.

Harada Kinjiro 原田謹次郎, comp. *Chūgoku meiga hōkan* 中國名畫寶鑑. Tokyo: Otsuka kogeisha, 1959.

Hashimoto shūzō Min Shin ga mokuroku 橋本所藏明淸畫目錄 Tokyo: Kadokawa shoten, 1972.

Hirayama Kangetsu 平山觀月. *Shin Chūgoku shodo-shi* 新中國書道史. Tokyo: Yuhodo, 1962.

Hiroshi Higuchi 樋口弘. *Chūgoku hanga shūsei* 中國版畫集成. Tokyo: Mito sho-oku, 1967.

Ho Liang-chün 何良俊. *Ssu-yu-chai hua-lun* 四友齋畫論. Mei-shu ts'ung-shu, vol. 12, no. III/3. Taipei: I-wen yin-shu-kuan, n. d.

Ho, Ping-ti. "The Salt Merchants of Yang-chou: A Study of Commercial Capitalism in Eighteenth-century China." *Harvard Journal of Asiatic Studies* 17 (1954): 130-68.

Ho T'ien-chien 賀天健. *Hsüeh-hua shan-shui kuo-ch'eng tzu-shu* 學畫山水過程自述. Peking: Jen-min mei-shu ch'u-pan-she, 1962.

Ho Tung-hsü 何東序 and Wang Shang-ning 汪尙寧, comps. *Hui-chou fu-chih* 徽州府志. Chung-kuo shih-hsüeh ts'ung-shu, no. 15. Reprinted., Taipei: Hsüeh-sheng shu-chü, 1965.

Hsiao Ch'ih-mu shan-shui Wang Yü-yang tui-t'i shuang-chüeh ts'e 蕭尺木山水王漁洋對題雙絕冊. Shanghai: Shen-chou kuo-kuang-she, 1916.

Hsiao Yün-ts'ung 蕭雲從. *T'ai-ping shan-shui shih-hua* 太平山水詩畫. Preface dated 1648. Reprint ed., Osaka: Tombo-sha, 1931.

Hsieh Chao-chih 謝肇淛. *Wu-tsa-tsu* 五雜俎. Shanghai: Chung-yang shu-tien, 1935.

Hsieh Chih-liu 謝稚柳, *Chu Ta*朱耷. Shanghai: Jen-min mei-shu ch'u-pan-she, 1958.

————. *T'ang Wu-tai Sung Yüan ming-chi* 唐五代宋元名跡. Shanghai: Ku-tien Wen-hsüeh ch'u-pan-she, 1957.

Hsieh Kuo-chen 謝國楨. *Nan-Ming shih-lüeh* 南明史略 Shanghai: Jen-min ch'u-pan-she, 1957.

Hsieh Lan-sheng 謝蘭生. *Ch'ang-hsing-hsing-chai shu-hua t'i-pa* 常惺惺齋書畫題跋. Macao: Wen-chi t'u-shu kung-ssu, n.d.

Hsin-an-p'ai ming-hua-chi 新安派名畫集. Shanghai: Shen-chou kuo-kuang-she, 1924.

Hsin-an ming-hua shan-ts'e 新安名畫扇冊. Shanghai: Shen-chou kuo-kuang-she, 1931.

Hsin-an ming-hua chi-chin ts'e 新安名畫集錦冊. Shanghai: Shen-chou kuo-kuang-she, 1920.

Hsü Ch'eng-yao 許承堯, comp. *She-hsien-chih* 歙縣志. Shanghai: Lü-hu t'ung-hsiang-hui, 1937; reprinted., Taipei: Ch'eng-wen ch'u-pan-she, 1975.

Hsü Hsia-k'o 徐霞客. *Hsü Hsia-k'o yu-chi* 徐霞客遊記. Edited by Ting Wen-chiang 丁文江, 20 vols. Reprinted., Taipei: Ting-wen shu-chü, 1972.

Hsü Pang-ta 徐邦達. *Li-tai liu-ch'uan shu-hua tso-p'in pien-nien-piao* 歷代流傳書畫作品編年表 Shanghai: Jen-min mei-shu ch'u-pan-she, 1963.

Hsü Piao-jan徐表然, comp. *Wu-i-chih* 武夷志. Chin-chiang, Fukien:, n. p., 1619.

Hsü Yün-ch'iu 徐澐秋. "Hsiao Ch'ih-mu yün-tai shu-shu t'u-chüan" 蕭尺木雲台疏樹圖卷. In *I-lin ts'ung-lu* 藝林叢錄, 8. 191-93. Hong Kong: Commercial Press, 1962.

Hu Chi-t'ang 胡積堂. *Pi-hsiao-hsien shu-hua-lu* 筆嘯軒書畫錄 N.p., n.d.

Hu I 胡藝. "Hua-chia Hsiao Yün-ts'ung" 畫家蕭雲從. *An-hui shih-hsüeh* 安徽史學, no. 3 (June 1960), pp. 35-39.

————. "Hsiao Yün-ts'ung nien-p'u" 蕭雲從年譜. *Mei-shu yen-chiu* 美術研究, no. 1 (1960), pp. 48-55.

Huang Pin-hung 黃賓虹. *Ku-hua-wei* 古畫微. Shanghai: Commercial Press, 1925.

————. *Ku-hua-wei fu Huang-shan hua-yüan-lüeh* 古畫微附黃山畫苑略 Hong Kong: Commercial Press, 1961.

Huang-shan 黃山. Hong Kong: Feng-kuang ch'u-pan-she, n.d.

Huang-shan tao-yu shou-ts'e 黃山導遊手冊. Hong Kong: T'u-i ch'u-pan kung-ssu, n.d.

Huang-shan t'ien-hsia-ch'i 黃山天下奇. Hong Kong: San Yau Publishing Co., 1979.

Huang Yen-p'ei 黃炎培 and Lü I-shou 呂頤壽. *Huang-shan* 黃山. Shanghai: Commercial Press, 1917.

Hummel, Arthur W., ed. *Eminent Chinese of the Ch'ing Period (1644-1912)*. Reprinted., New York: Paragon Book Reprint Corp., 1964.

I-lin hsün-k'an 藝林旬刊, no. 25 (September 1, 1928).

The Individualists: Chinese Painting and Calligraphy of the 17th Century from the Collection of John M. Crawford, Jr. Providence, Rhode Island: Bell Gallery, List Art Center, Brown University, 1980.

I-yüan chen-shang-chi 藝苑眞賞集. Shanghai: I-yüan chen-shang-she, 1914-20.

I-yüan chi-chin 藝苑集錦. T'ien-tsin: T'ien-tsin mei-shu ch'u-pan-she, 1959.

Jao Tsung-i 饒宗頤. "Fang I-chih chih hua-lun" 方以智之畫論. *The Journal of the Institute of Chinese Studies at the Chinese University of Hong Kong* 7.1 (1974): 113-31.

————. "Wan-Ming hua-chia yü hua-lun" 晚明畫家與畫論. *The Journal of the Institute of Chinese Studies at the Chinese University of Hong Kong*. 9.1 (1978): 39-64.

Ko Chin-lang 葛金烺. *Ai-jih-yin-lu shu-hua-lu* 愛日吟廬書畫錄. Shanghai: n.p., 1910.

Ko-an shan-shui ho-chin chüan 格庵山水合錦卷. Shanghai: Shen-chou kuo-kuang-she, 1936.

Kokka 國華, no. 329 (October, 1917).

Ku-kung ming-hua san-pai-chung 故宮名畫三百種. Taichung: National Palace Museum, 1959.

Ku-kung shu-hua-chi 故宮書畫集, 48 vols. Peking: Palace Museum, 1939-36.

Ku-kung shu-hua-lu 故宮書畫錄, rev. ed. Taipei: National Palace Museum, 1965.

Ku Yen-wu 顧炎武. *T'ien-hsia chün-kuo li-ping-shu* 天下郡國利病書. Ssu-pu ts'ung-k'an, no. 238. Shanghai: Commercial Press, 1935-36.

K'uan-yü 寬予. "Yü-ch'ing-chai fa-t'ieh k'ao-lüeh" 餘清齋法帖考略. In *I-lin ts'ung-lu* 藝林叢錄, 6.204-208. Hong Kong: Commercial Press, 1966.

Kung Pan-ch'ien shan-shui ts'e 龔半千山水冊. Shanghai: Wen-ming shu-chü, 1909.

Kuo-hua yüeh-k'an 國畫月刊, 1.8 (June 1935).

Kuo Wei 郭威, comp. *Sung-Yüan i-lai ming-hua-chi* 宋元以來名畫集. Taipei: Ya-yün-t'ang, n.d.

————, Comp. *Pa-ta, Shih-t'ao, Chien-chiang, Shih-ch'i ssu-seng hua-chi* 八大石濤漸江石谿四僧畫集. Taipei: Ya-yün-t'ang, 1972.

Kuo Wei-ch'ü 郭味渠. *Chung-kuo pan-hua shih-lüeh* 中國版畫史略. Peking: Chao-hua mei-shu ch'u-pan-she, 1962.

————. *Sung Yüan Ming Ch'ing shu-hua-chia nien-piao* 宋元明清書畫家年表. Peking: Jen-min mei-shu ch'u-pan-she, 1962.

Laing, Ellen Johnston. "The State of Ming Painting Studies." *Ming Studies*, no. 3 (Spring 1977), pp. 9-25.

Landscape Paintings by Kwangtung Masters During the Ming & Ch'ing Periods.

Hong Kong: Art Gallery, Institute of Chinese Studies, Chinese University of Hong Kong, 1973.

Lao T'ien-pi 勞天庇, ed. *Chih-lo-lou shu-hua-lu: Ming i-min chih-pu* 至樂樓書畫錄: 明遺民之部 Hong Kong: Ho-shih Chih-lo-lou, 1973.

Ledderose, Lothar. *Mi Fu and the Classical Tradition of Chinese Calligraphy*. Princeton: Princeton University Press, 1979.

Lee, Sherman E. *The Colors of Ink*. New York: The Asia Society, 1974.

————. *A History of Far Eastern Art*, rev. ed. New York: Harry N. Abrams, 1973.

Li, Chi. *The Travel Diaries of Hsü Hsia-k'o*. Hong Kong: The Chinese University of Hong Kong, 1974.

Li Chia 李佳. *Tso-an i-te-lu* 左庵一得錄. N.p., 1908.

Li, Chu-tsing. *A Thousand Peaks and Myriad Ravines: Chinese Paintings in the Charles A. Drenowatz Collection,* 2 vols. Ascona: Artibus Asiae, 1974.

————. *The Autumn Colors on the Ch'iao and Hua Mountains: A Landscape by Chao Meng-fu*. Ascona: Artibus Asiae, 1965.

Lien Ch'üan 廉泉, ed. *Shan-mien ta-kuan* 扇面大觀. Kobe: Shan-mien-kuan, 1915.

Liu Hai-su 劉海粟, ed. *Tsin-T'ang Sung-Yüan Ming-Ch'ing ming-hua pao-chien* 晉唐宋元明清名畫寶鑑. Shanghai: Shen-pao-she, n.d.

————, ed. *Tsin-T'ang Sun-Yüan Ming-Ch'ing ming-hua ta-kuan* 晉唐宋元明清名畫大觀, 4 vols. Shanghai: Chung-hua shu-chü, 1936.

Liu-yin-t'ang ts'ang-hua 六瑩堂藏畫. N.p.: Liu-yin-t'ang, 1940.

Loehr, Max. "Chinese Landscape Painting and Its Real Content." *Rice University Studies* 59.4 (Fall 1973): 67–96.

Lu Hsin-yüan 陸心源. *Jan-li-kuan kuo-yen-lu* 穰梨館過眼錄. N.p. 1892.

Ma Pin 馬彬. "Ch'ien-tan Hsin-an hua-p'ai" 淺談新安畫派. *I-yüan to-ying* 藝苑掇英, no. 20 (April, 1983), pp. 35–37.

Metropolitan Annual Report: 1971. New York: Metropolitan Museum of Art, 1971.

Min-shin sansui meiga-hen 明清山水名畫選. N.p., n.d.

Ming-Ch'ing chih-chi ming-hua t'e-chan 明清之際名畫特展. Taipei: National Palace Museum, 1970.

Ming-Ch'ing shan-mien-hua hsüan-chi 明清扇面畫選輯. Shanghai: Jen-min mei-shu ch'u-pan-she, 1959.

Ming-hua chi-ts'e ching-p'ing 名畫集冊精品 Shanghai: Wen-ta shu-she, 1920.

Ming-hua fa-shu chi-miao 名畫法書集妙 Shanghai: Shen-chou kuo-kuang-she, 1935.

Ming-k'an ming-shan-t'u pan-hua-chi 明刊名山圖版畫集. Shanghi: Jen-min mei-shu ch'u-pan-she, 1958.

Ming-mo ssu-seng hsüan-chi 明末四僧選輯. Hong Kong: Pan-tao Chung-kuo shu-hua ch'u-pan-she, n.d.

Miyazaki Ichisada 宮崎市定. "Chō Fu to sono jidai" 張溥とその時代. *Tōyōshi kenkyū* 東洋史研究, 33.3 (December 1974): 1-42.

Mu Hsiao-t'ien 穆孝天. *Anhui wen-fang ssu-pao-shih* 安徽文房四寶史. Shanghai: Jen-min mei-shu ch'u-pan-she, 1962.

Naitō Torajirō 內藤虎次郎, comp. *Shincho shogafu* 清朝書畫譜. Osaka: Hakubundō, 1922.

Nanking po-wu-kuan ts'ang-hua-chi 南京博物館藏畫集 2 vols. Peking: Wen-wu ch'u-pan-she, 1966.

Ni Tsan 倪瓚. *Ch'ing-pi-ko ch'üan-chi* 淸閟閣全集. Reprint ed., Taipei: Hsüeh-sheng shu-chü, 1969.

1000 Jahre chinesische Malerei. Munich: Haus der Kunst, 1959.

Pai-mei-chi 百梅集. 2 vols. Shanghai: Commercial Press, 1927.

Pao Yu-wen 鮑幼文. "Hui-chou te wen-fang ssu-pao" 徽州的文房四寶. *An-hui shih-hsüeh t'ung-hsün* 安徽史學通訊, nos. 12-13 (October 1959), pp. 160-66.

P'an T'ien-shou 潘天壽. *Chung-kuo hui-hua-shih* 中國繪畫史. Shanghai: Commercial Press, 1926.

Pang, Mae Ann Quan. "Wang Yüan-ch'i (1642-1715) and Formal Construction in Chinese Landscape Painting." Ph.D. dissertation, University of California, Berkeley, 1976.

Peintures chinoises et calligraphies anciennes: XIIIe-XIXe siècles de la collection Chiang Er-shih. Paris: Musée Cernuschi, 1971.

Peintures chinoises Ming et Ts'ing: XVe-XIXe siècles de la collections Mu-fei. Paris: Musée Cernuschi, 1967.

Peterson, Willard J. *Bitter Gourd: Fang I-chih and the Impetus for Intellectual Change*. New Haven: Yale University Press , 1979.

The Richard Bryant Hobart Collection of Chinese Ceramics & Paintings. New York: Park-Bernet Galleries, Inc., 1969.

Rosenzweig, Daphne Lange. "A Landscape Handscroll by Hsiao Yün-ts'ung." *Allen Memorial Art Museum Bulletin* 32.1 (1974-75): 34-55.

Seize peintures de maitres chinois XIIe-XVIIIe siècles de la collection Chiang Er-shih. Paris: Musée Cernuschi, 1959-60

Shan-mien 扇面. Shanghai: Yu-chen shu-chü, n.d.

Shang Yüeh 尙鉞. *Chung-kuo tzu-pen chu-i kuan-hsi fa-sheng chi yen-pien te ch'u-pu yen-chiu.* 中國資本主義關係發生及演變的初步研究 Peking: San-lien shu-tien, 1956.

Shen-chou kuo-kuang-chi 神州國光集. 18 vols. Shanghai: Shen-chou kuo-kuang-she, 1908-10.

Shen-chou ta-kuan 神州大觀. 16 vols. Shanghai: Shen-chou kuo-kuang-she, 1912-21.

Shen-chou ta-kuan hsü-pien 神州大觀續編. 11 vols. Shanghai: Shen-chou kuo-kuang-she, 1927-31.

Shih-pai-chai-chu-jen 十百齋主人 [Chin Ai 金瑷] . *Shih-pai-chai shu-hua-lu* 十百齋書畫錄. 11 vols. N.p., n.d.

Shimada Shūiirō 島田修次郎. "Ippin gafū ni tsuite" 逸品畫風について, *Bijutsu kenkyū* 美術研究, no. 116 (1950), pp. 20-46. Translated by James Cahill as "Concerning the I-p'in Style of Painting." *Oriental Art* 7.2 (Summer 1961): 66-74; 8.3 (Autumn 1962): 130-37; 10.1 (Spring 1964): 19-26.

Shina meiga-sen 支那名畫選. Tokyo: Geisōdō, n.d.

Shina nanga taisei 支那南畫大成. 22 vols. Tokyo: Kobunsha, 1935-37.

Silbergeld, Jerome Leslie. "Political Symbolism in the Landscape Painting and Poetry of Kung Hsien (c. 1620-1689)." Ph. D. dissertation, Stanford University, 1974.

Sirén, Osvald, *A History of Later Chinese Painting*, 2 vols. London: The Medici Society, 1938.

————. *Chinese Painting: Leading Masters and Principles*, 7 vols. New York: The Ronald Press Company, 1956-58.

————. *The Chinese on the Art of Painting*. New York: Schoken Books, 1963.

Sickman, Laurence, ed. *Chinese Calligraphy and Painting in the Collection of John M. Crawford, Jr.* New York: The Pierpont Morgan Library, 1962.

Speiser, Werner. *Ausstellung chinesische Malerci*. Köln: Museum für Ostasiatische Kunst, 1955.

Struve, Lynn A. "Ambivalence and Action: Some Frustrated Scholars of the K'ang-hsi Period." In *From Ming to Ch'ing: Conquest, Regions, and Continuity in Seventeenth-Century China,* pp. 321 – 65. Edited by Jonathan D. Spence and John E. Wills, Jr. New Haven: Yale University Press, 1979.

Style Transformed. Taipei: National Palace Museum, 1977.

Su-chou po-wu-kuan ts'ang-hua-chi 蘇州博物館藏畫集. Peking: Wen-wu ch'u-pan-she, 1963.

Su Tsung-jen 蘇宗仁, comp. *Huang-shan ts'ung-k'an* 黃山叢刊. Peking: Pai-i-yen-chai, 1937.

Sung Chao Kou chen-ts'ao ch'ien-tzu-wen 宋趙構眞草千字文. Shanghai: Shanghai Museum, n.d.

Ta-feng-t'ang ming-chi 大風堂名蹟. 4 vols. Kyoto: Benridō, 1955.

T'ai-shan ts'an-shih-lou ts'ang-hua 泰山殘石樓藏畫. Shanghai: Hsi-ling yin-she, 1926-29.

Tanaka Kanrō 田中乾郎. *Sōgen irai meiga shoshū* 宋元以來名畫小集. Tokyo: Hunkyōdō, 1947.

T'ang Sung Yüan Ming Ch'ing hua-hsüan 唐宋元明淸畫選. Kuangchou: I-shu hua-pao-she, 1963.

Teng Chih-ch'eng 鄧之誠, ed. *Ch'ing-shih chi-shih ch'u-pien* 淸詩紀事初編. Reprint ed., Taipei: Ting-wen shu-chü, 1971.

T'ien-tsin-shih i-shu po-wu-kuan ts'ang-hua-chi 天津市藝術博物館藏畫集. 2 vols. Peking: Wen-wu ch'u-pan-she, 1959-63.

T'ien-yin-t'ang ming-hua-hsüan 天隱堂名畫選. Tokyo: n.p., 1963.

Ting-ching-t'ang ts'ang Chung-kuo Ming-Ch'ing shu-hua t'u-lu 定靜堂藏中國明淸書畫圖錄. Tokyo: Fei-niao hua-lang ch'u-pan-pu, 1968.

Tomita, Kojiro, and Tseng, Hsien-ch'i, eds. *Portfolio of Chinese Paintings in the Museum: Yüan to Ch'ing Periods*. Boston : Museum of Fine Arts, 1961.

Tōsō Genmin meiga taikan 唐宋元明名畫大觀. 2 vols. Tokyo, 1929.

Tsin-T'ang Wu-tai Sung-Yüan Ming-Ch'ing ming-chia shu-hua-chi 晉唐五代宋元明淸名家書畫集. Shanghai: Commercial Press, 1937.

Vanderstappen, Harrie A."The Style of Some Seventeenth-Century Chinese Paintings." In *Artists and Traditions: Uses of the Past in Chinese Culture*, pp. 149-68. Edited by Christin F. Murck. Princeton: The Art Museum, Princeton University, 1976.

Vannotti, Franco. *Peinture chinoise de la dynastie Ts'ing (1644-1912): Collection Vannotti Muzzano (Lugano)*. Geneva: Collections Baur, 1974.

Wang, Chi-chen. "Notes on Chinese Ink." *Metropolitan Museum Studies* 3 (1930-31): 114-33.

Wang Chi-ch'ien 王季遷. "Ni Yün-lin chih-hua" 倪雲林之畫. *National Palace Museum Quarterly*, 1.3 (January 1967): 15-38.

———— and Contag, Victoria. *Seals of Chinese Painters and Collectors of the Ming and Ch'ing Periods*. Shanghai: Commercial Press, 1940.

Wang Chieh 王杰 et al. *Shih-ch'ü pao-chi hsü-pien* 石渠寶笈續編 Reprinted., Taipei: National Palace Museum, 1971.

Wang Hung-tu 汪洪度. *Huang-shan ling-yao-lu* 黃山領要錄. Chih-pu-tsu-chai ts'ung-shu, vol. 150. Reprinted., Shanghai: Commercial Press, 1921.

Wang Po-min 王伯敏. *Chung-kuo pan-hua-shih* 中國版畫史. Shanghai: Jen-min mei-shu ch'u-pan-she, 1961.

Wang Shih-chen 王世貞. *Ku-pu-ku lu* 觚不觚錄. Shanghai: Commercial Press, 1937.

Wang Shih-ch'ing 汪世清. "Kung Hsien te Ts'ao-hsiang-t'ang chi" 龔賢的《草香堂集》 *Wen-wu* 文物 no. 5 (1978), pp. 45-49.

———— and Wang Ts'ung 汪聰. *Chien-chiang tzu-liao-chi* 漸江資料集. Hofei, Anhui: Jen-min ch'u-pan-she, 1964; rev. ed., 1984.

Wang Shih-hung 汪士鋐, comp. *Huang-shan-chih* 黃山志 An-hui ts'ung-shu 安徽叢書. Shanghai: Anhui ts'ung-shu pien-yin-ch'u, 1935.

Wang Shun-min 汪舜民 and P'eng Tse 彭澤, comp. *Hui-chou fu-chih* 徽州府志. Chung-kuo shih-hsüeh ts'ung-shu, no. 15. Reprinted., Taipei: Hsüeh-sheng shu-chü, 1965.

Wang Ts'ai-pai 汪采白. *Hsin-an hua-p'ai* 新安畫派. Nanking: Chin-ling ta-hsüeh Chung-kuo wen-hua yen-chiu-so, 1948.

Wang Wei-an 王維龕 (Wang Chuang-wei 王壯爲). "Mei-ho ti-i hua-ts'e" 梅壑第一畫冊. *Chung-kuo shu-hua* 中國書畫, no. 9 (December 1969), pp. 9-19.

Wen Chao-t'ung 溫肇桐. *Yüan-chi ssu-ta hua-chia* 元季四大畫家. Hong Kong: Hsing-fu ch'u-pan-she, 1960.

Weng, Wan-go. *Chinese Painting and Calligraphy: A Pictorial Survey.* New York: Dover Publications, Inc., 1978.

Whitfield, Roderick. *In Pursuit of Antiquity: Chinese Painting of the Ming and Ch'ing Dynasties from the Collection of Mr. and Mrs. Earl Morse.* Princeton: The Art Museum, Princeton University, 1967.

Wu Ch'i-chen 吳其貞. *Shu-hua-chi* 書畫記. Shanghai: Jen-min mei-shu ch'u-pan-she, 1963.

Wu, K. T. "Color Printing in the Ming Dynasty." *T'ien-hsia Monthly* 11 (1940): 30-44.

————. "Ming Print and Printer." *Harvard Journal of Asiatic Studies* 7 (1943): 203-60.

Wu, William, "Kung Hsien. " Ph. D. dissertation, Princeton University, 1979.

Wu Yin-ming 吳因明. "Wan-Ming chiang-nan fo-hsüeh feng-ch'i yü wen-jen-hua." 晚明江南佛風氣與文人畫, *Hsin-ya shu-yüan hsüeh-shu nien-k'an* 新亞書院學術年刊, no. 2 (September 1960), pp. 1-22.

Wu-jan 旡染. "Hsiao Yün-ts'ung T'ai-p'ing shan-shui t'u." 蕭雲從太平山水圖. In *I-lin ts'ung-lu* 藝林叢錄, 10. 218-23. Hong Kong: Commercial Press, 1962.

Yao Weng-wang 姚翁望. *An-hui hua-chia hui-pien* 安徽畫家匯編. Hofei, Anhui: Anhui Provincial Museum, 1979.

Young, M, W., ed. *The Eccentric Painters of China.* Ithaca, New York: Cornell University, 1965.

Yü Chien-hua 俞劍華. *Chung-kuo hui-hua-shih* 中國繪畫史. Shanghai: Commercial Press, 1937.

————, comp. *Chung-kuo hua-lun lei-pien* 中國畫論類編. Peking: Jen-min mei-shu ch'u-pan-she, 1957.

Yü-hsiang 予向 (Huang Pin-hung 黃賓虹). "Chien-chiang ta-shih shih-chi i-wen" 漸江大師事蹟佚聞. *Chung-ho yüeh-k'an* 中和月刊 1 (1940): 26-33, 46-65.

Yü Ta-fu 郁達夫. *Yü Ta-fu yu-chi* 郁達夫遊記. Shanghai: Shanghai tsa-chih kung-ssu, 1948.

Yuhas, Louise R. "The Landscape Art of Lu Chih (1496-1576)." Ph.D. dissertation, The University of Michigan, 1979.

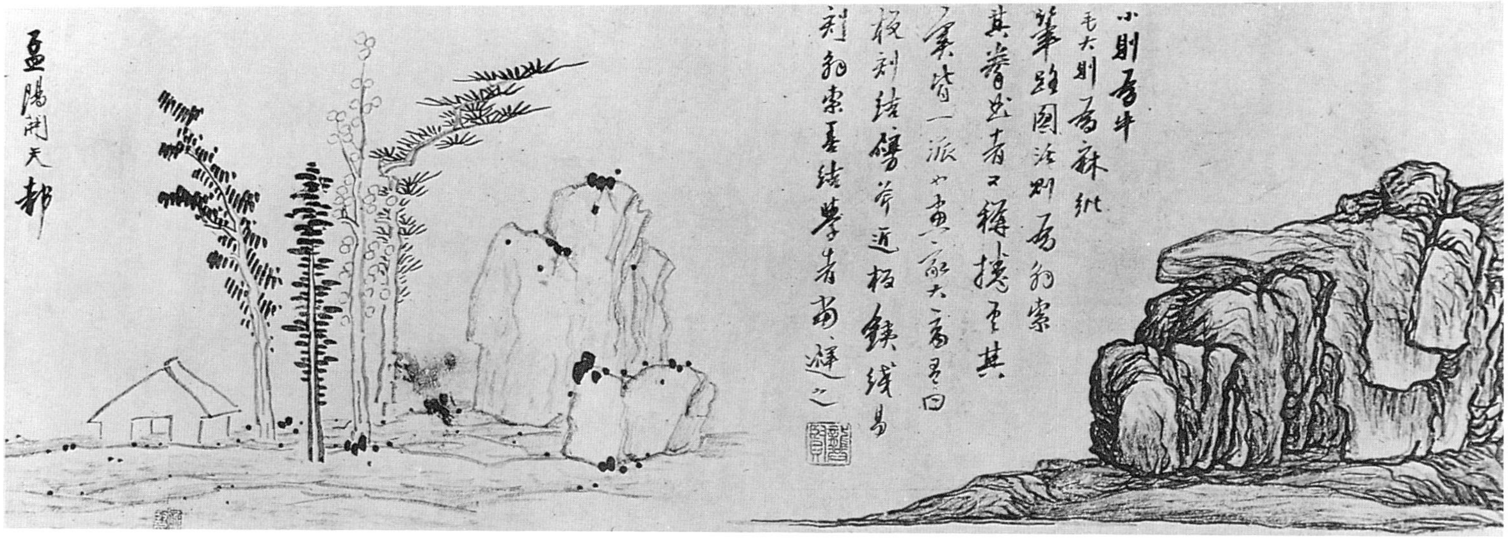

1. Kung Hsien: "Landscape." Handscroll, ink on paper, height 26.9 cm. Fogg Art Museum, Harvard University. Section 5.

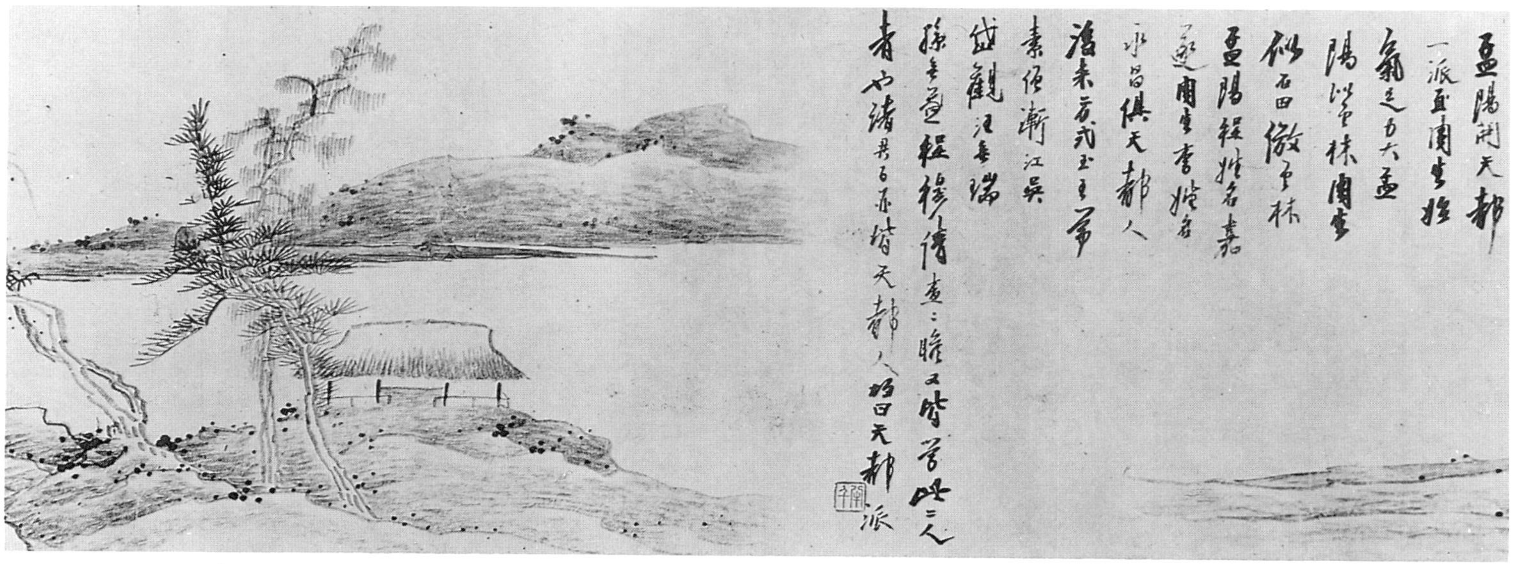

2. Kung Hsien: "Landscape." Section 6.

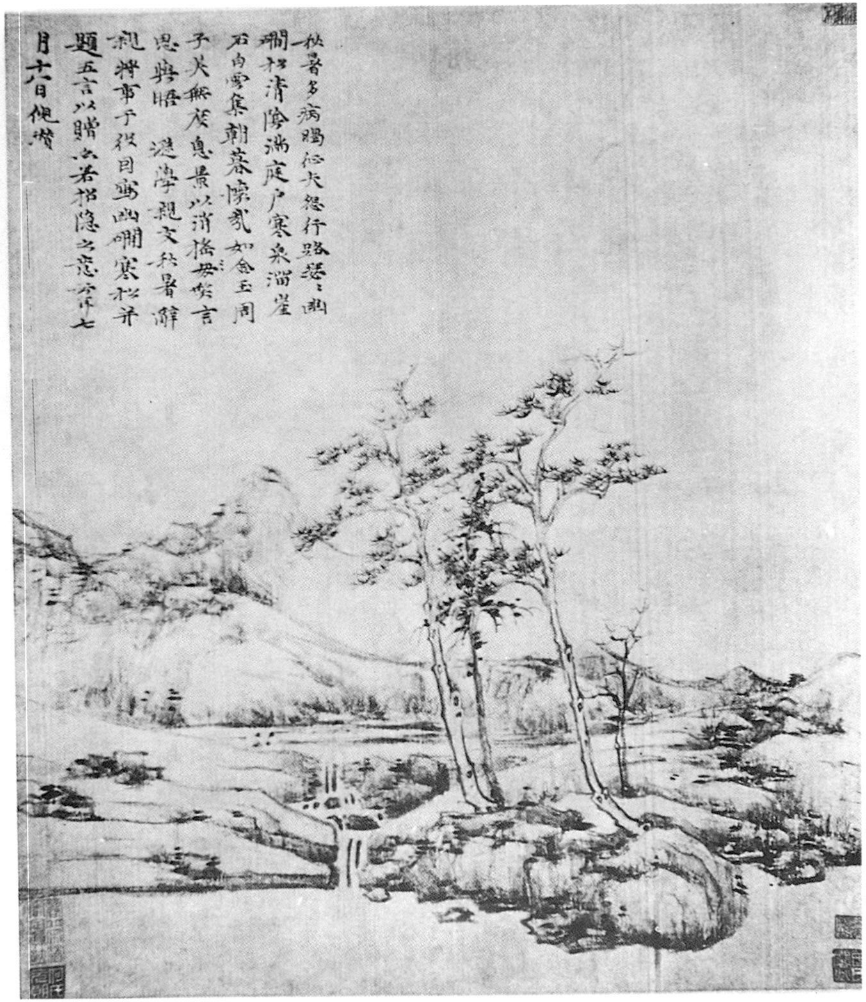

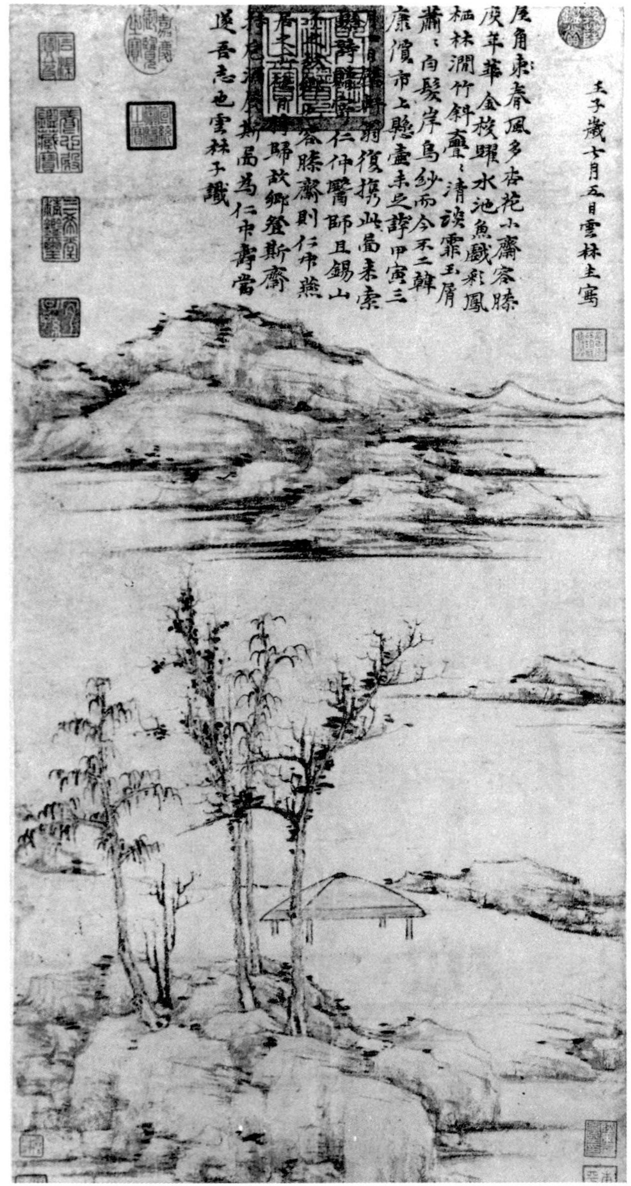

3. Ni Tsan: "Wintry Pines in a Quiet Valley." Hanging scroll , ink on paper. Palace Museum, Peking.

4. Ni Tsan: "The Jung-hsi Studio," 1372. Hanging scroll, ink on paper, 74.7×35.5 cm. National Palace Museum, Taipei.

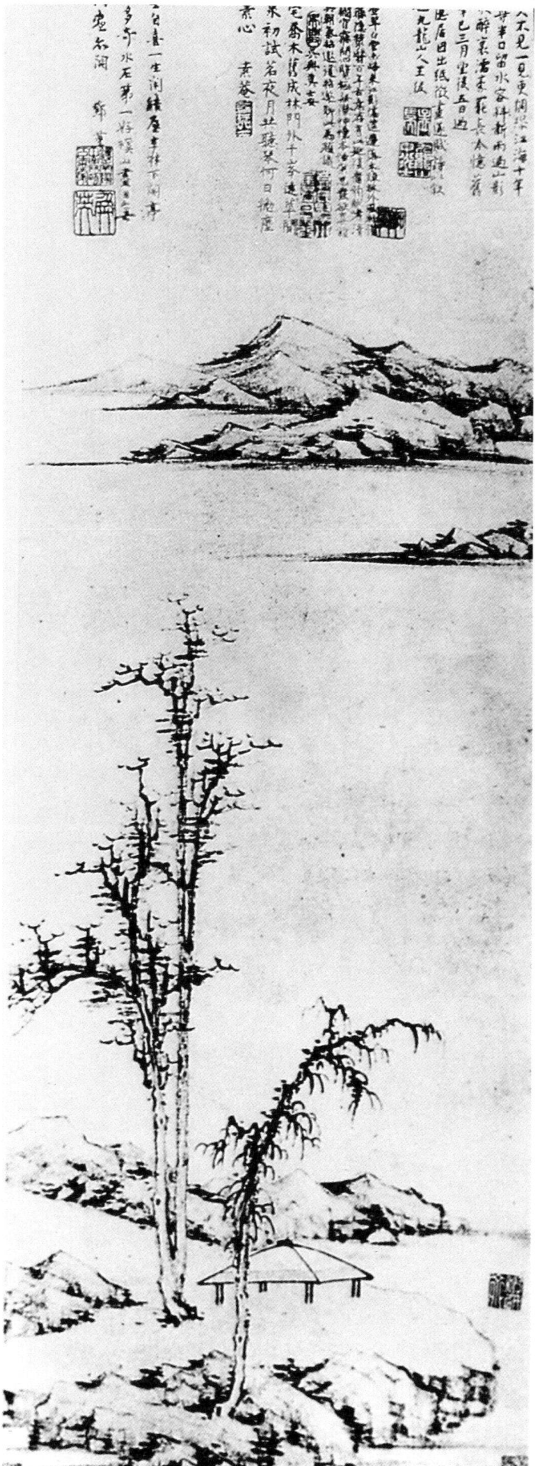

5. Wang Fu: "River Landscape in the Manner of Ni Tsan," 1401. Hanging scroll, ink on paper, 95×32 cm. Private collection, Tokyo.

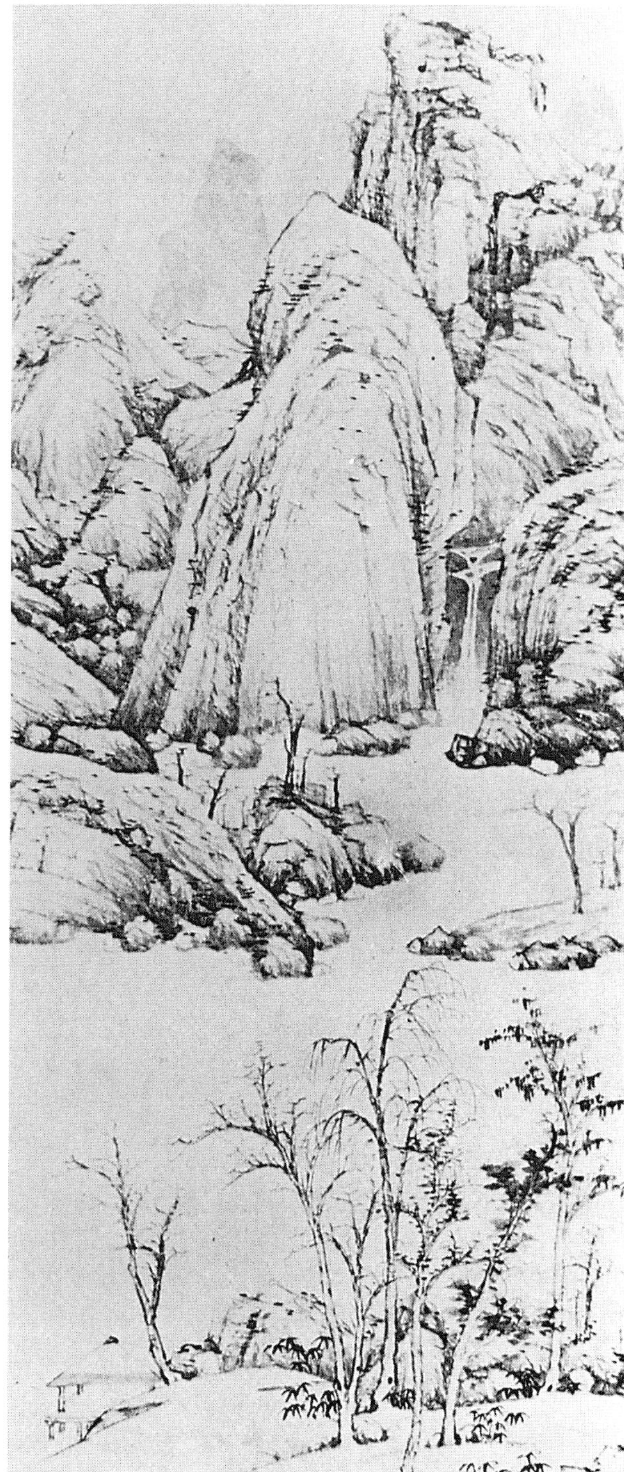

6. Liu Chüeh: "Landscape." Hanging scroll, ink on paper, 148.9×54.9 cm. Musée Guimet, Paris.

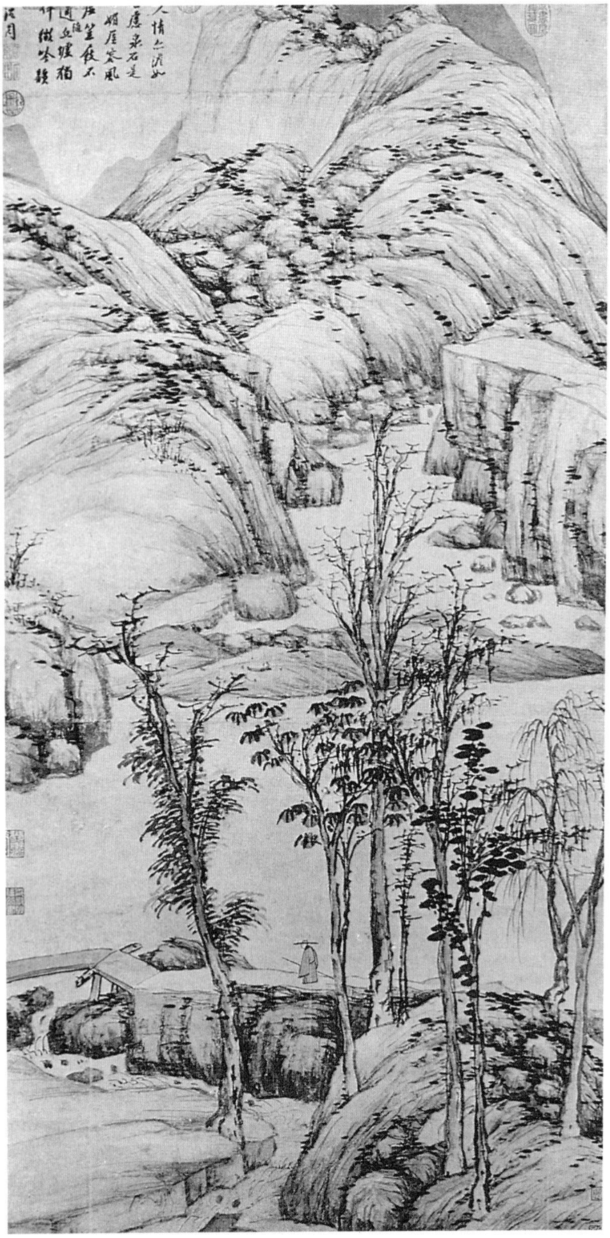

7. Shen Chou: "Walking with a Staff." Hanging scroll, ink on paper, 159.1×72.2 cm. National Palace Museum, Taipei.

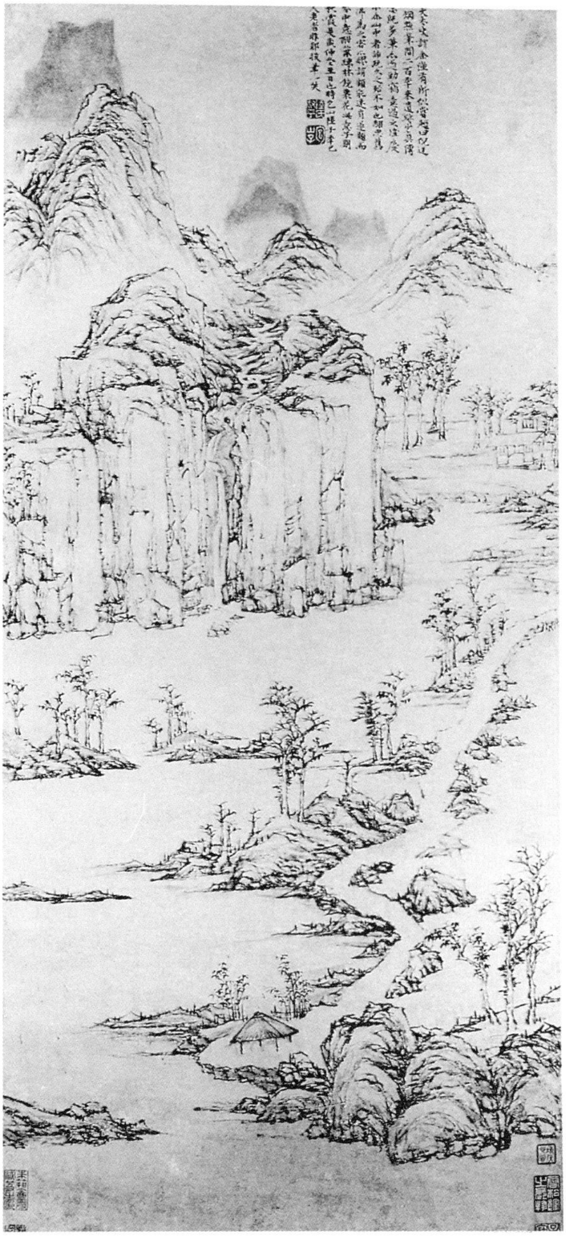

8. Lu Chih: "Taoist Retreat in Mountain and Stream: An Autumn Landscape after Ni Tsan," 1567. Hanging scroll, ink on paper, $42\frac{15}{16}\times 18$ in. Cleveland Museum of Art.

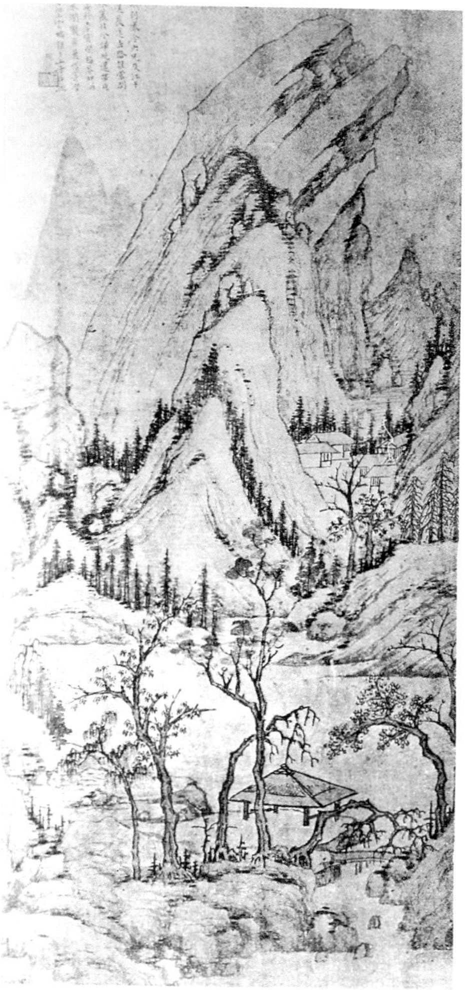

9. Ting Yün-p'eng: "Parting at End of the Year," 1585. Hanging scroll, ink and light colors on paper. Whereabouts unknown.

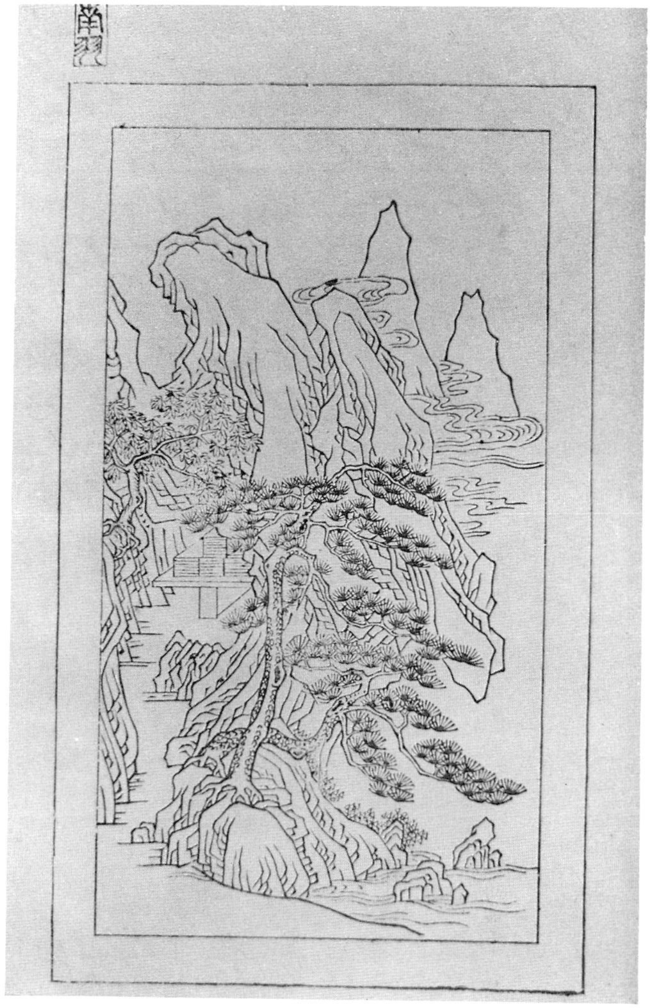

10. Ting Yün-p'eng: Design for ink-cake. From *Fang-shih mo-p'u,* 7:14a.

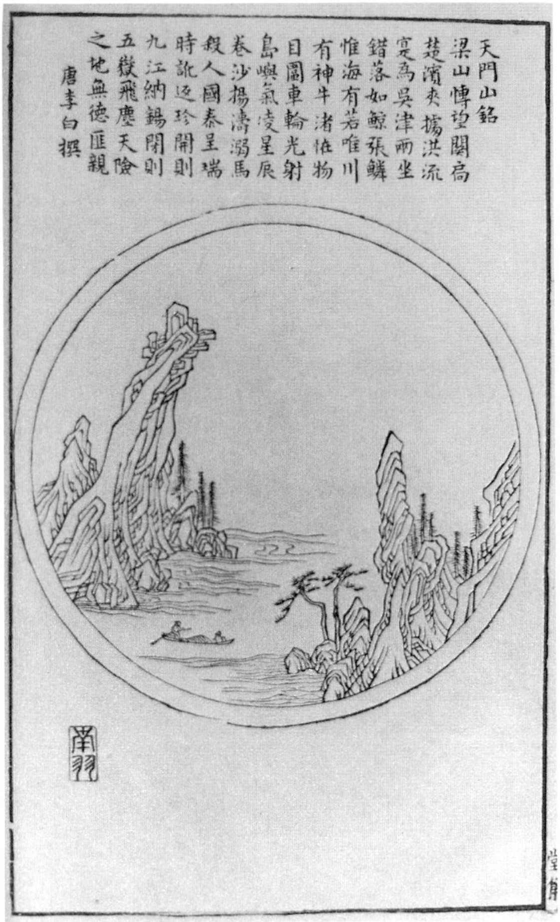

11. Ting Yün-p'eng: Design for ink-cake. From *Fang-shih mo-p'u,* 3:31b.

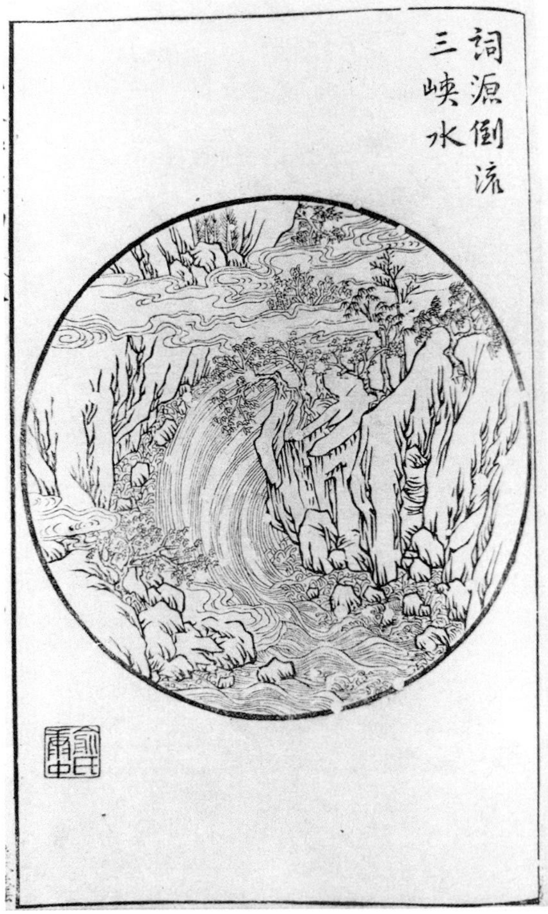

12. Yü Chia-shen: Design for ink-cake. From *Fang-shih mo-p'u,* 4:41a.

13. Hsiao Yün-ts'ung: "Small and Large Ching Mountains." Woodcut. From *T'ai-ping shan-shui shih-hua* album.

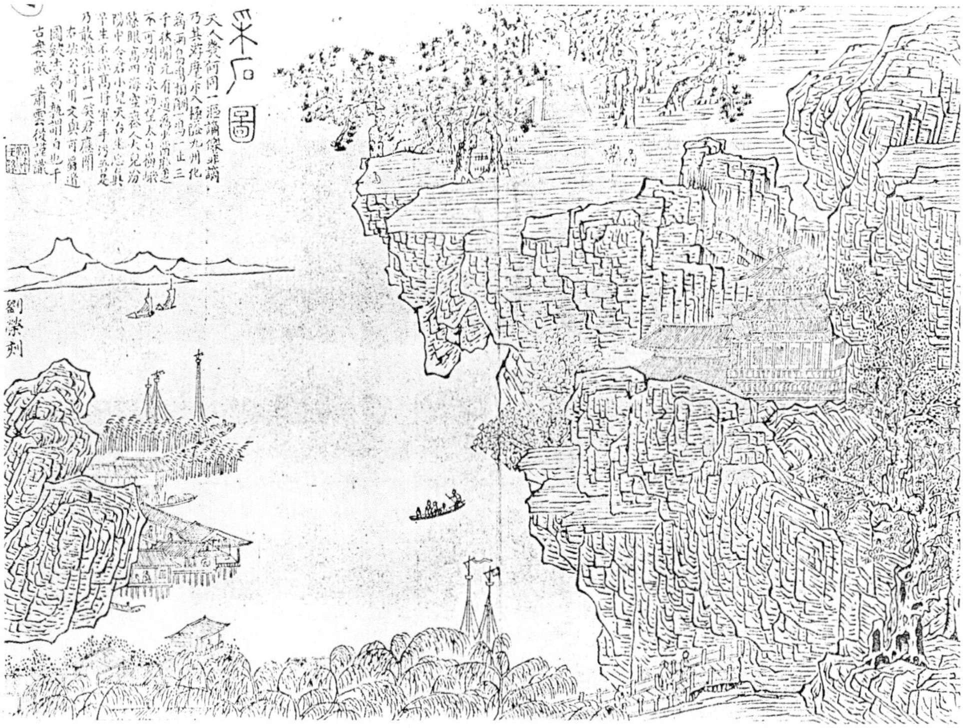

14. Hsiao Yün-ts'ung: "Ts'ai-shih Cliff." Woodcut. From *T'ai-ping shan-shui shih-hua* album.

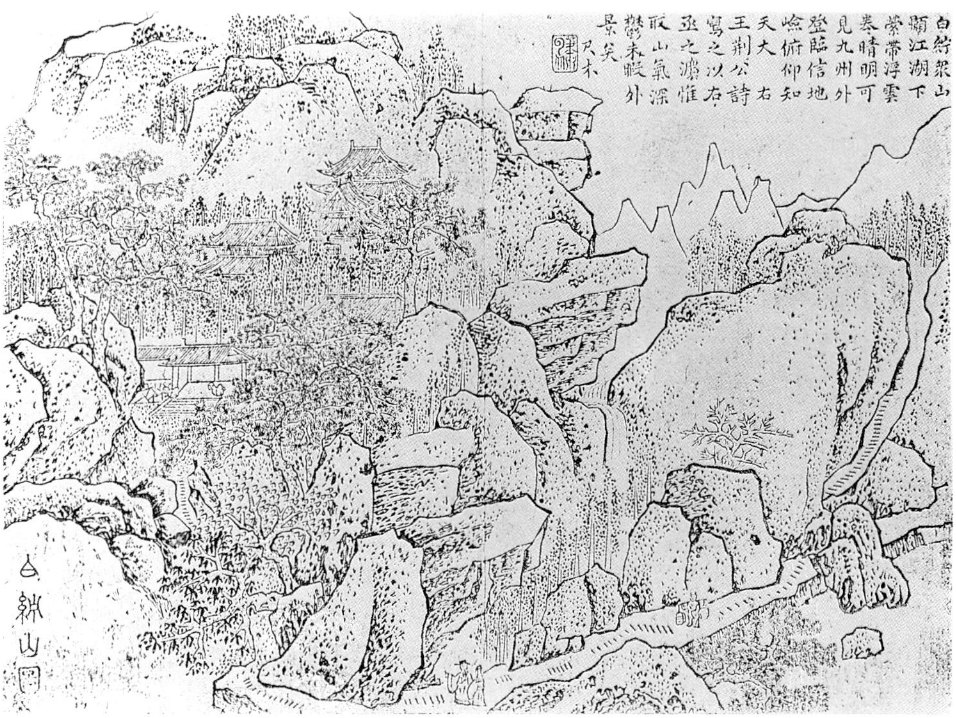

15. Hsiao Yün-ts'ung: "Mt. Pai-chu." Woodcut. From *T'ai-p'ing shan-shui shih-hua* album.

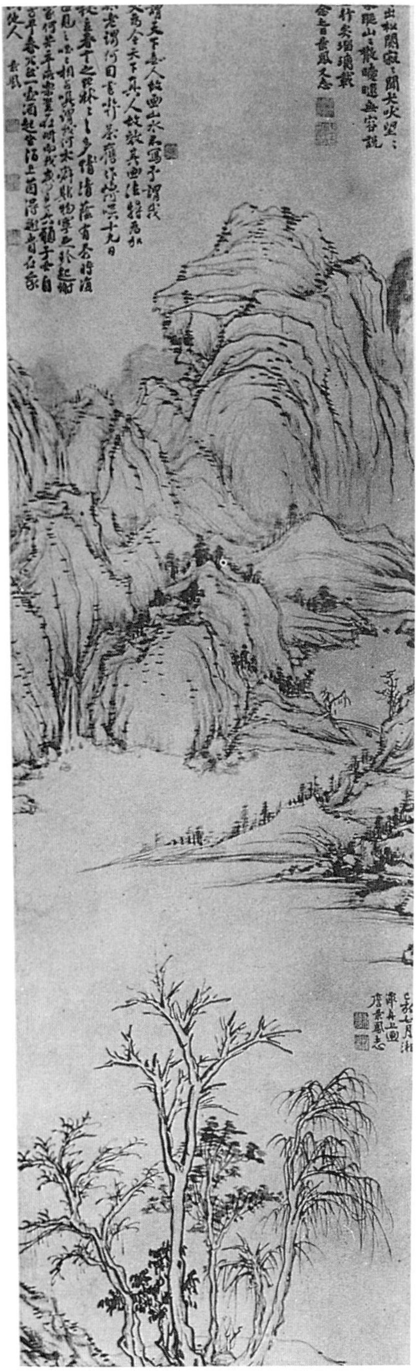

16. Chan Ching-feng: "Landscape," 1597. Hanging scroll, ink on paper. Whereabouts unknown.

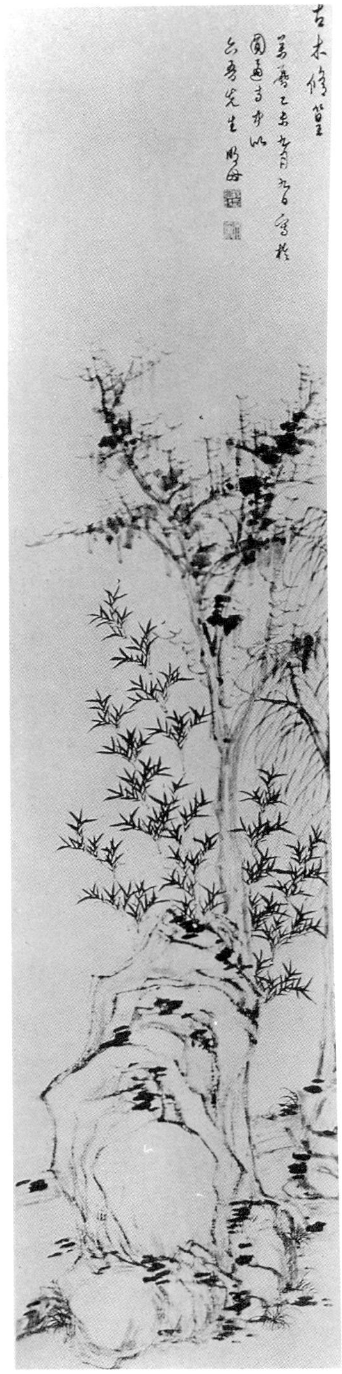

17. Yang Ming-shih: "Old Tree and Bamboo." Hanging scroll, ink on paper, 136×32 cm. Palace Museum. Peking.

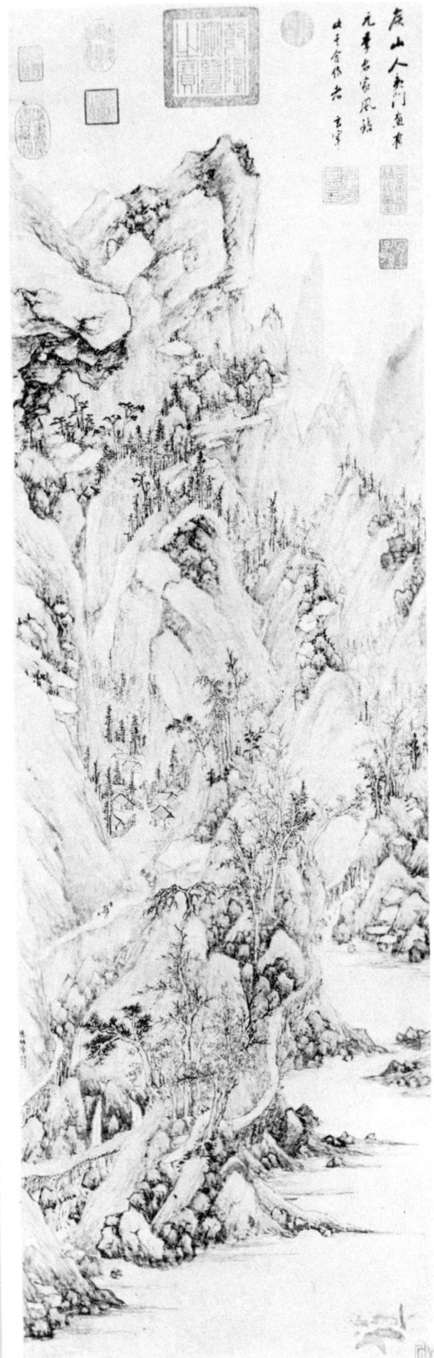

18. Hou Mou-kung: "Landscape." Hanging scroll, ink and light colors on paper, 112×34.9 cm. National Palace Museum, Taipei.

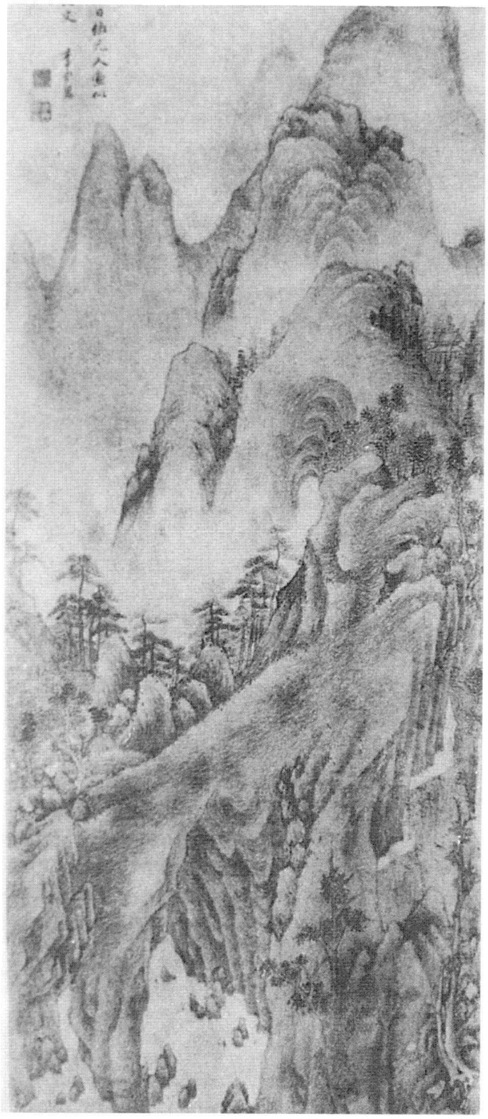

19. Li Yung-ch'ang: "In the Manner of Yüan Masters," 1634. Hanging scroll. Whereabouts unknown.

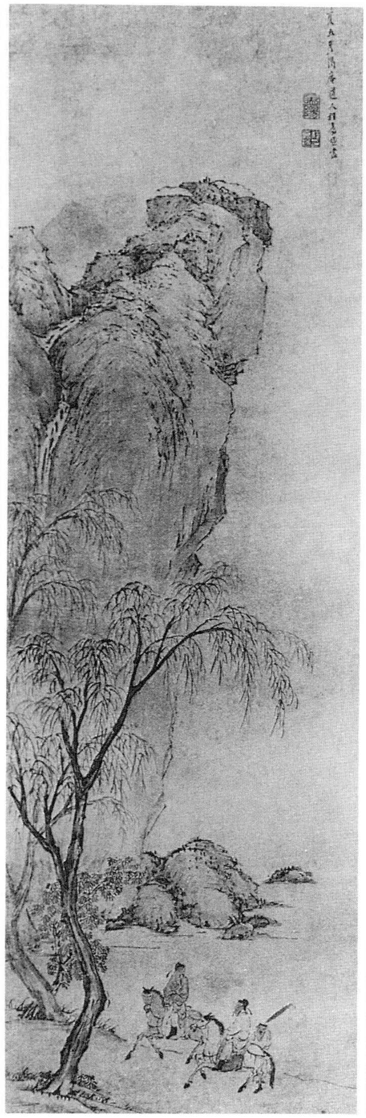

20. Ch'eng Chia-sui: "Horse-Riders on the Willow Bank," 1639. Hanging scroll ink and colors on paper, 108×32 cm. T'ien-tsin Museum.

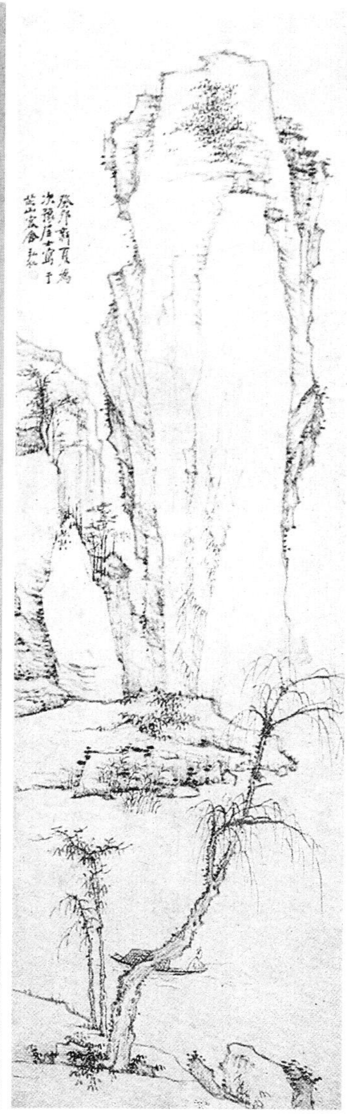

21. Hung-jen: "Landscape," 1663. Hanging scroll. Liaoning Provincial Museum.

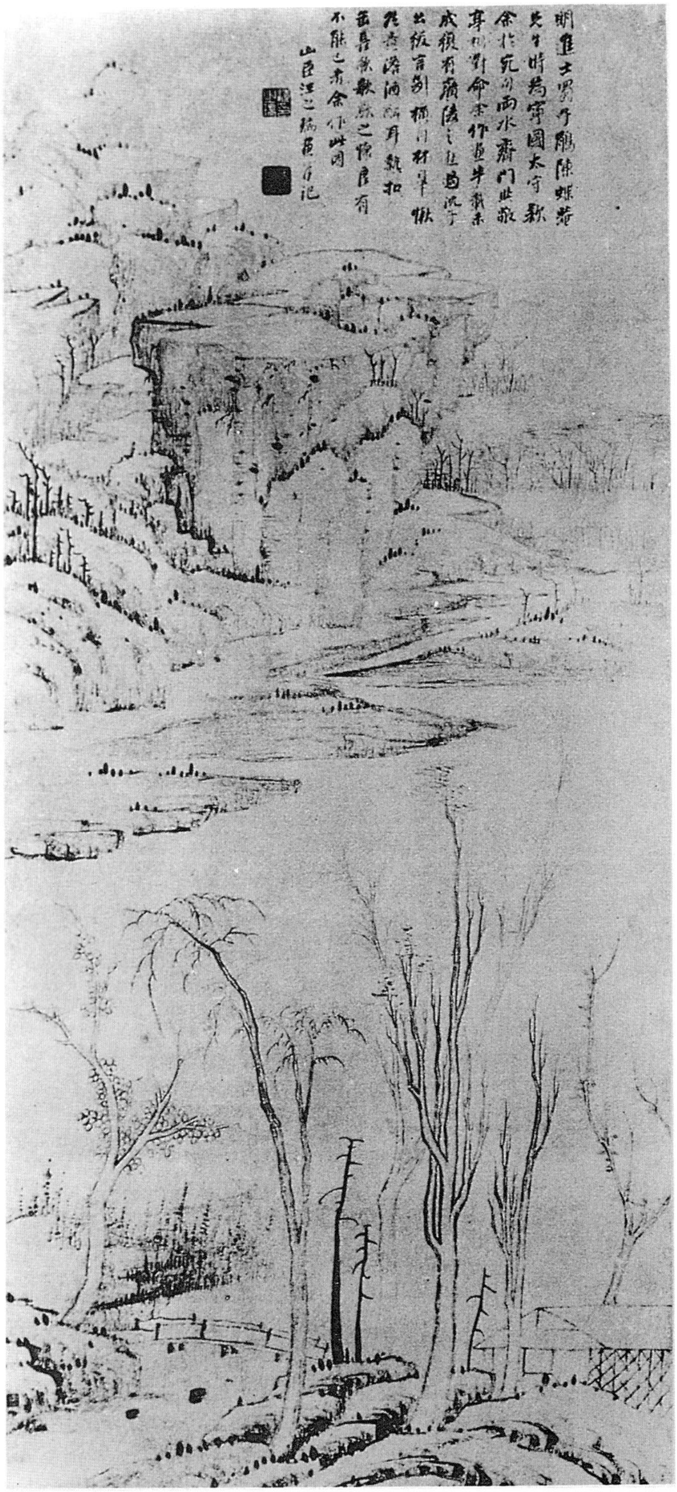

22. Wang Chih-jui: "Landscape." Hanging scroll. Whereabouts unknown.

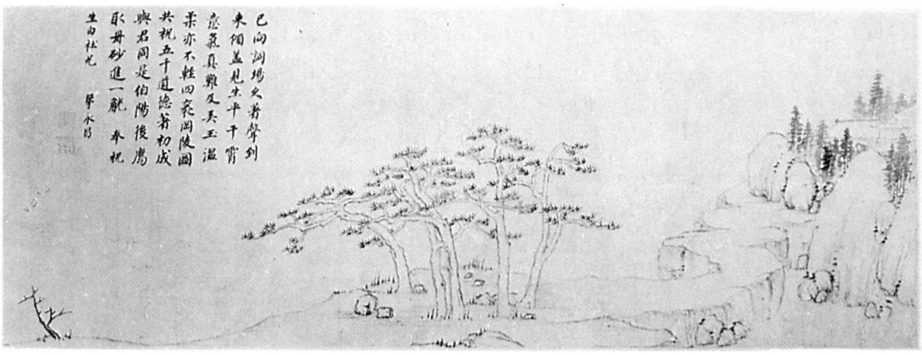

A

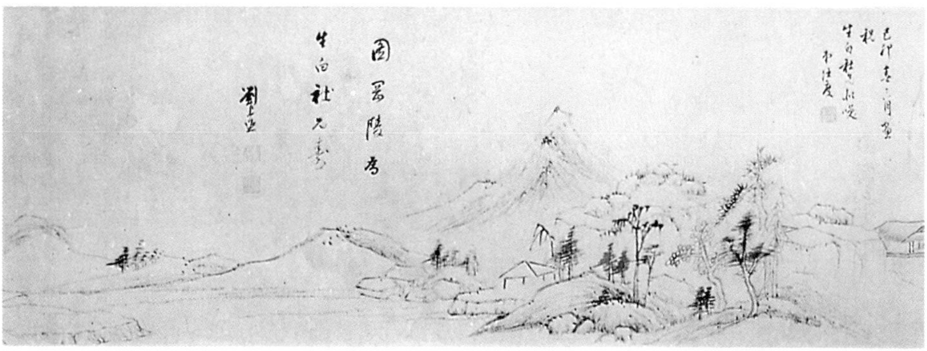

B

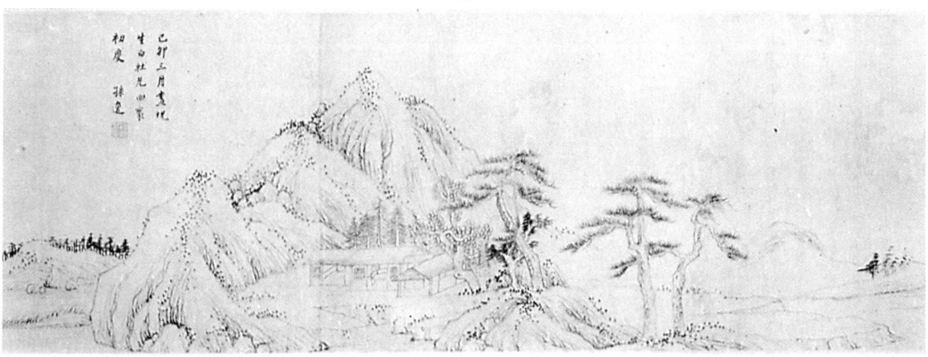

C

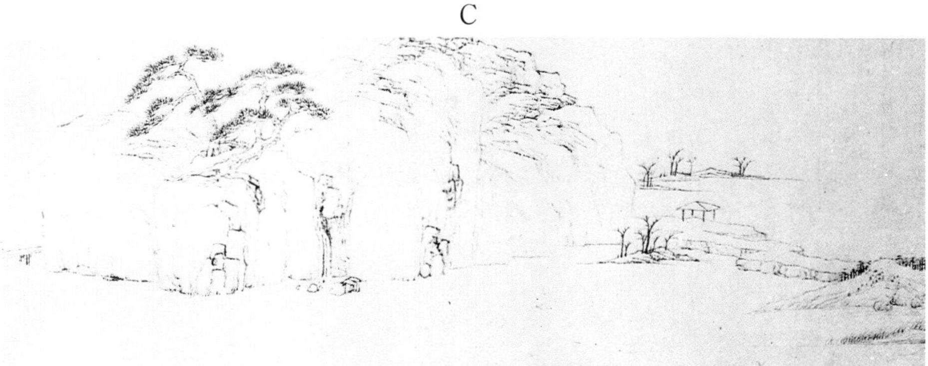

D

23. Liu Shang-yen and others: "Joint Handscroll of Landscape for Li Sheng-pai," 1639. Shanghai Museum.

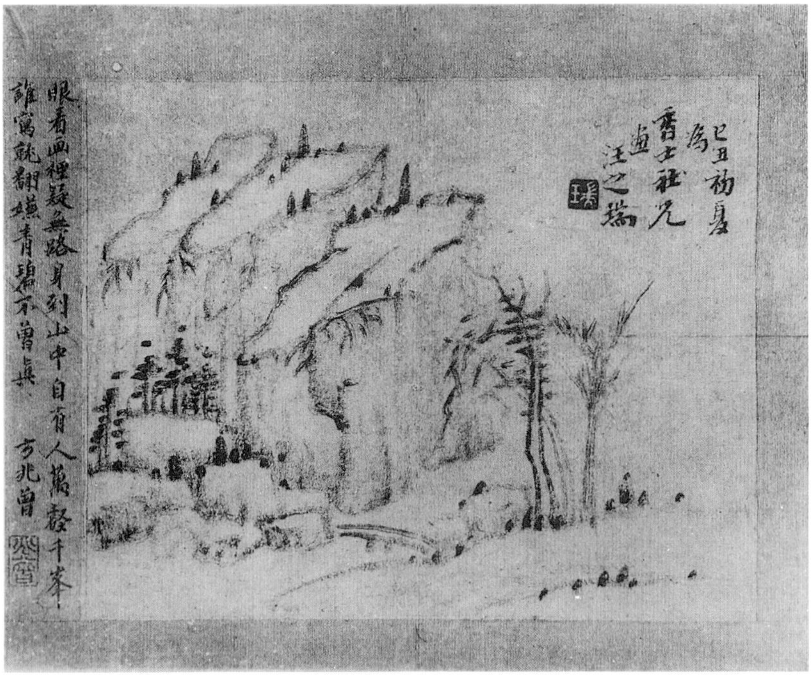

24. Wang Chih-jui: "Landscape," 1649. Album leaf. Whereabouts unknown.

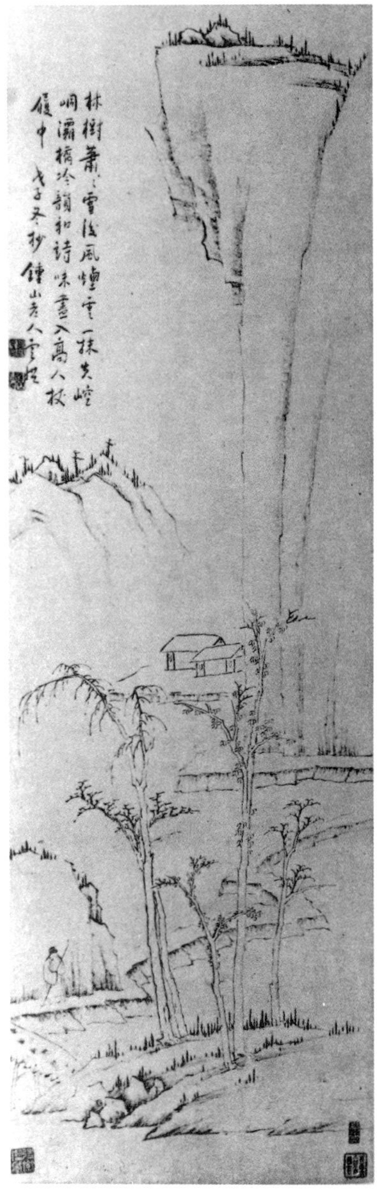

26. Hsiao Yün-ts'ung: 'Walking with a Staff in a Sparse Forest," 1648. Hanging scroll, ink on paper, 107.2×31.5 cm. T'ien-tsin Museum.

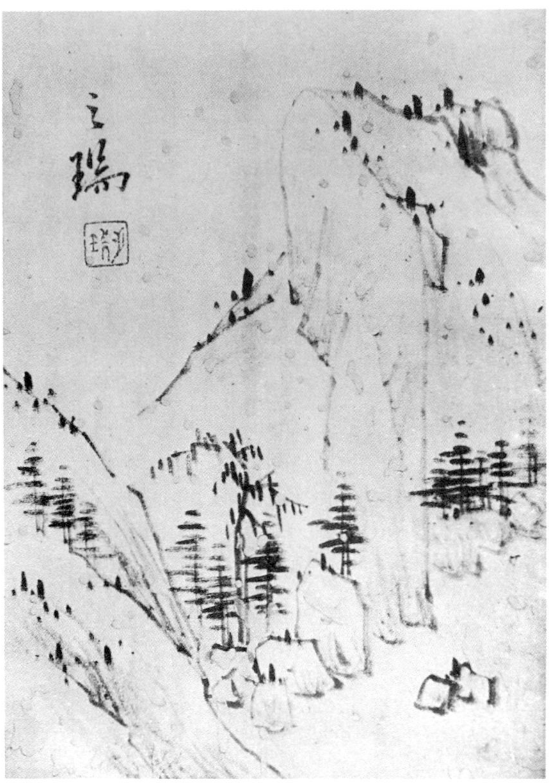

25. Wang Chih-jui: "Landscape." Album leaf. Whereabouts unknown.

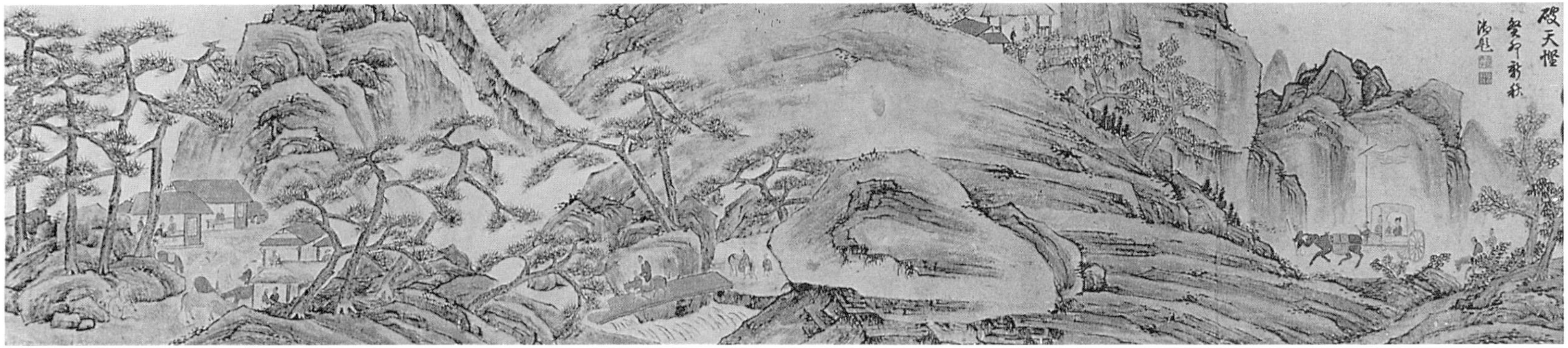

27. Hsiao Yün-ts'ung: "Travellers in Autumn Mountains," 1657 . Handscroll, ink and colors on paper, 25.4×250.4 cm. Tokyo National Museum. Section 2.

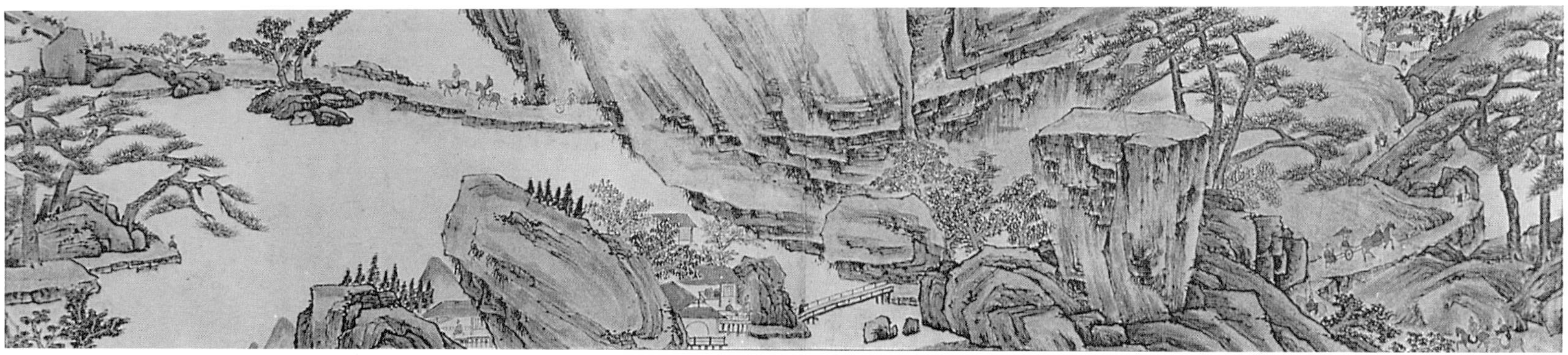

28. Hsiao Yün-ts'ung: "Travellers in Autumn Mountains," Section 3.

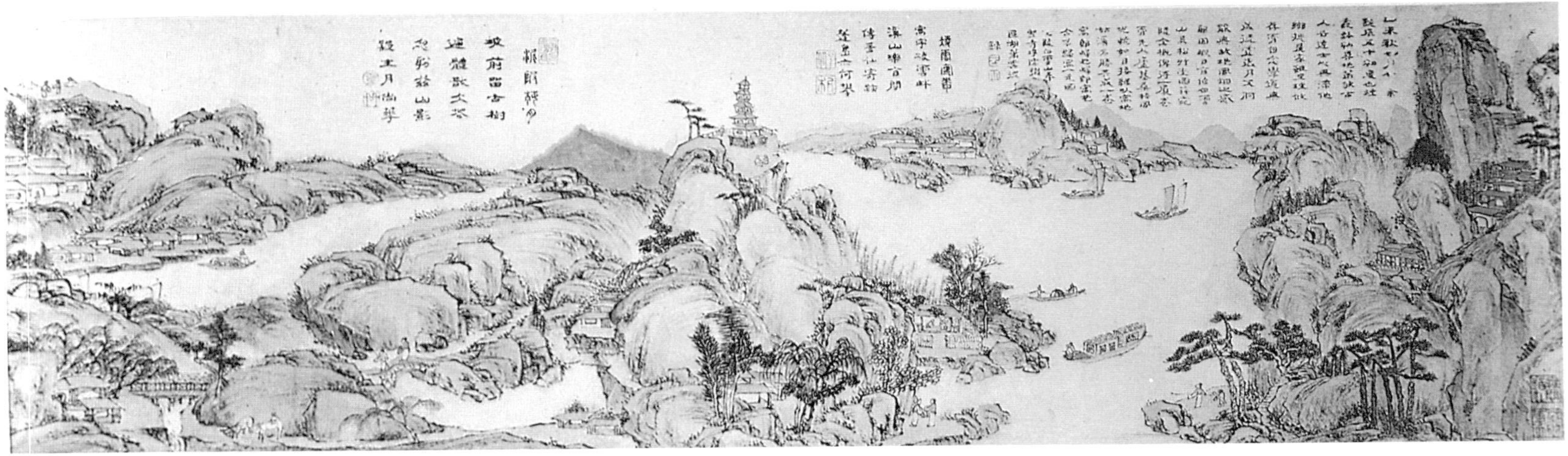

29. Hsiao Yün-ts'ung: "Going Home and Living Abroad Are the Same Thing," 1656. Handscroll, ink and colors on paper, 23.5×1,302 cm. Drenowatz collection, Zürich. Section 1.

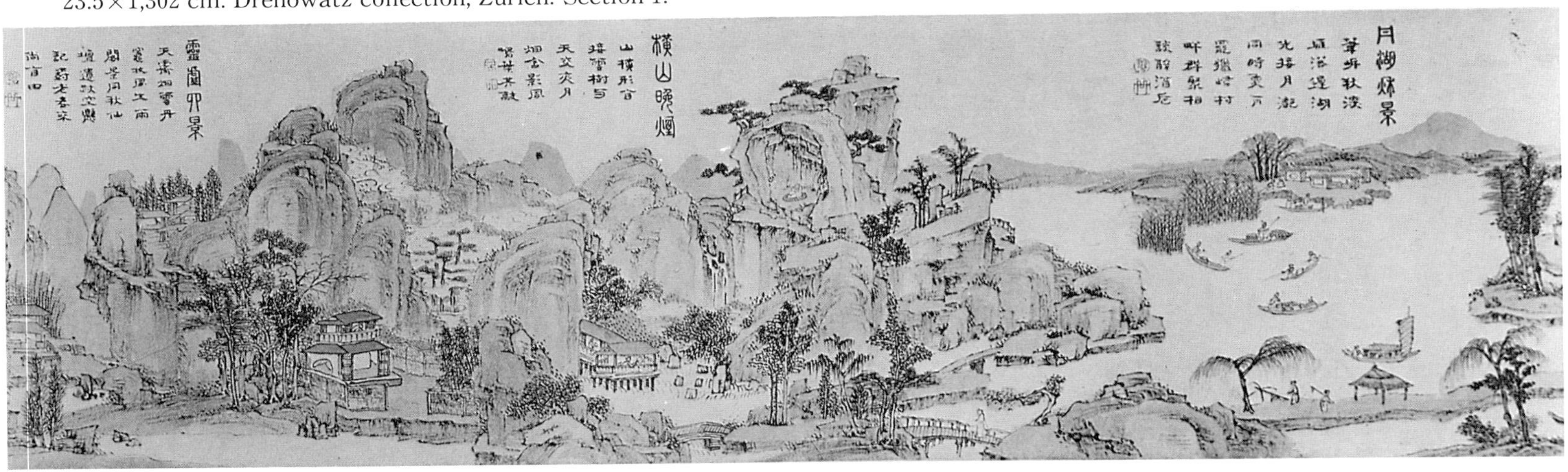

30. Hsiao Yün-ts'ung: "Going Home and Living Abroad Are the Same Thing." Section 4.

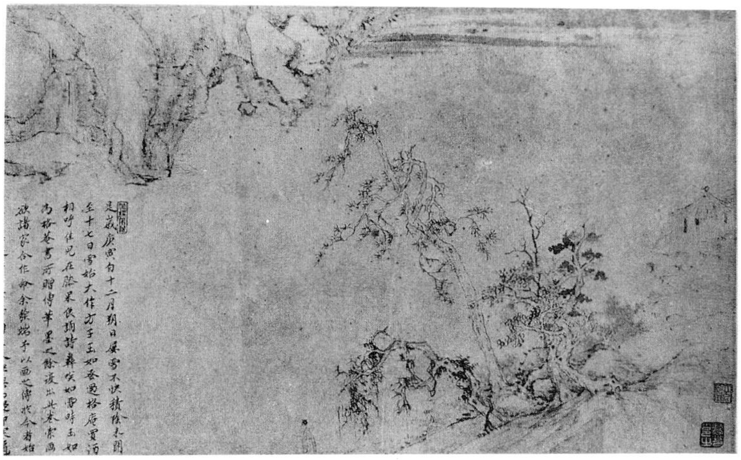

31. "Joint Handscroll of Landscape for Ko-an," 1650-51. Whereabouts unknown. Section 1.

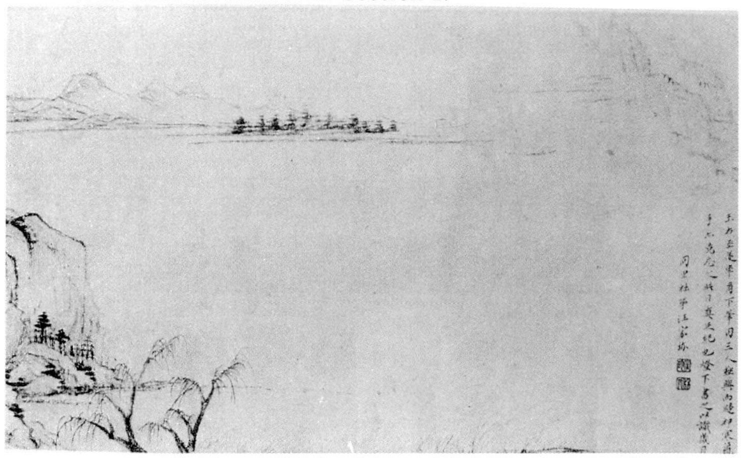

32. "Joint Handscroll of Landscape for Ko-an." Section 2.

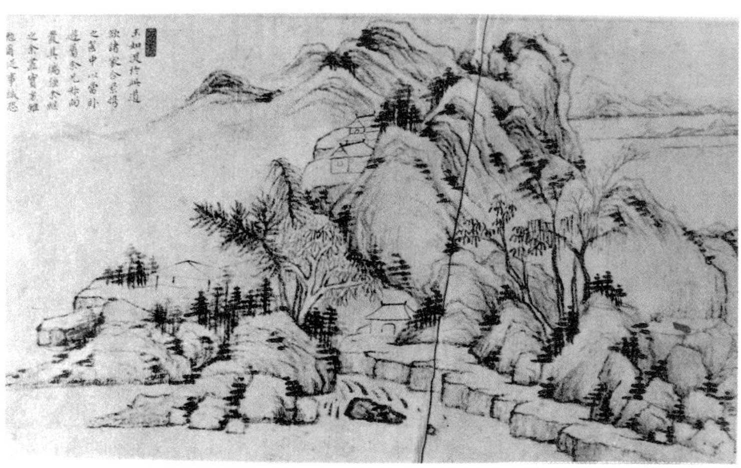

33. "Joint Handscroll of Landscape for Ko-an." Section 3.

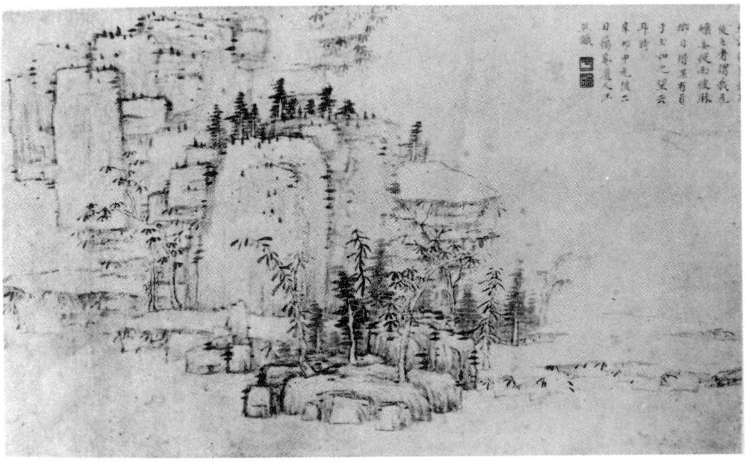

34. "Joint Handscroll of Landscape for Ko-an." Section 4.

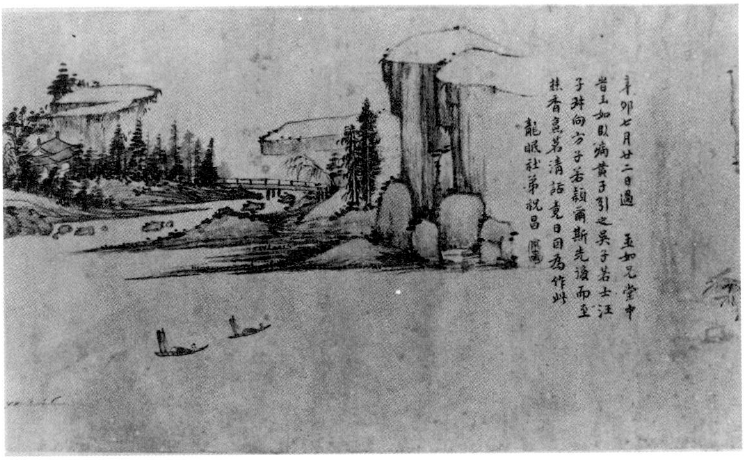

35. "Joint Handscroll of Landscape for Ko-an." Section 5.

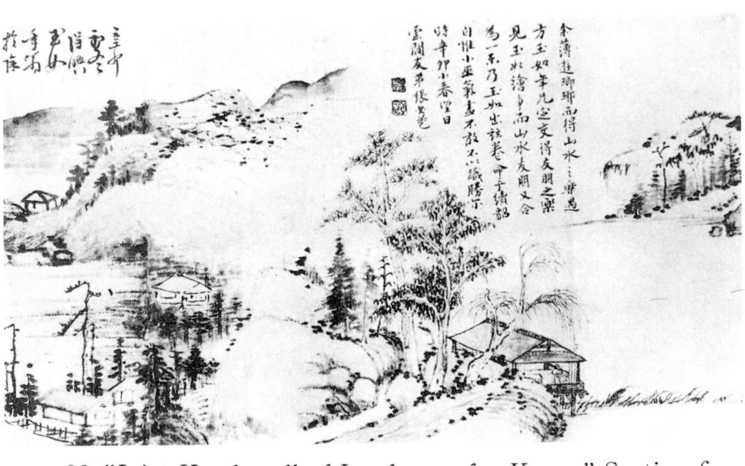

36. "Joint Handscroll of Landscape for Ko-an." Section 6.

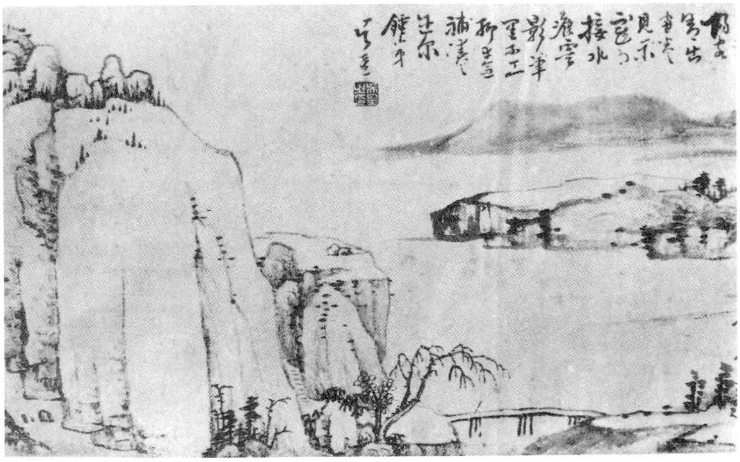

37. "Joint Handscroll of Landscape for Ko-an." Section 7.

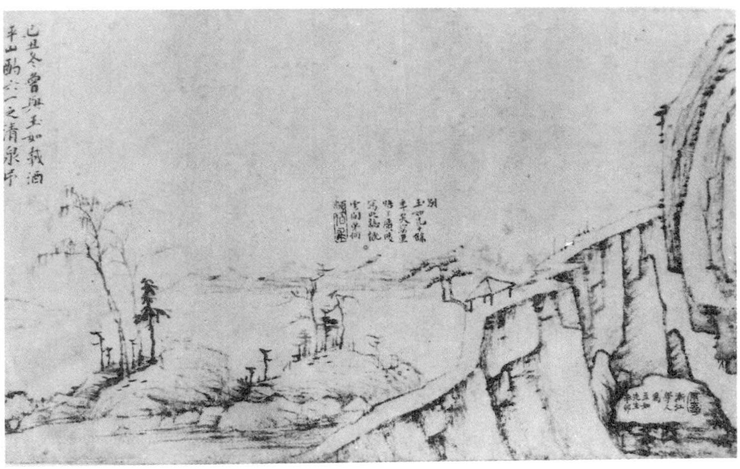

38. "Joint Handscroll of Landscape for Ko-an." Section 8.

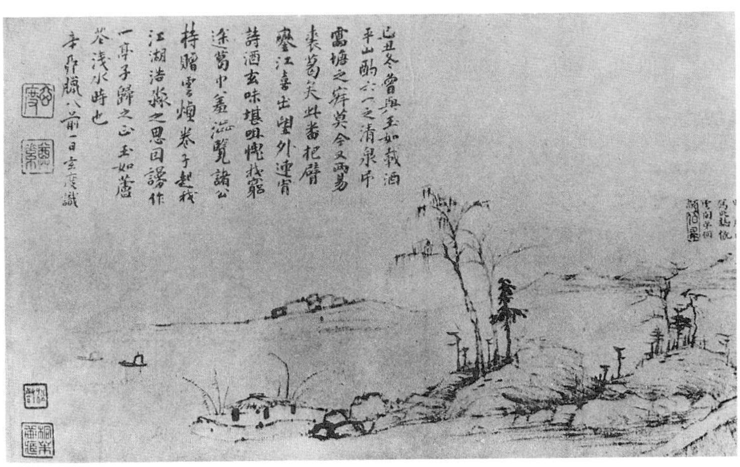

39. "Joint Handscroll of Landscape for Ko-an." Section 9.

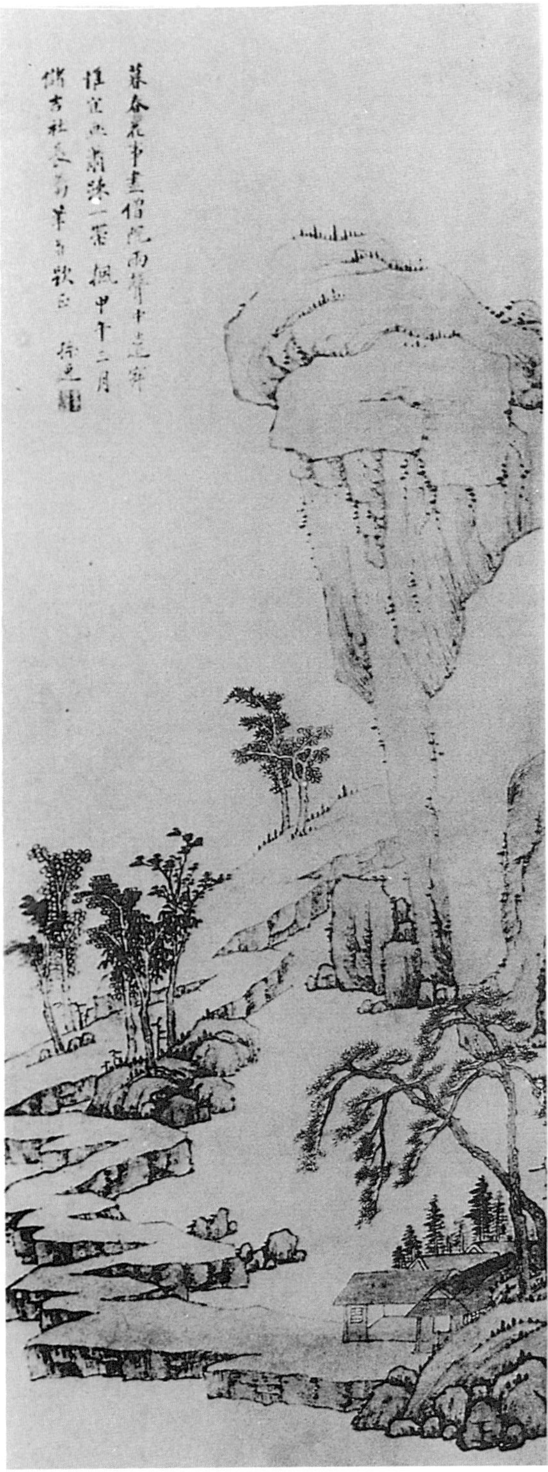

40. Sun I: "Flowers at End of Spring," 1654. Hanging scroll. Whereabouts unknown (formerly Huang Pin-hung collection).

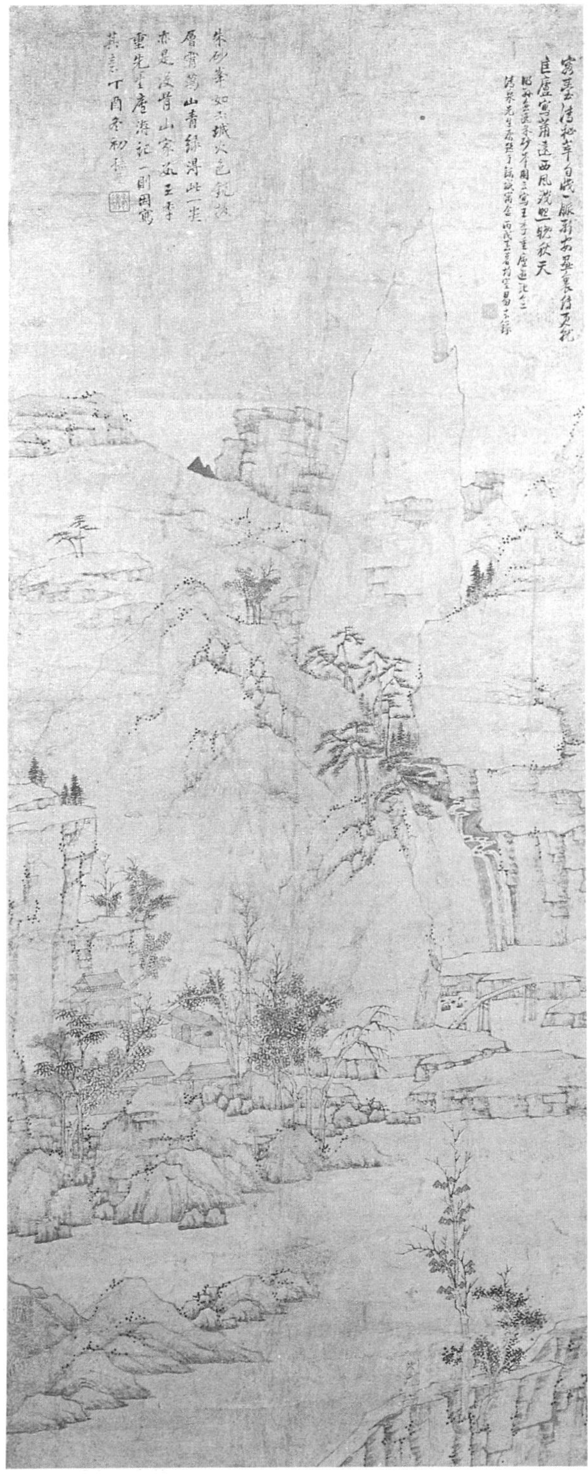

41. Sun I: "Cinnabar Peak," 1657. Hanging scroll, ink and colors on paper, 3 ft. $8\frac{7}{8}$ in. × 1 ft. $6\frac{1}{4}$ in. Chien-lu collection, Ann Arbor.

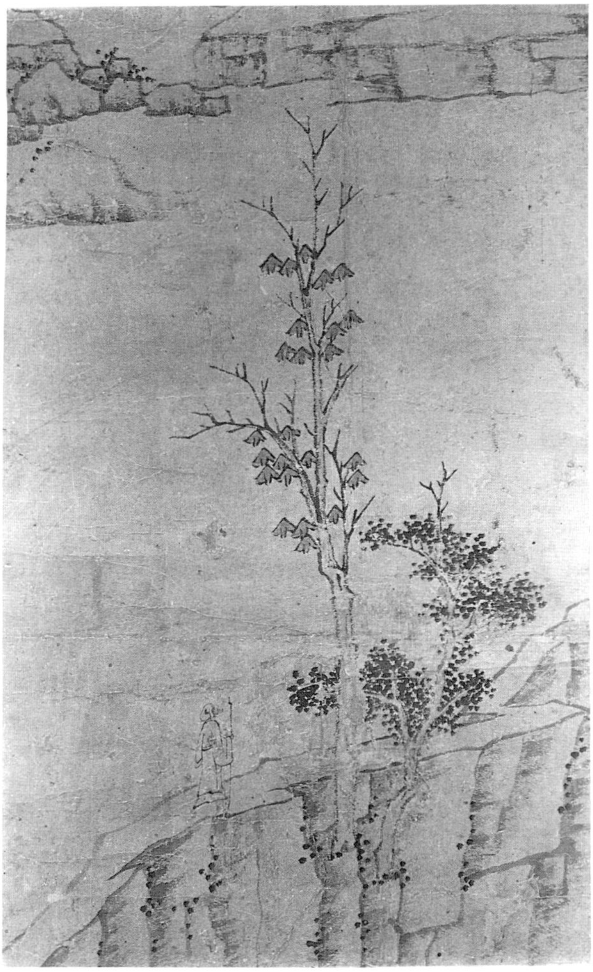

42. Sun I: "Cinnabar Peak." Detail.

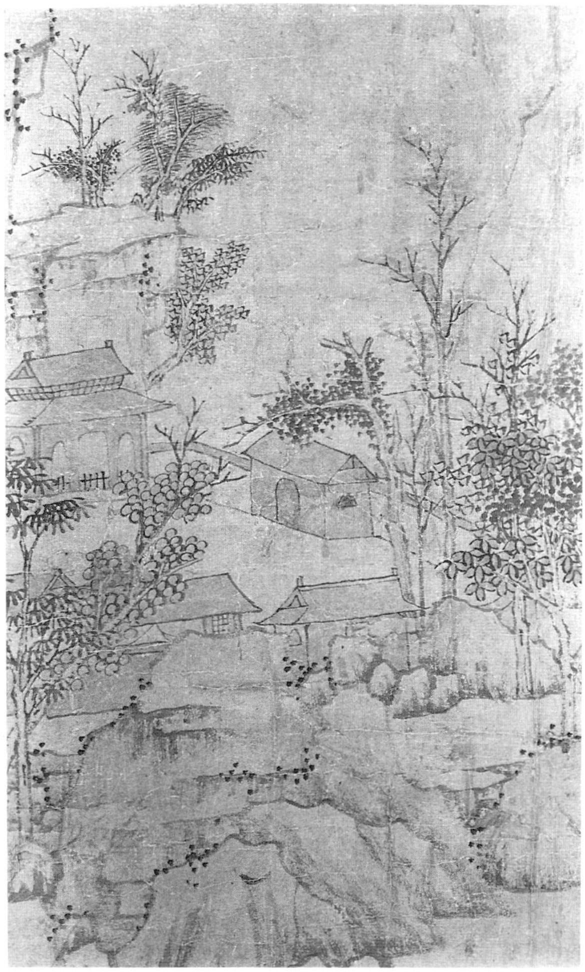

43. Sun I: "Cinnabar Peak." Detail.

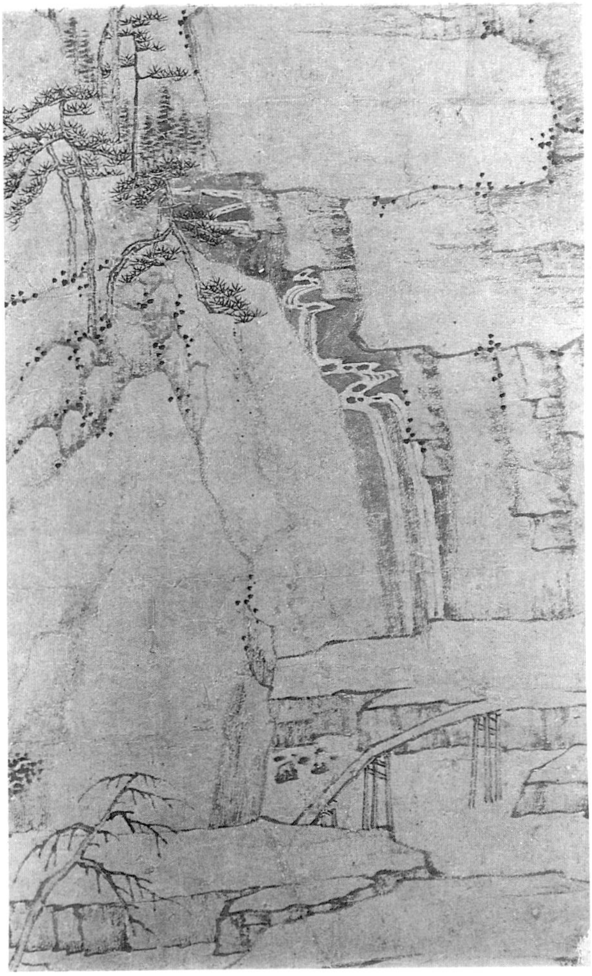

44. Sun I: "Cinnabar Peak." Detail.

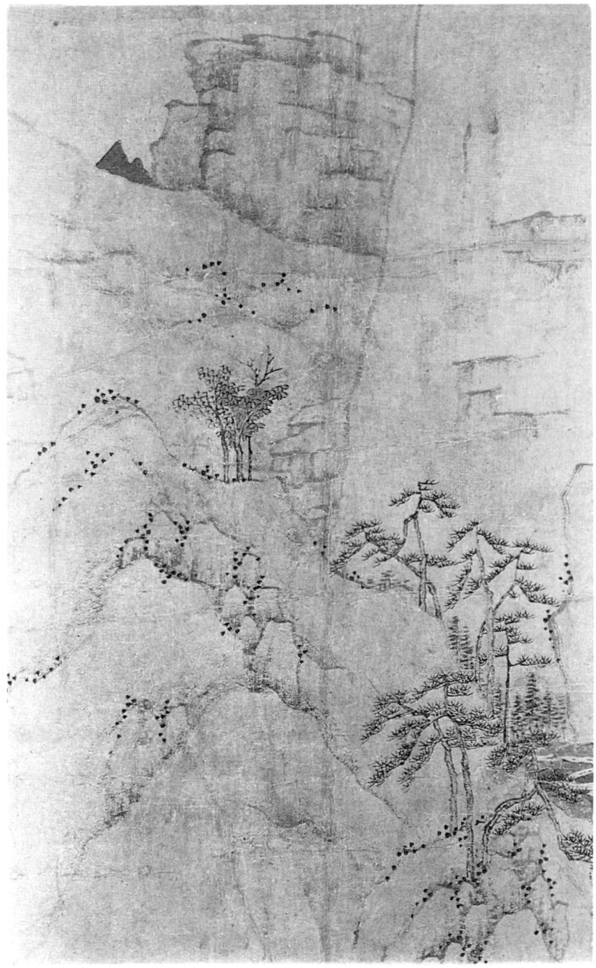

45. Sun I: "Cinnabar Peak." Detail.

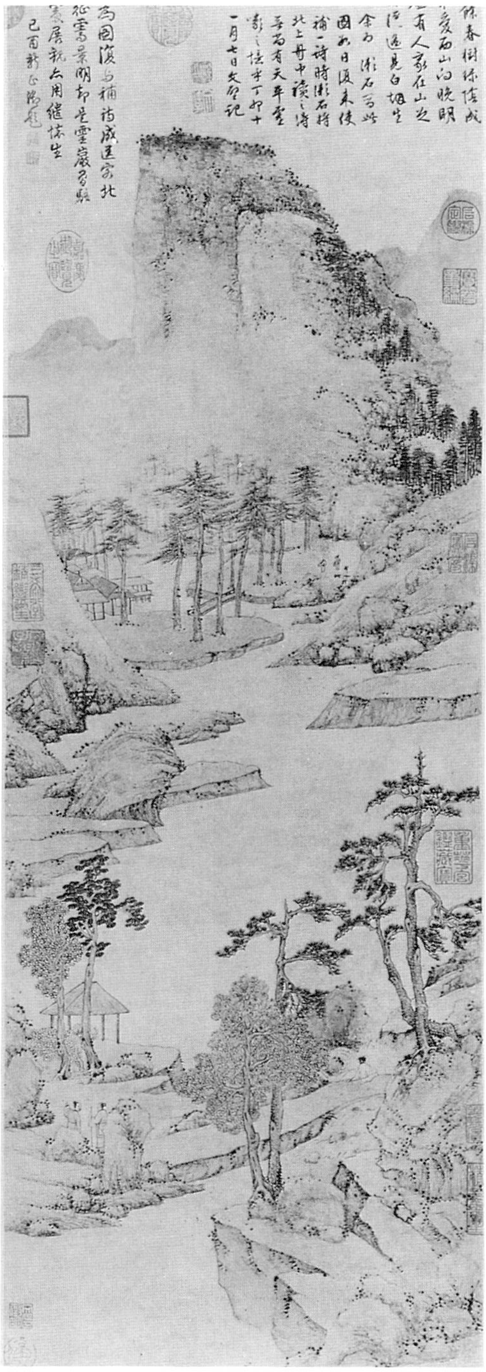

46. Wen Cheng-ming: "Spring Trees after Rain," 1507. Hanging scroll, ink and colors on paper, 94.3×33.3 cm. National Palace Museum, Taipei.

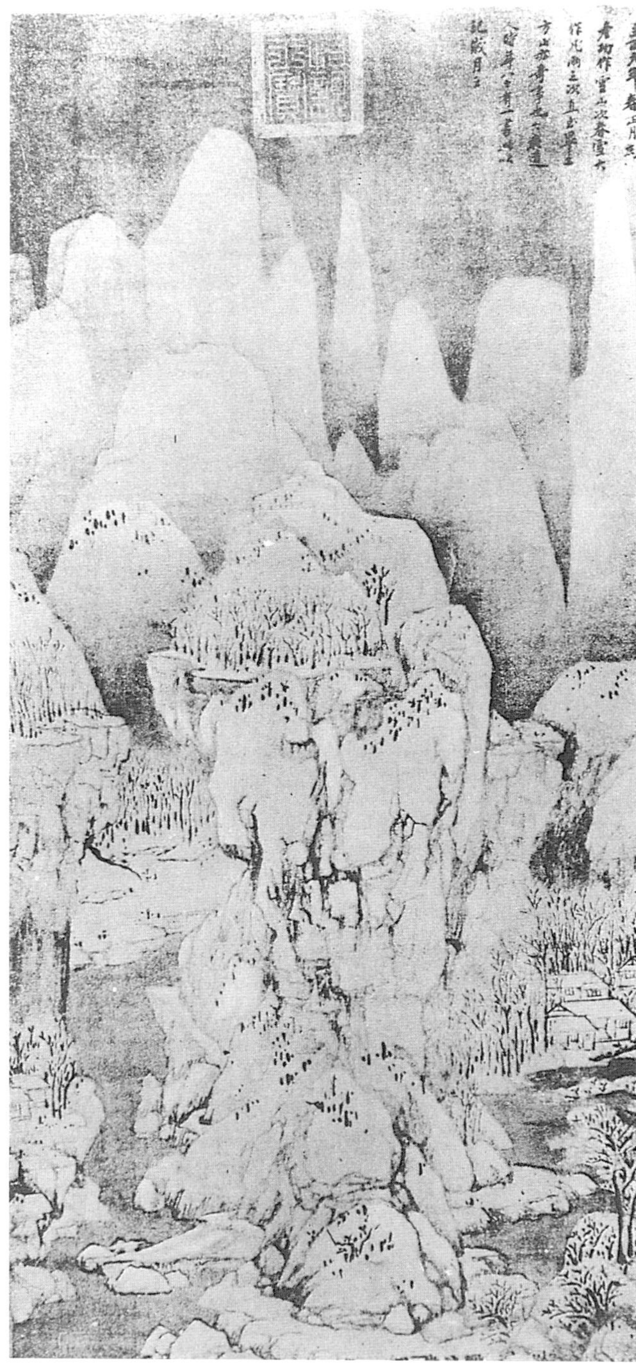

47. Huang Kung-wang: "Snow Clearing over Nine Peaks," 1349. Hanging scroll. Palace Museum, Peking.

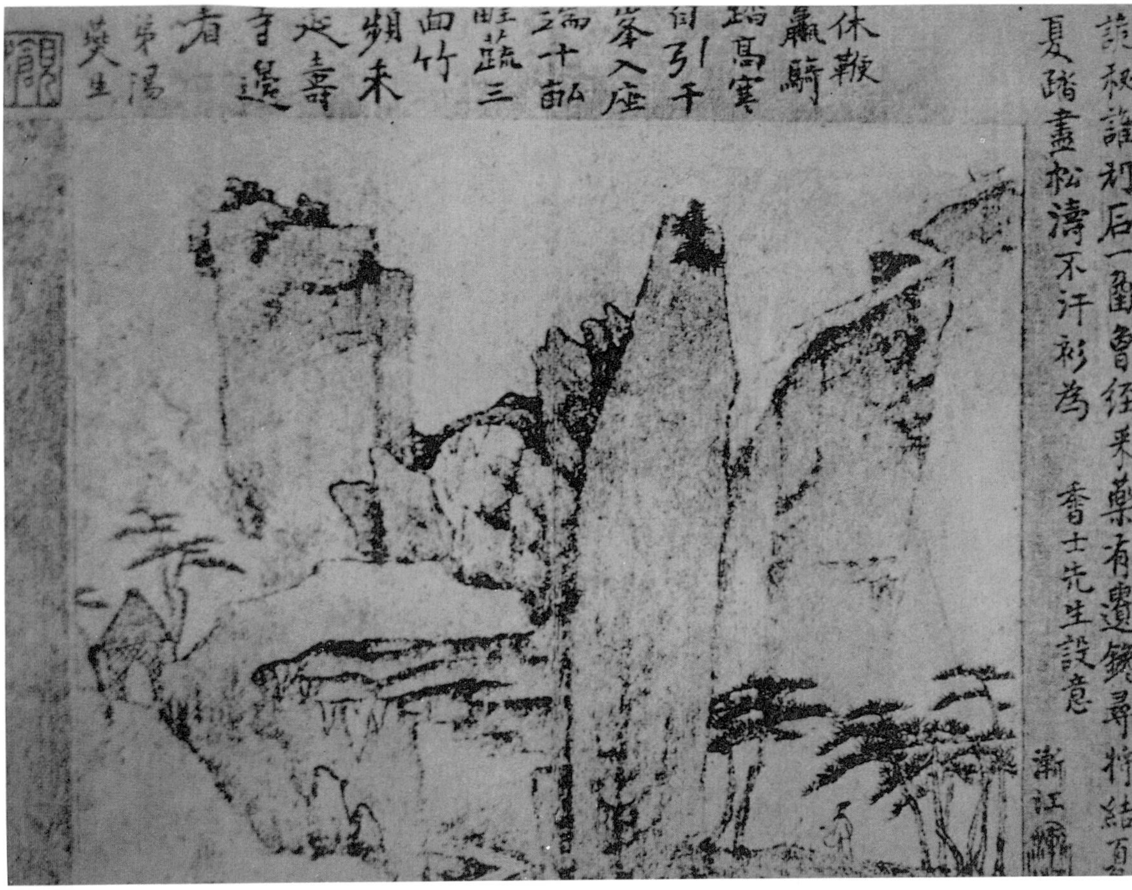

48. Hung-jen: "Landscape." Album leaf, ink and colors on paper. Whereabouts unknown.

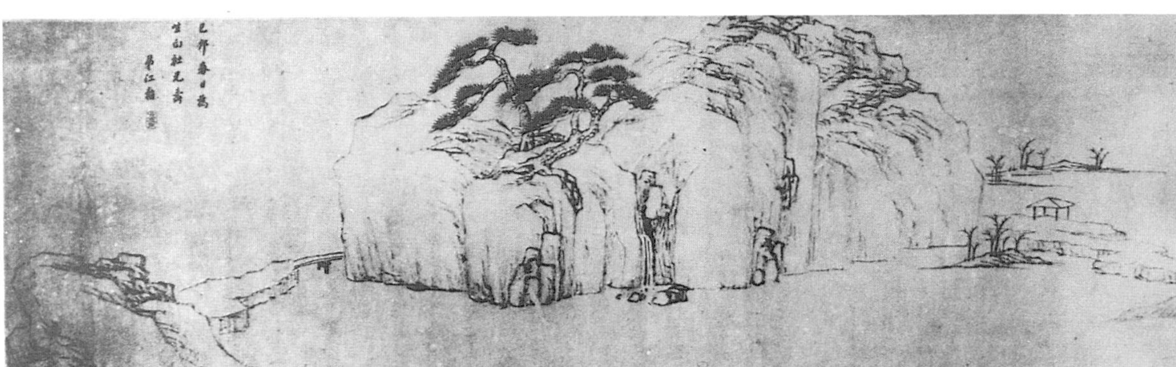

49. Hung-jen: "Landscape," 1639. Section of "Joint Handscroll of Landscape for Li Sheng-pai." Shanghai Museum.

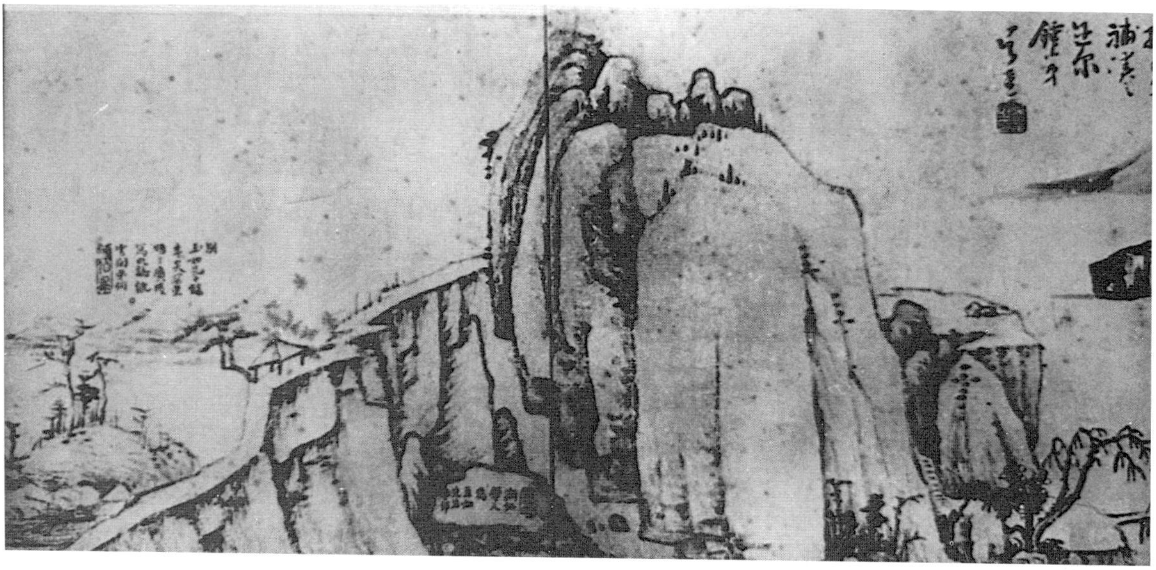

50. Hung-jen: "Landscape," 1651. Section of "Joint Handscroll of Landscape for Ko-an." Detail.

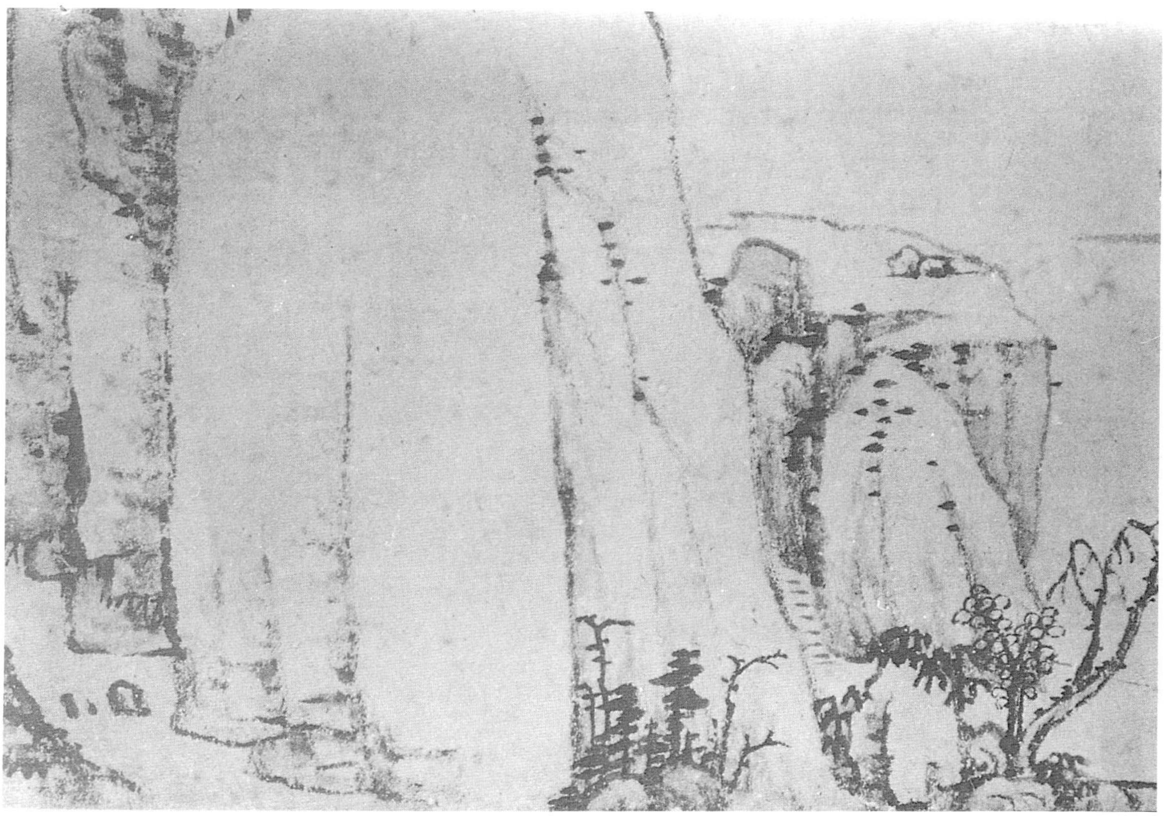

51. Hung-jen: "Landscape," 1651. Section of "Joint Handscroll of Landscape for Ko-an." Detail.

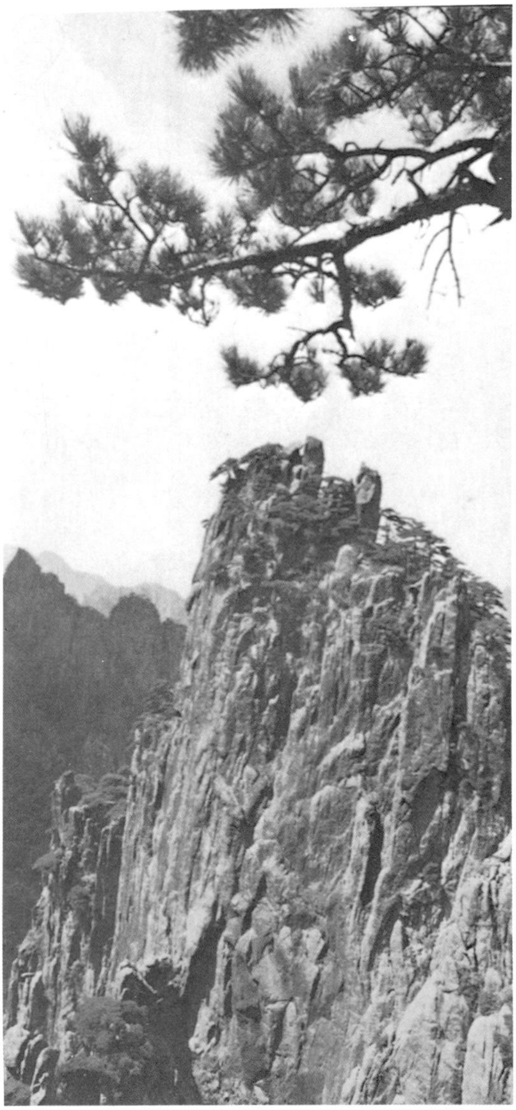

57. Rocks and Pines at Mt. Huang. From *Huang-shan tao-yu shou-ts'e,* p. 53.

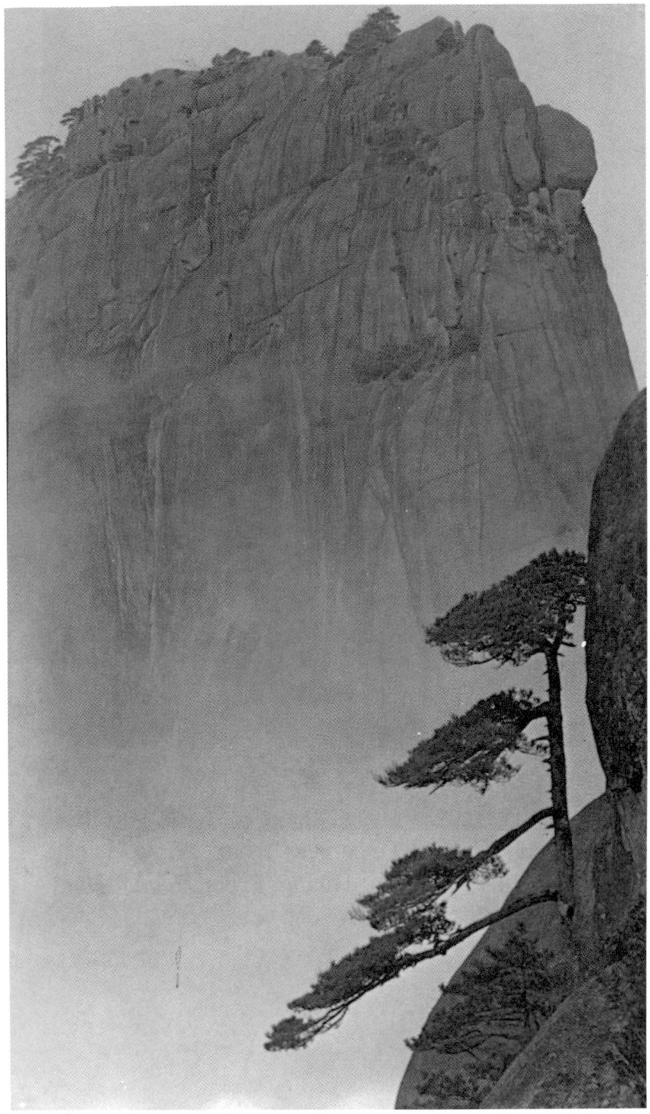

58. Rocks and Pines at Mt. Huang. From *Huang-shan,* p. 12.

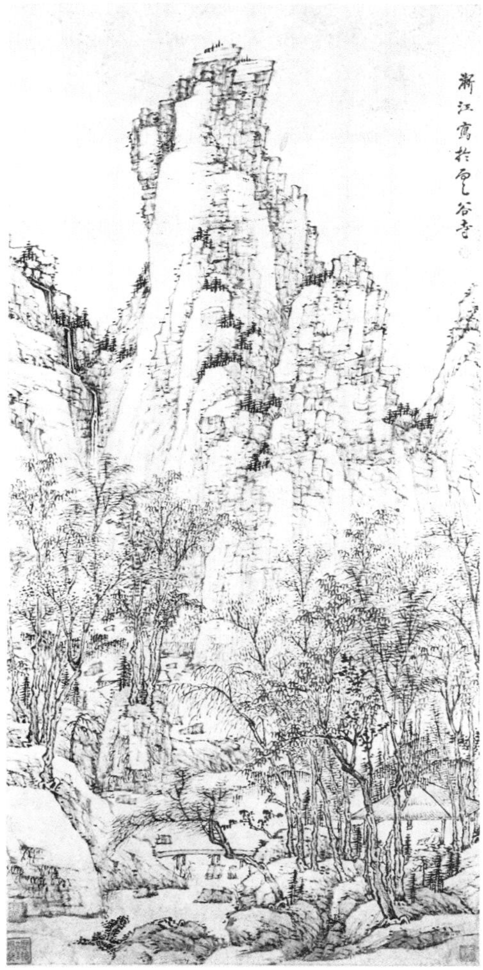

59. Hung-jen: "Landscape." Hanging scroll, ink on paper, $33\times16\frac{1}{2}$ in. Freer Gallery of Art, Washington, D.C.

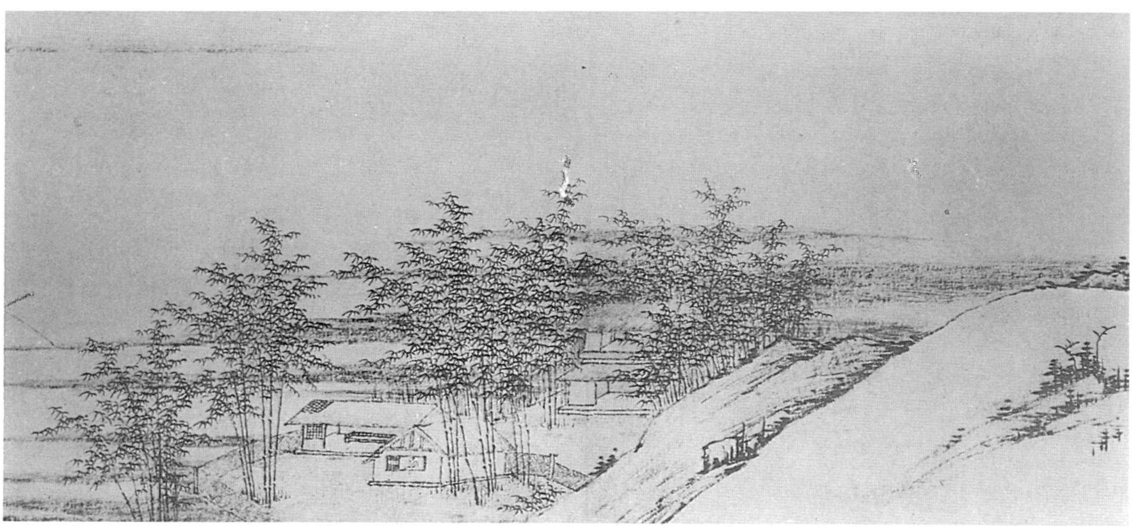

60. Hung-jen: "Bamboo Banks and Reed Shores," 1652. Handscroll, ink on paper, 24.9×240 cm. Sumitomo collection, Japan. Section 1.

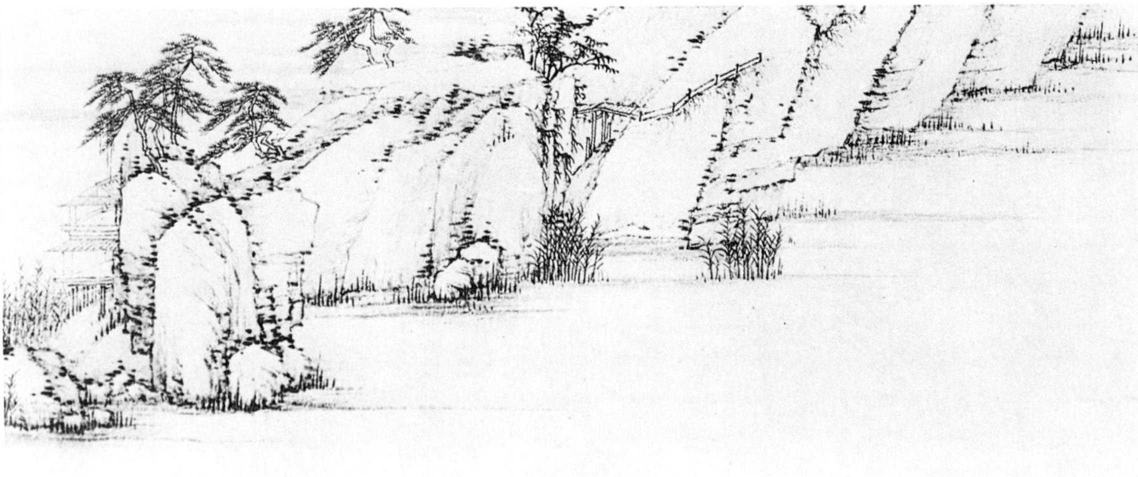

61. Hung-jen: "Bamboo Banks and Reed Shores." Section 2.

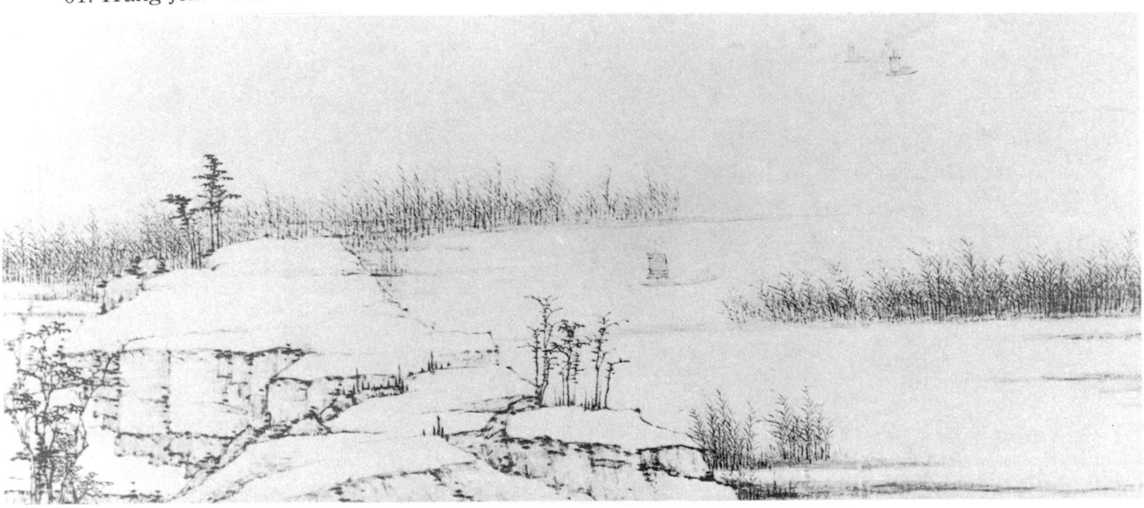

62. Hung-jen: "Bamboo Banks and Reed Shores." Section 3.

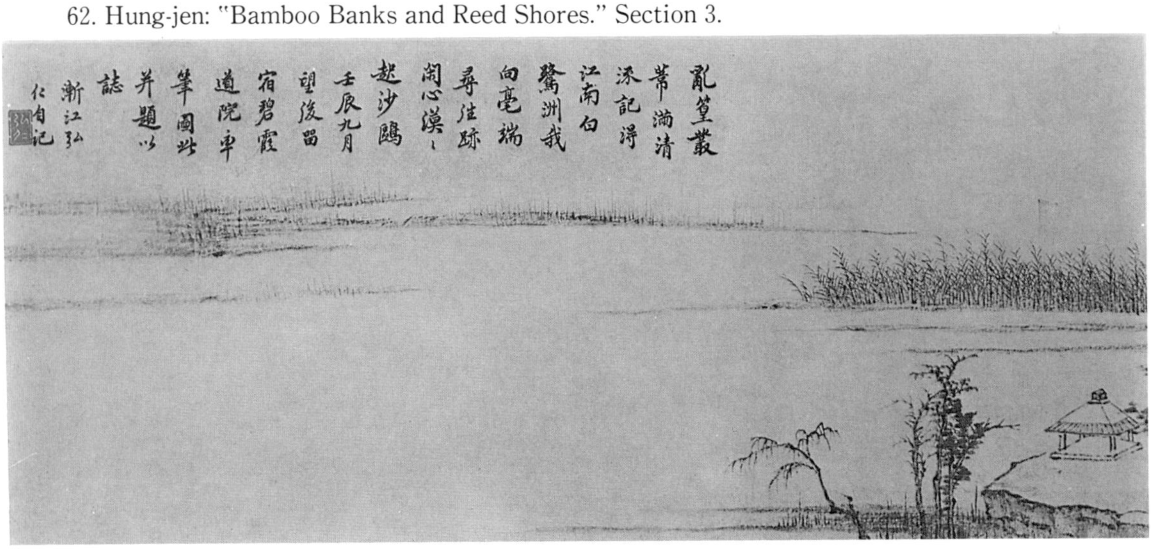

63. Hung-jen: "Bamboo Banks and Reed Shores." Section 4.

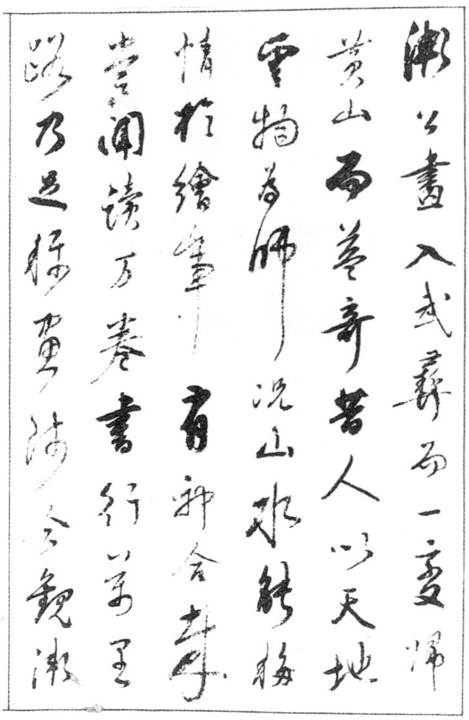

64. Cha Shih-piao: Colophon to Hung-jen's "Mt. Huang" Album. Palace Museum, Peking.

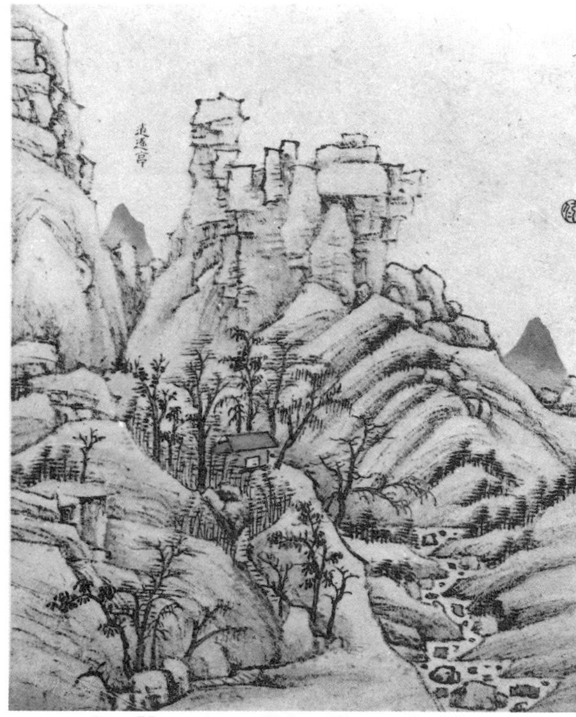

65. Hung-jen: "Mt. Huang." Album. Palace Museum, Peking. Leaf no. 1.

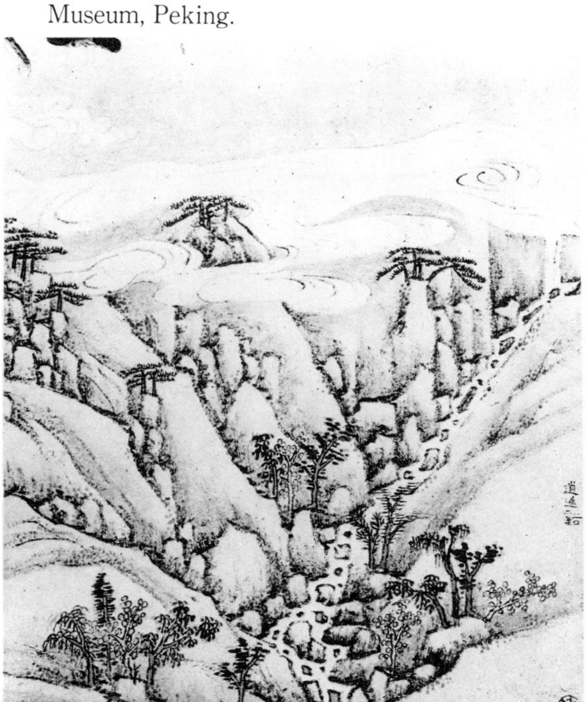

66. Hung-jen: "Mt. Huang." Leaf no. 2.

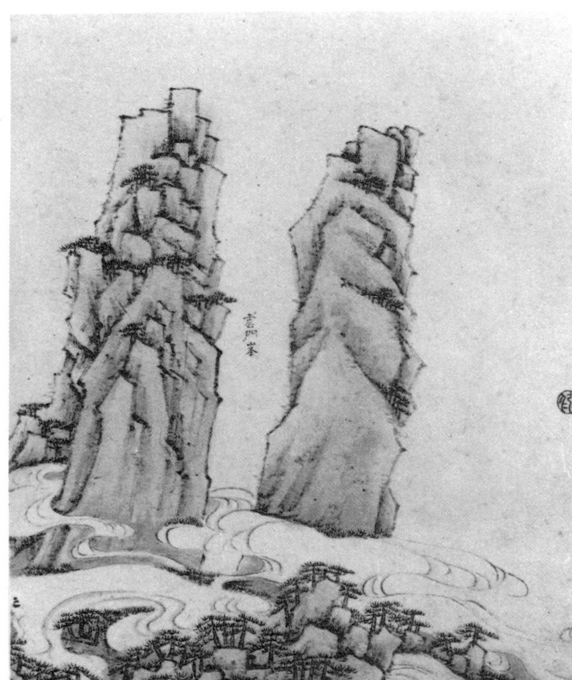

67. Hung-jen: "Mt. Huang." Leaf no. 3.

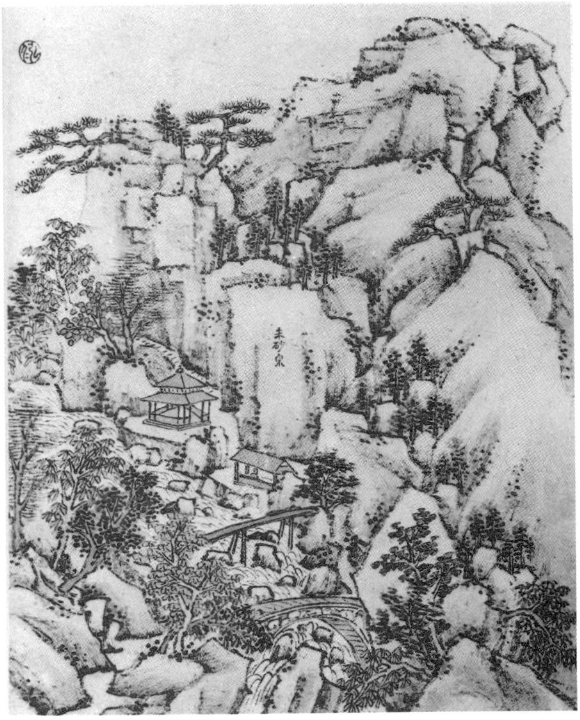

68. Hung-jen: "Mt. Huang." Leaf no. 4.

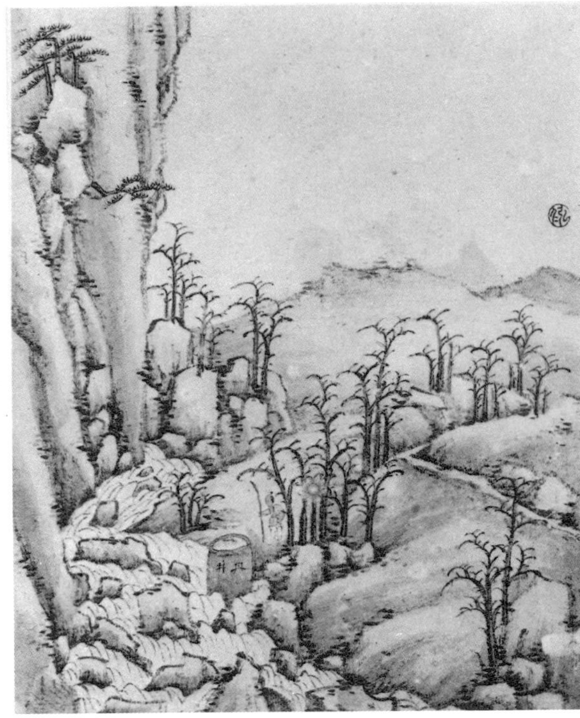

69. Hung-jen: "Mt. Huang." Leaf no. 5.

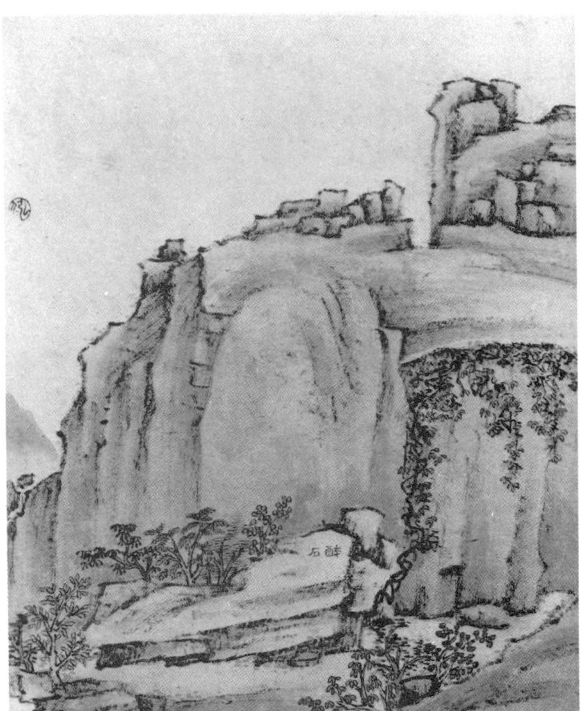

70. Hung-jen: "Mt. Huang." Leaf no. 6.

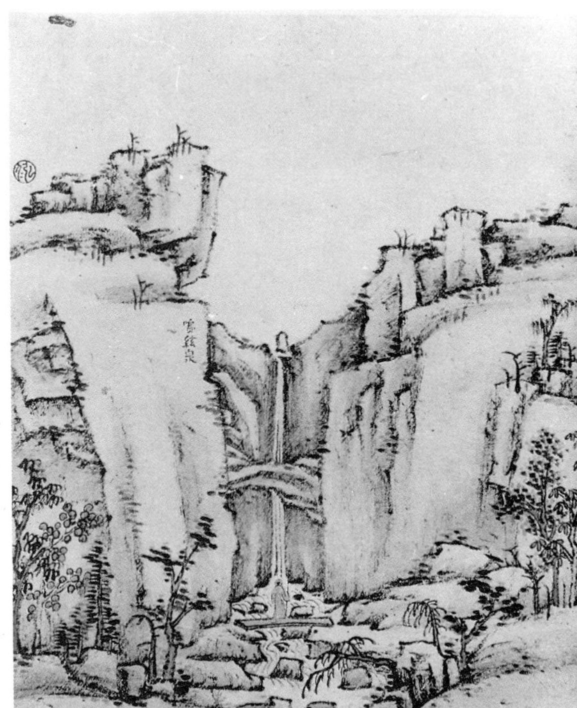

71. Hung-jen: "Mt. Huang." Leaf no. 7.

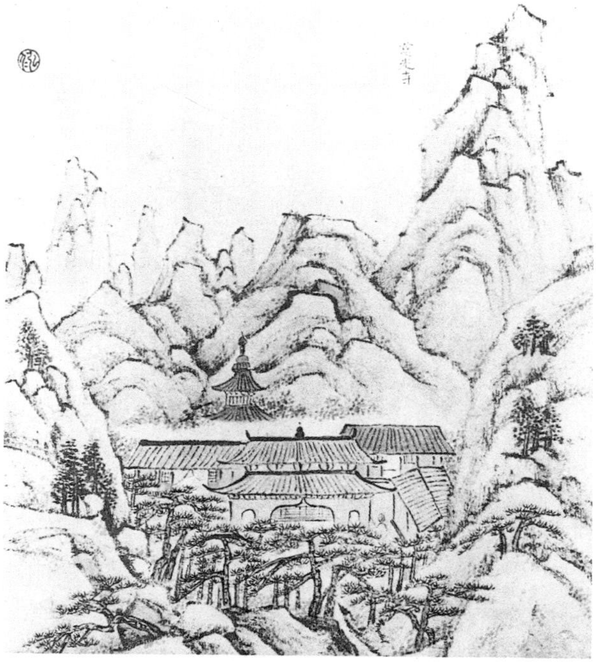

72. Hung-jen: "Mt. Huang." Leaf no. 8.

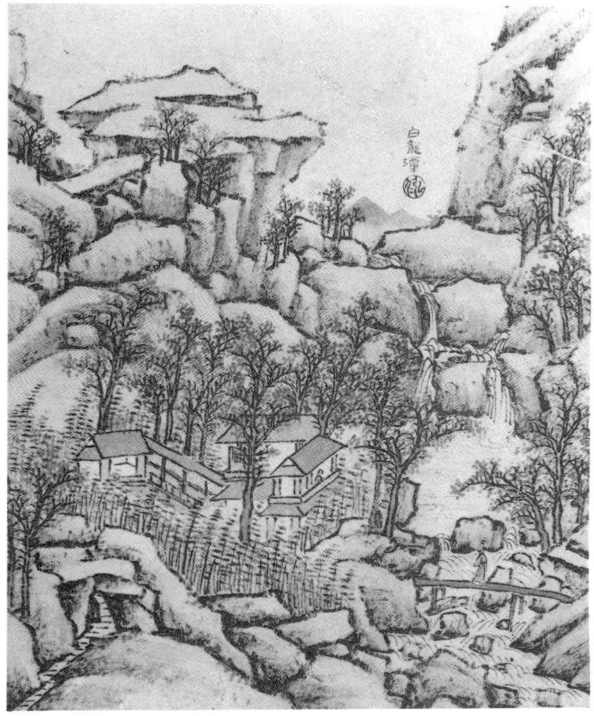

73. Hung-jen: "Mt. Huang." Leaf no. 9.

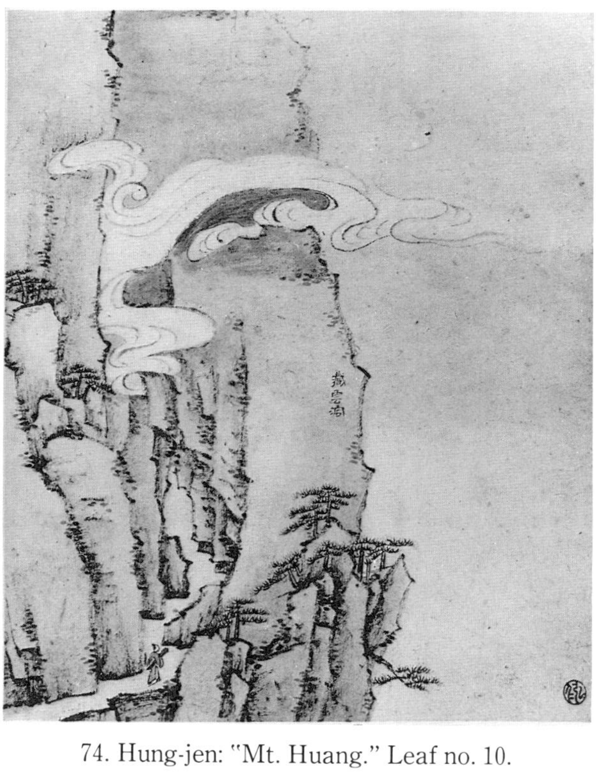

74. Hung-jen: "Mt. Huang." Leaf no. 10.

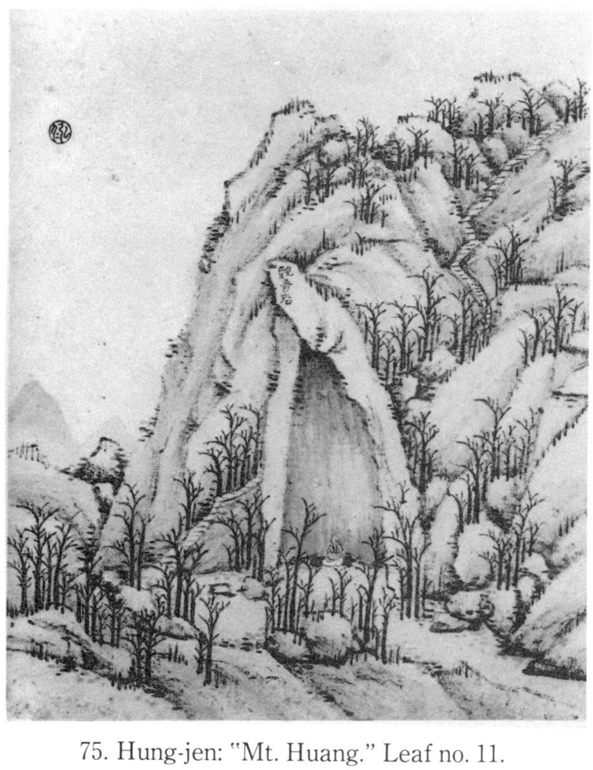

75. Hung-jen: "Mt. Huang." Leaf no. 11.

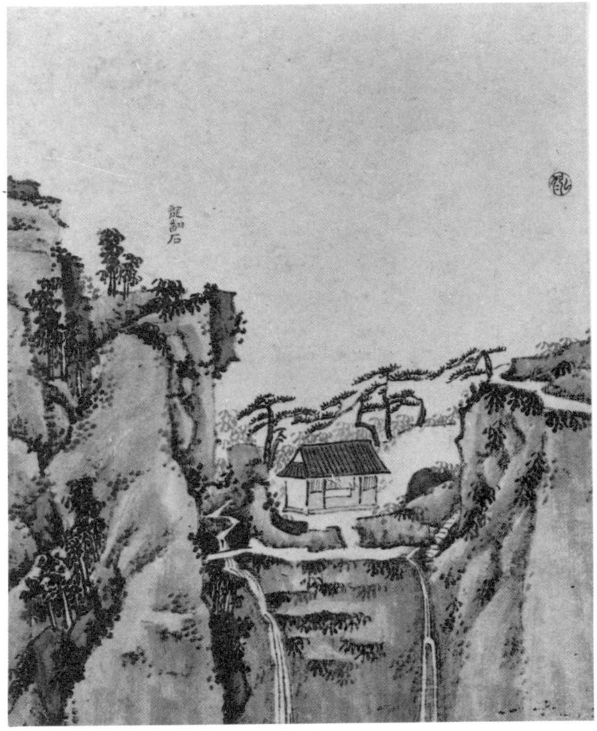

76. Hung-jen: "Mt. Huang." Leaf no. 12.

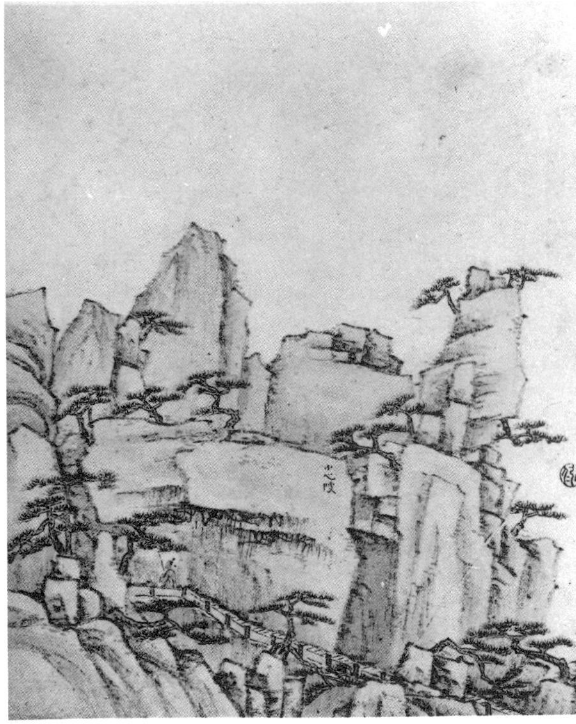

77. Hung-jen: "Mt. Huang." Leaf no. 13.

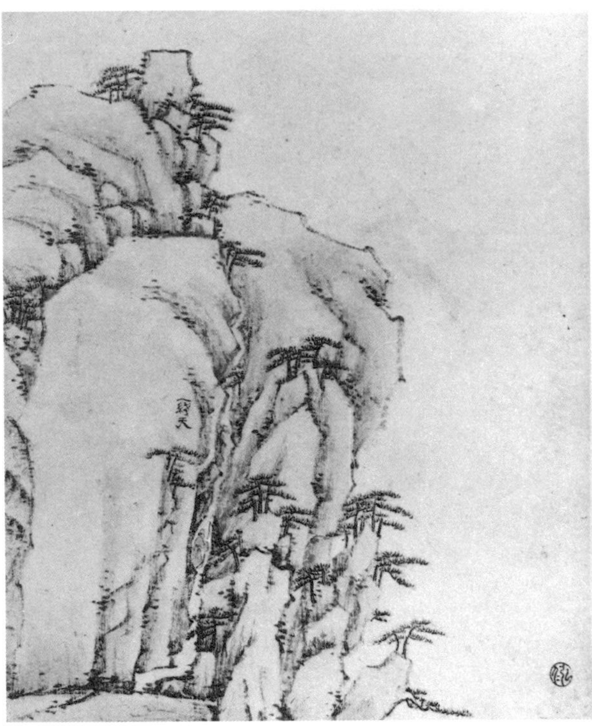

78. Hung-jen: "Mt. Huang." Leaf no. 14.

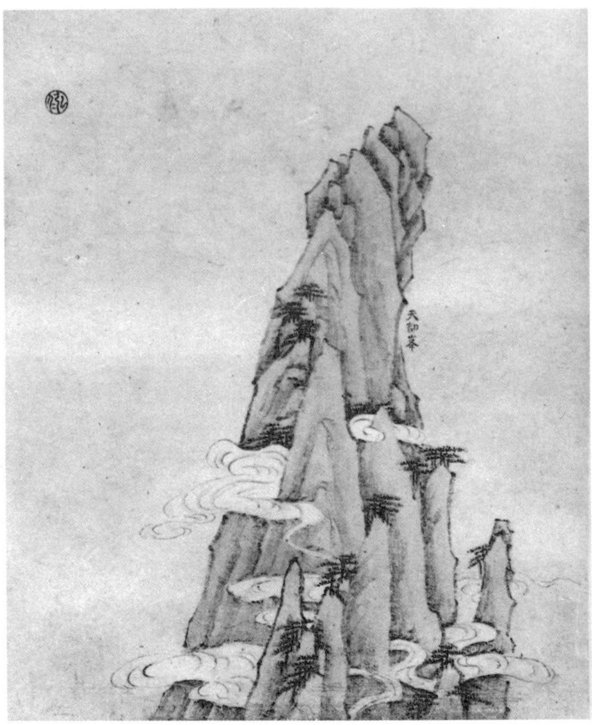

79. Hung-jen: "Mt. Huang." Leaf no. 15.

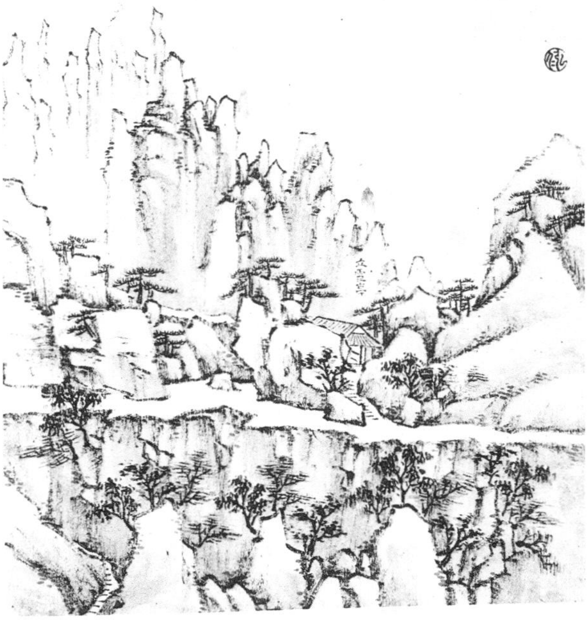

80. Hung-jen: "Mt. Huang." Leaf no. 16.

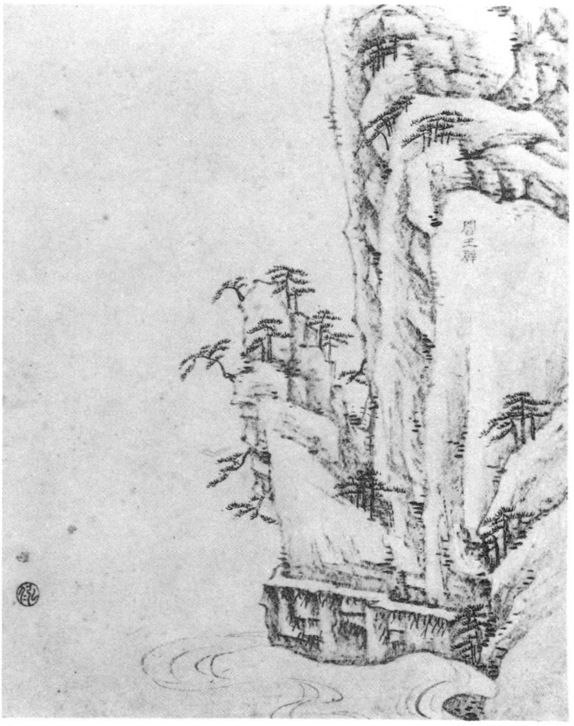

81. Hung-jen: "Mt. Huang." Leaf no. 17.

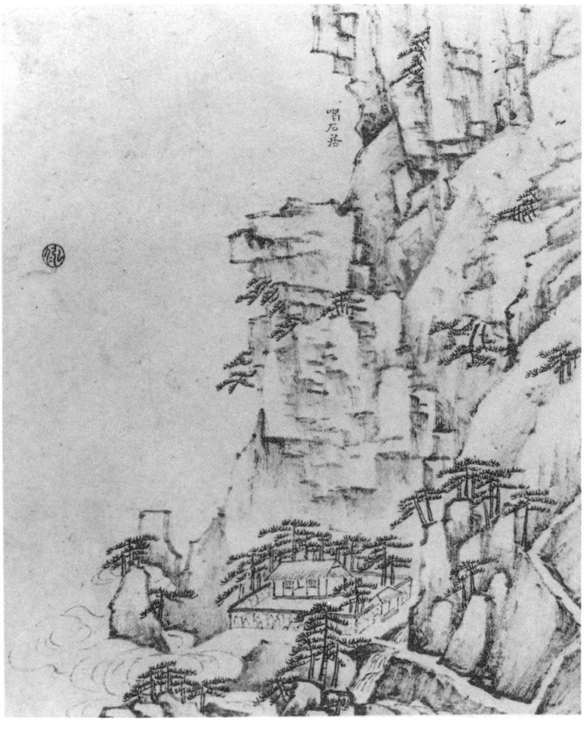

82. Hung-jen: "Mt. Huang." Leaf no. 18.

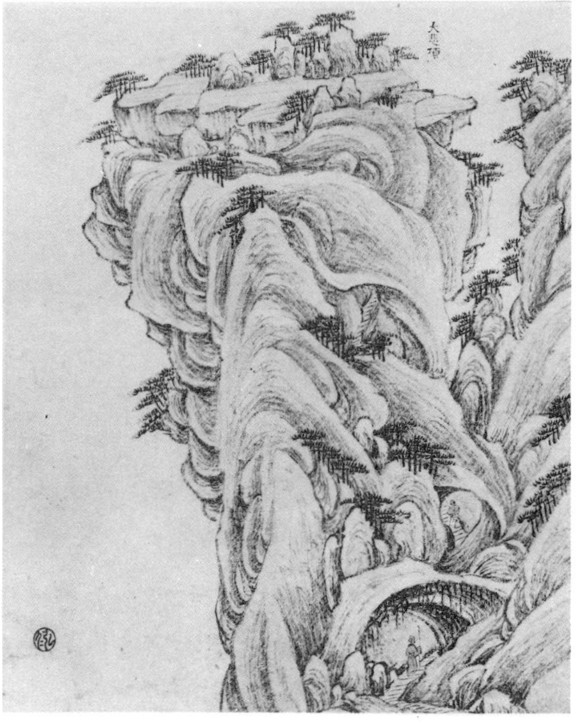

83. Hung-jen: "Mt. Huang." Leaf no. 19.

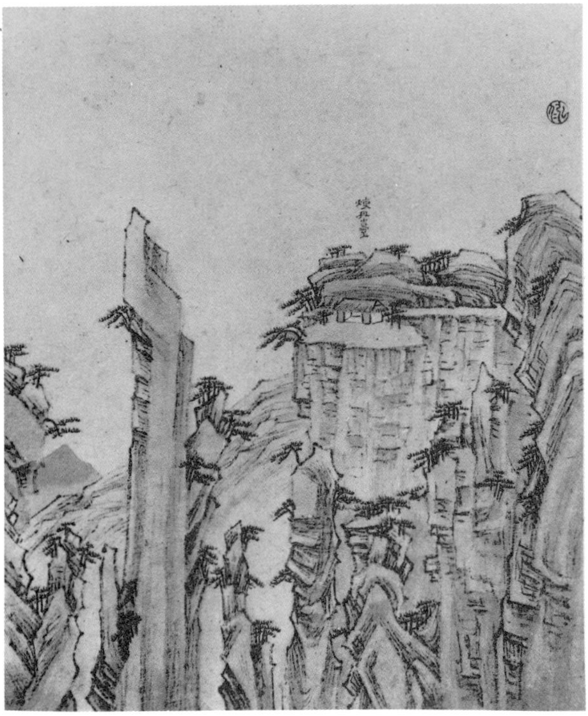

84. Hung-jen: "Mt. Huang." Leaf no. 20.

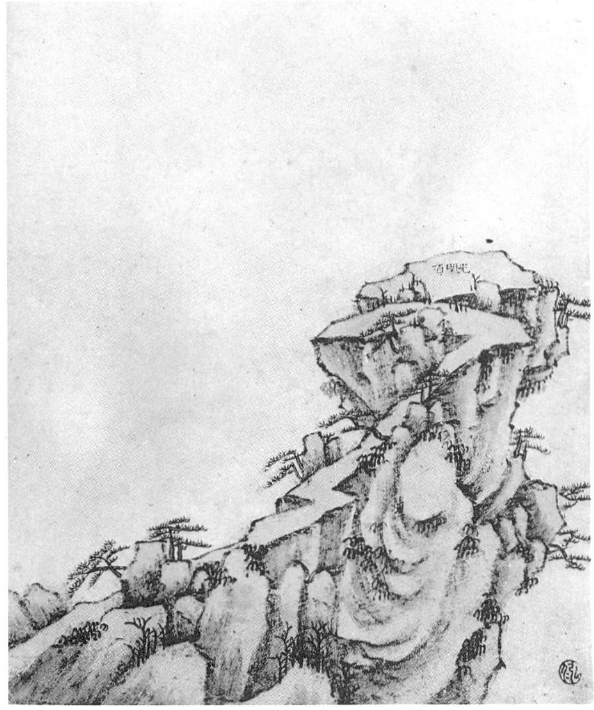

85. Hung-jen: "Mt. Huang." Leaf no. 21.

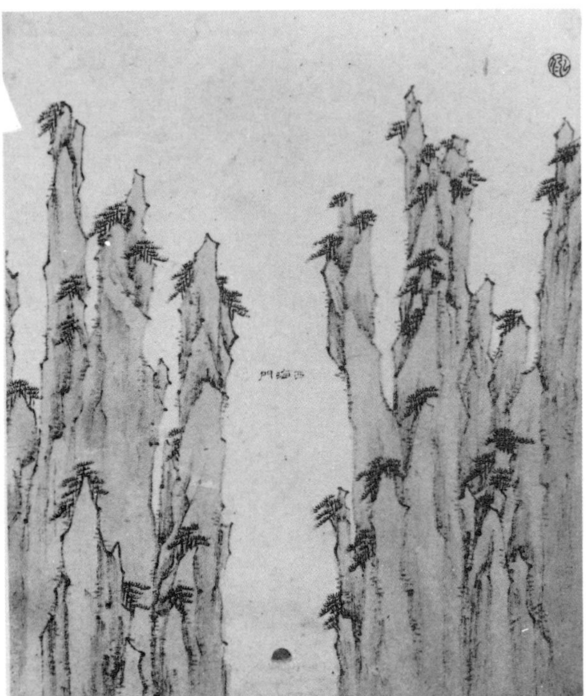

86. Hung-jen: "Mt. Huang." Leaf no. 22.

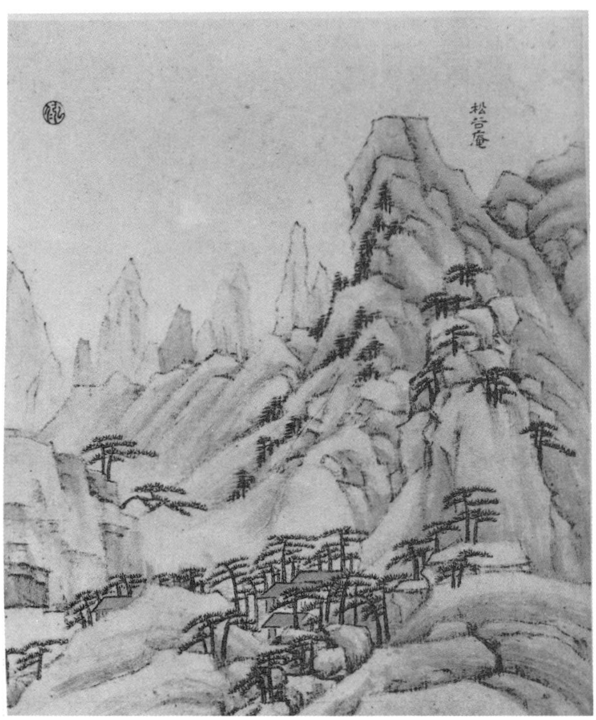

87. Hung-jen: "Mt. Huang." Leaf no. 23.

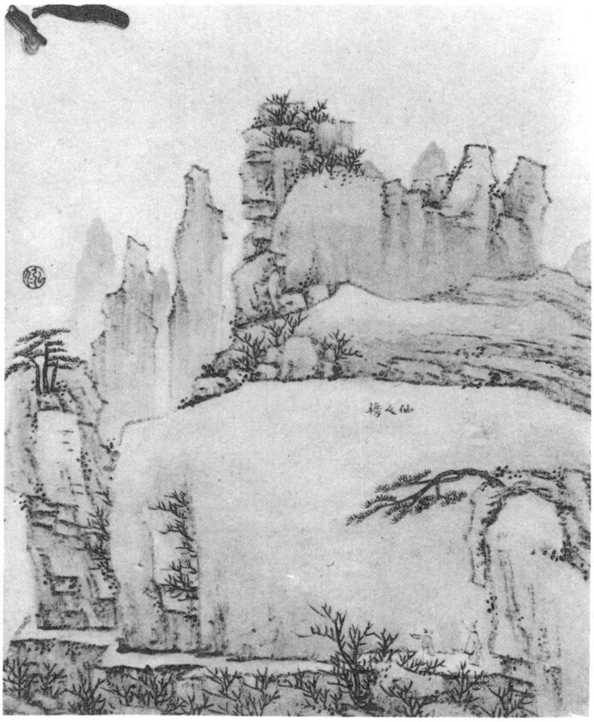

88. Hung-jen: "Mt. Huang." Leaf no. 24.

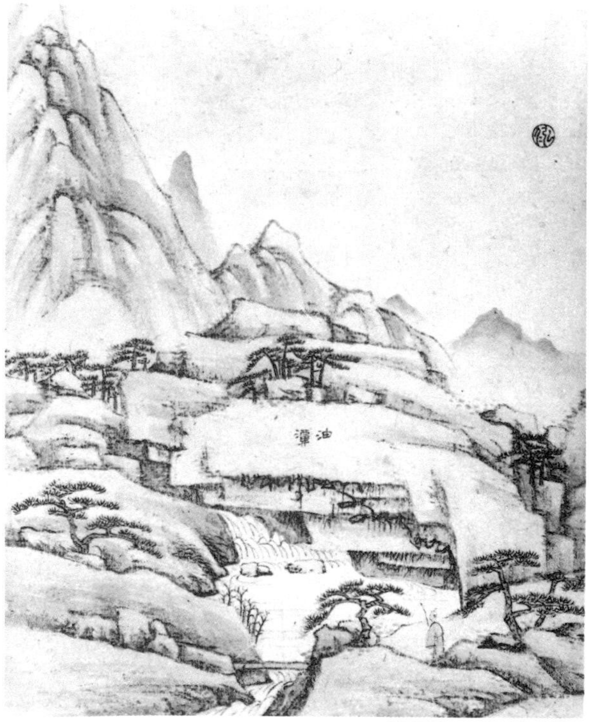

89. Hung-jen: "Mt. Huang." Leaf no. 25.

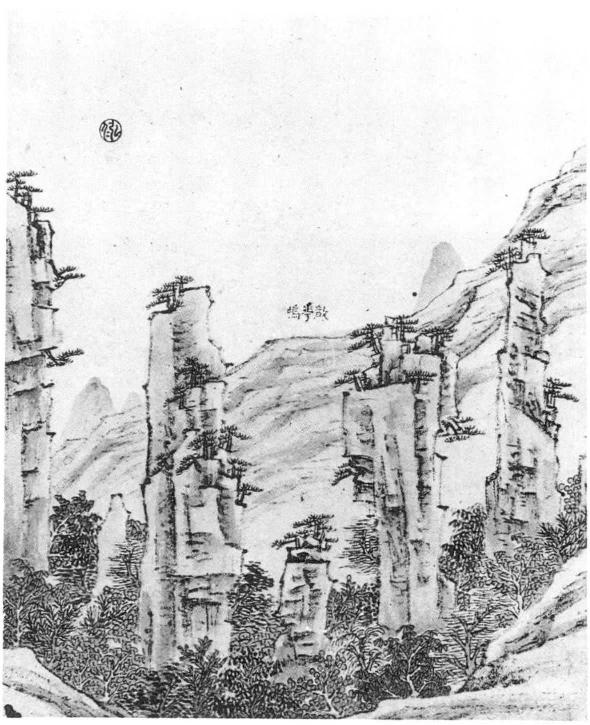

90. Hung-jen: "Mt. Huang." Leaf no. 26.

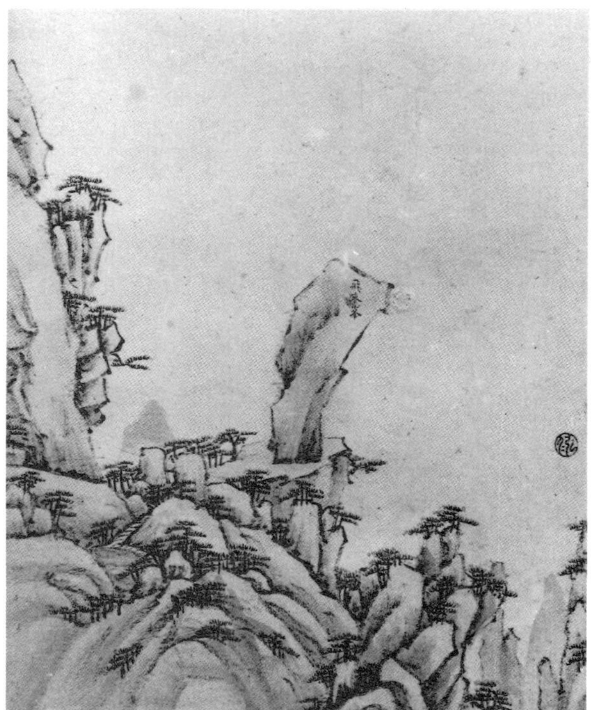

91. Hung-jen: "Mt. Huang." Leaf no. 27.

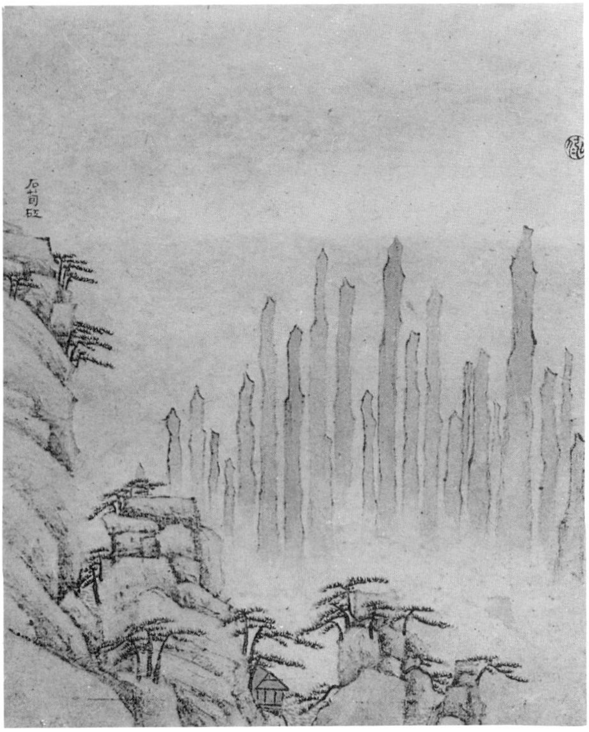

92. Hung-jen: "Mt. Huang." Leaf no. 28.

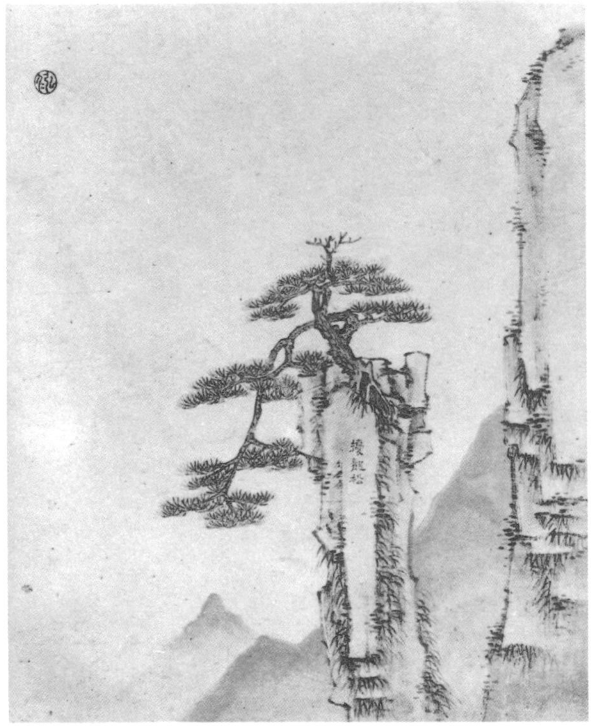

93. Hung-jen: "Mt. Huang." Leaf no. 29.

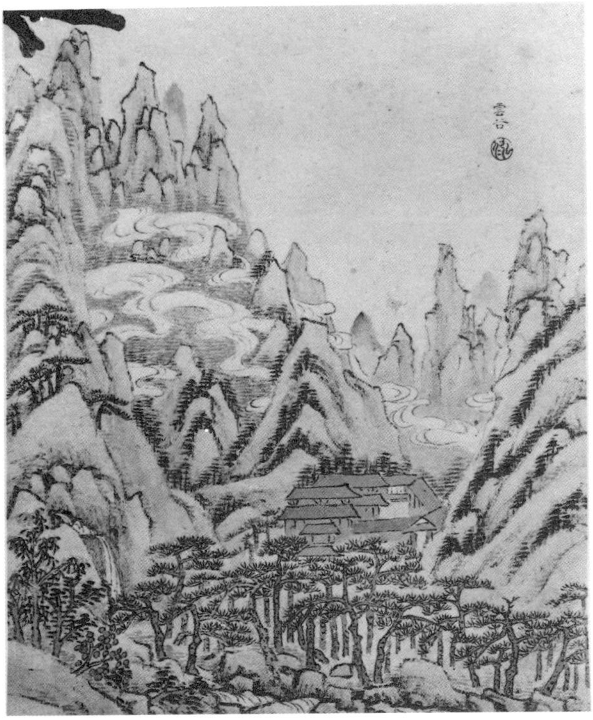

94. Hung-jen: "Mt. Huang." Leaf no. 30.

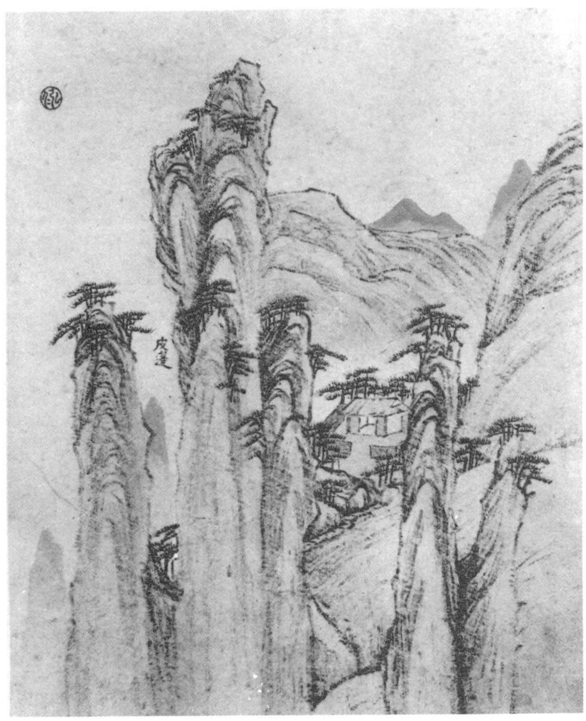

95. Hung-jen: "Mt. Huang." Leaf no. 31.

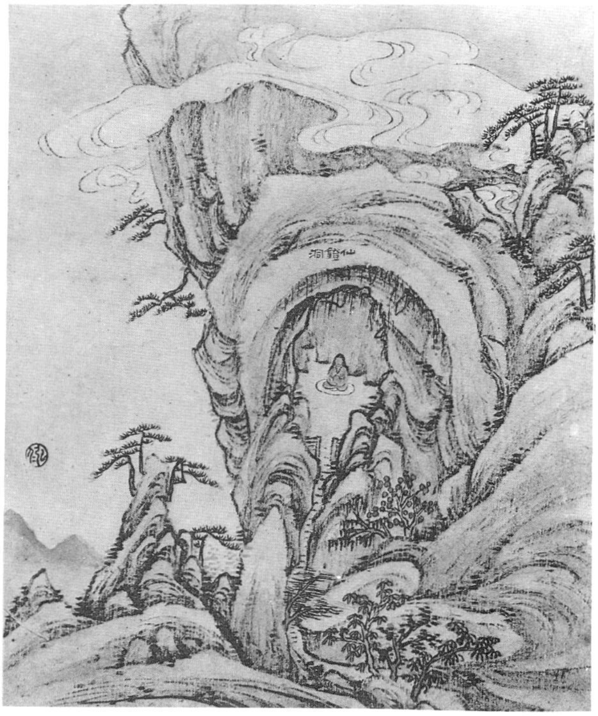

96. Hung-jen: "Mt. Huang." Leaf no. 32.

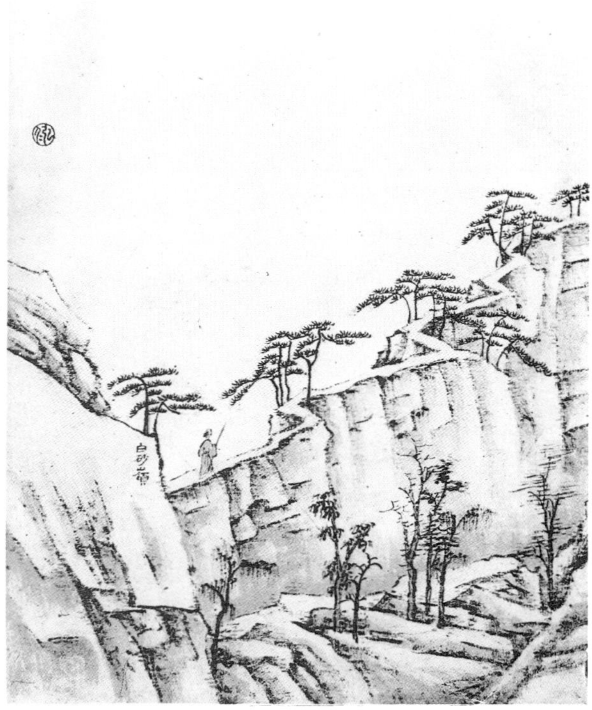

97. Hung-jen: "Mt. Huang." Leaf no. 33.

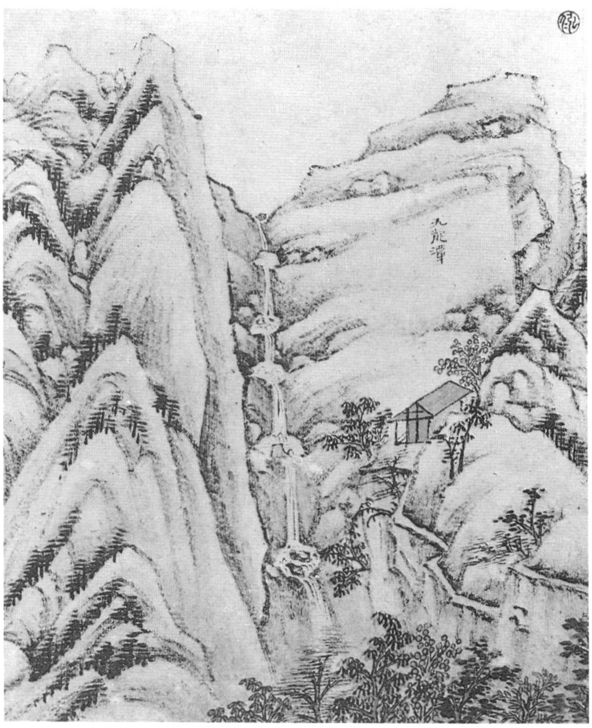

98. Hung-jen: "Mt. Huang." Leaf no. 34.

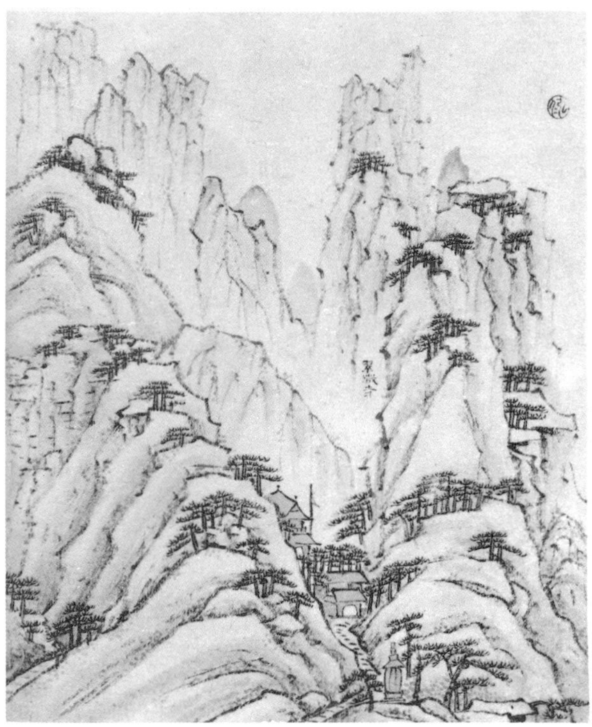

99. Hung-jen: "Mt. Huang." Leaf no. 35.

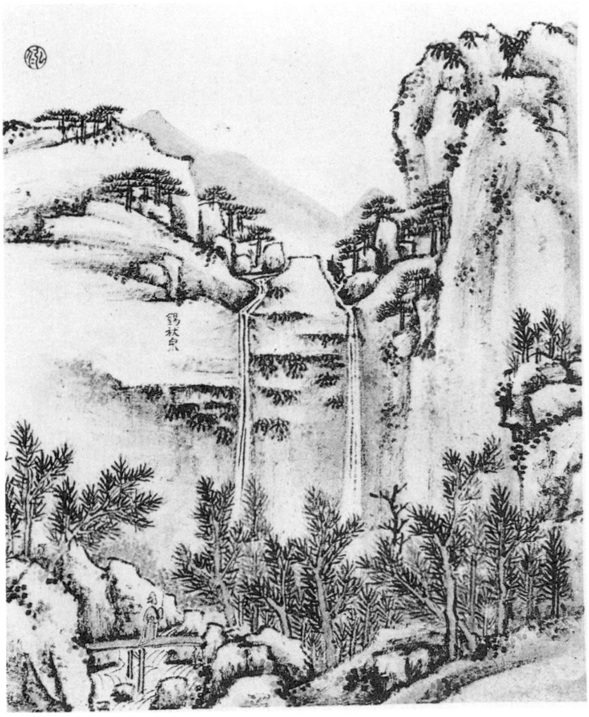

100. Hung-jen: "Mt. Huang." Leaf no. 36..

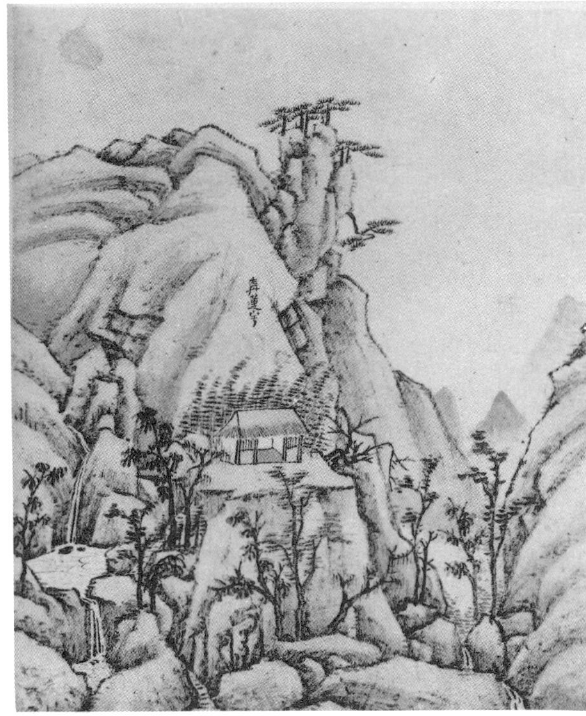

101. Hung-jen: "Mt. Huang." Leaf no. 37.

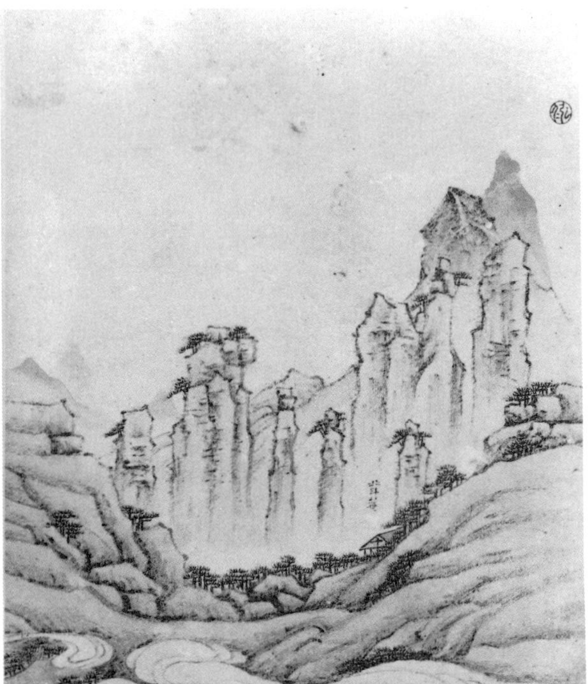

102. Hung-jen: "Mt. Huang." Leaf no. 38.

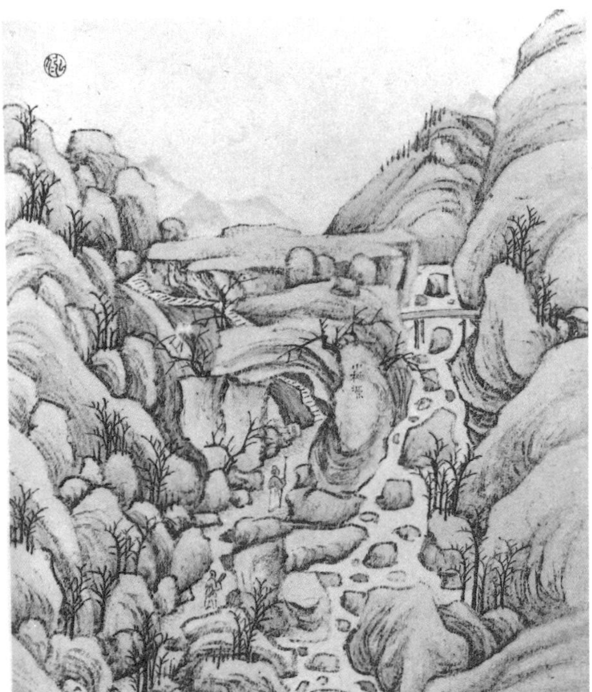

103. Hung-jen: "Mt. Huang." Leaf no. 39.

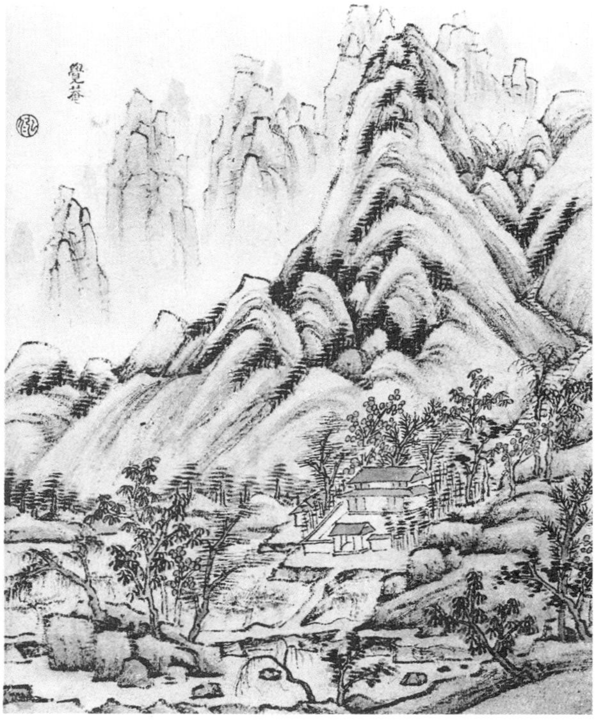

104. Hung-jen: "Mt. Huang." Leaf no. 40.

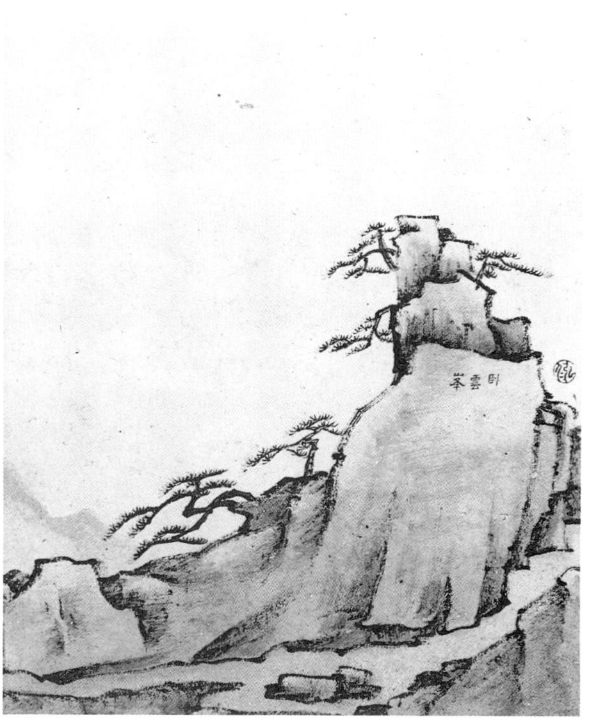

105. Hung-jen: "Mt. Huang." Leaf no. 41.

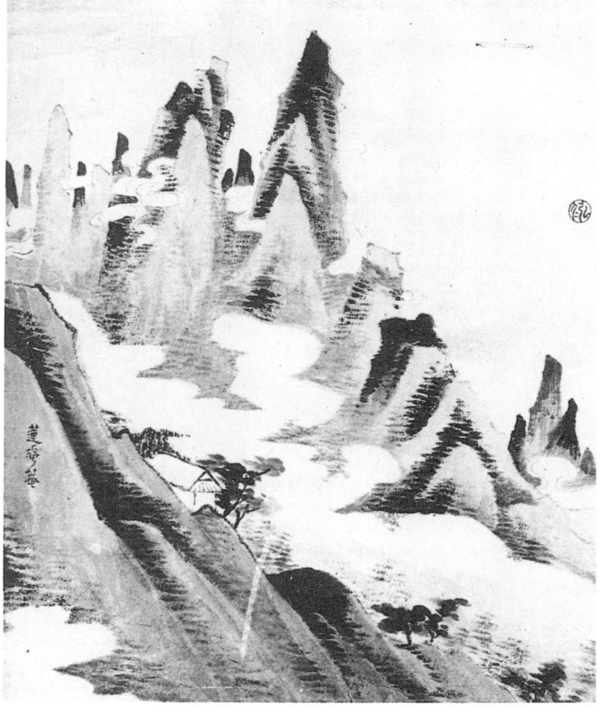

106. Hung-jen: "Mt. Huang." Leaf no. 42.

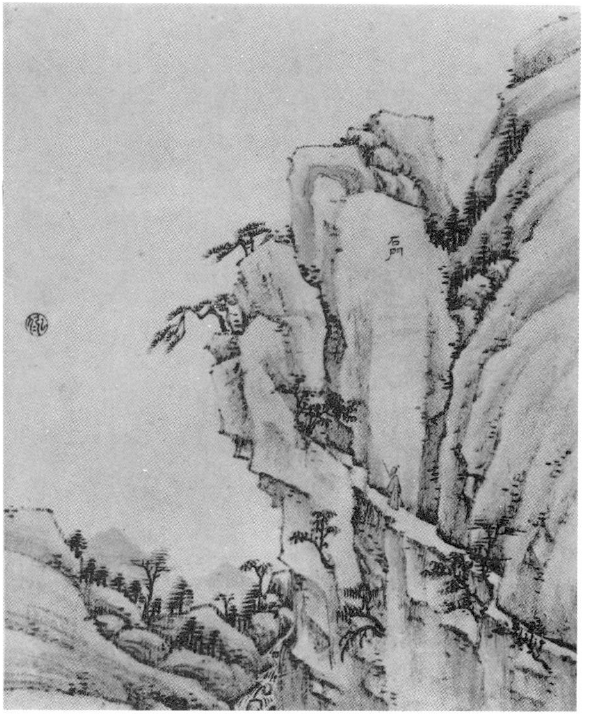

107. Hung-jen: "Mt. Huang." Leaf no. 43.

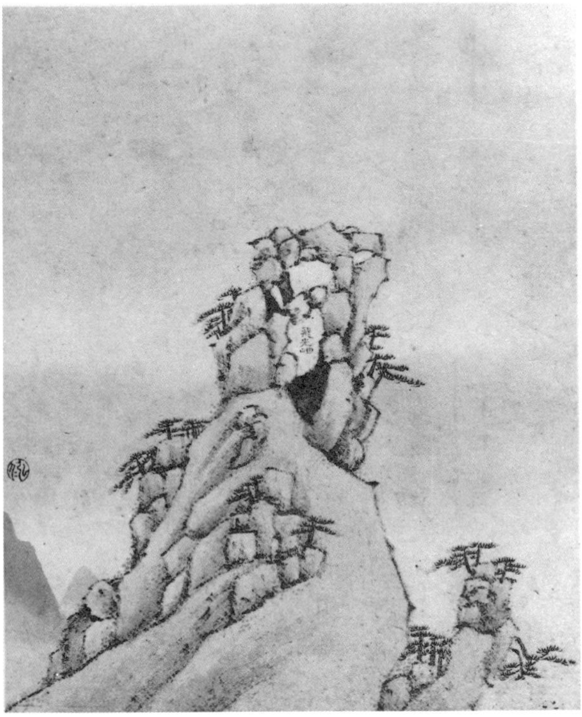

108. Hung-jen: "Mt. Huang." Leaf no. 44.

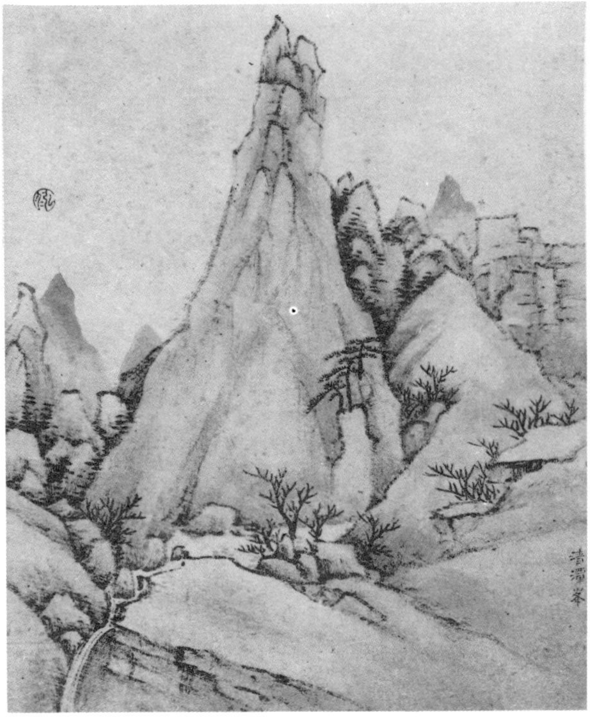

109. Hung-jen: "Mt. Huang." Leaf no. 45.

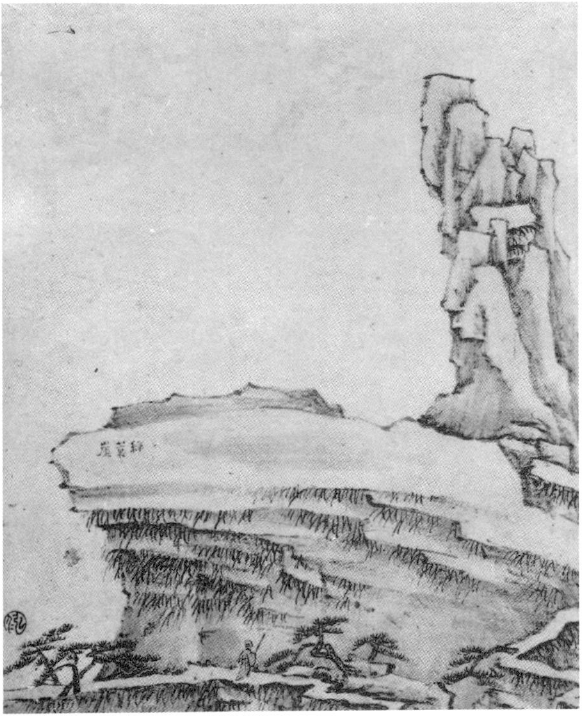

110. Hung-jen: "Mt. Huang." Leaf no. 46.

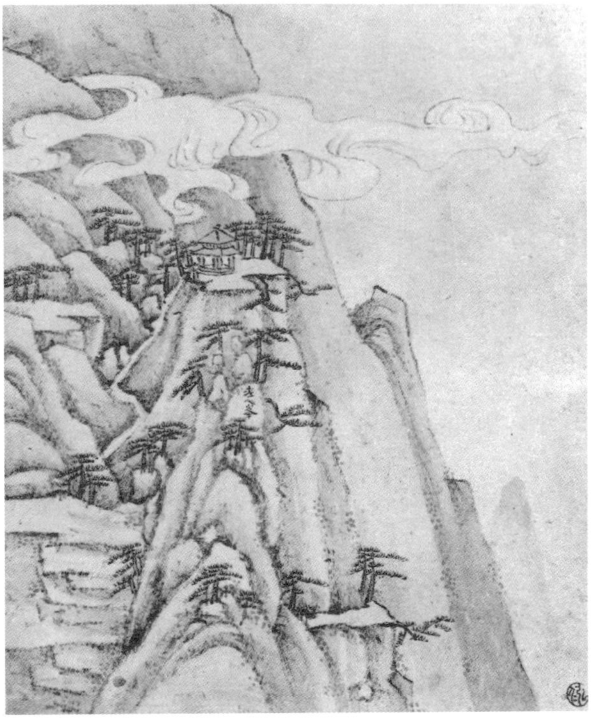

111. Hung-jen: "Mt. Huang." Leaf no. 47.

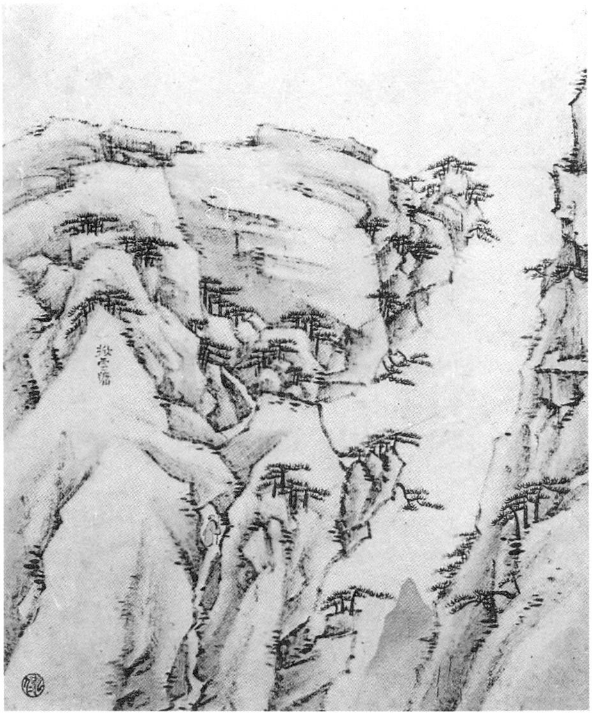

112. Hung-jen: "Mt. Huang." Leaf no. 48.

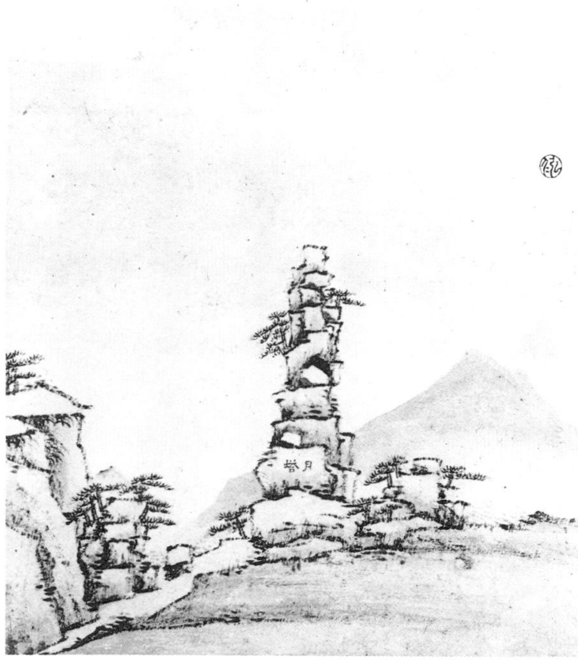

113. Hung-jen: "Mt. Huang." Leaf no. 49.

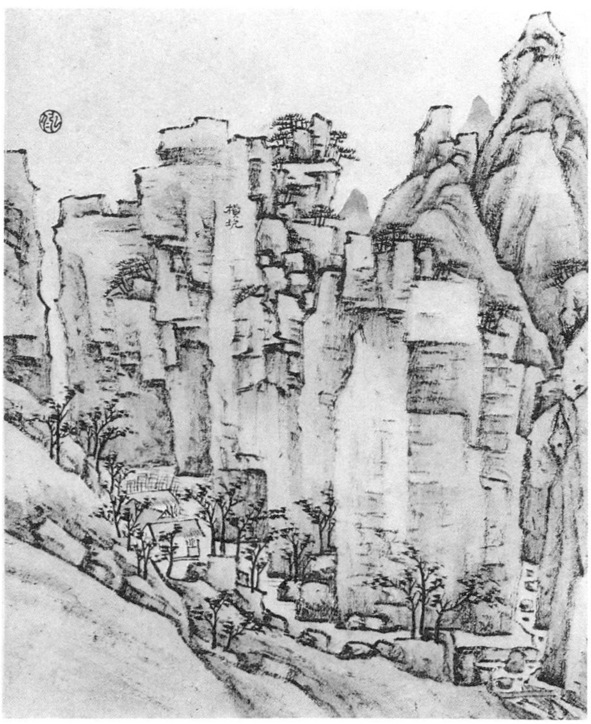

114. Hung-jen: "Mt. Huang." Leaf no. 50.

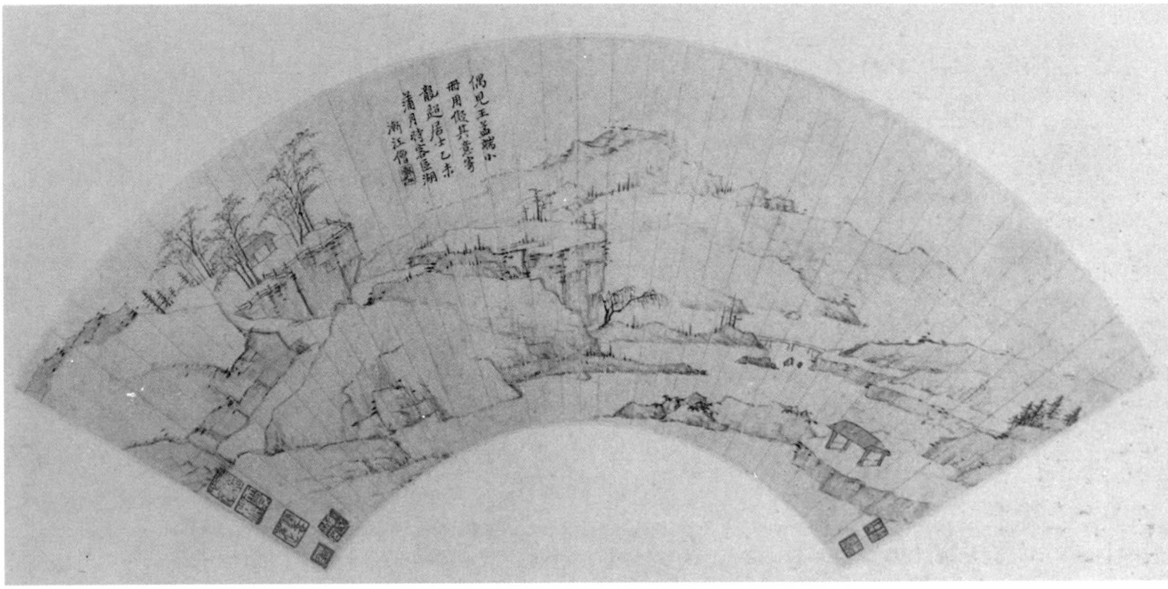

115. Hung-jen: "Landscape," 1655. Fan. Liu Tso-ch'ou collection, Hong Kong.

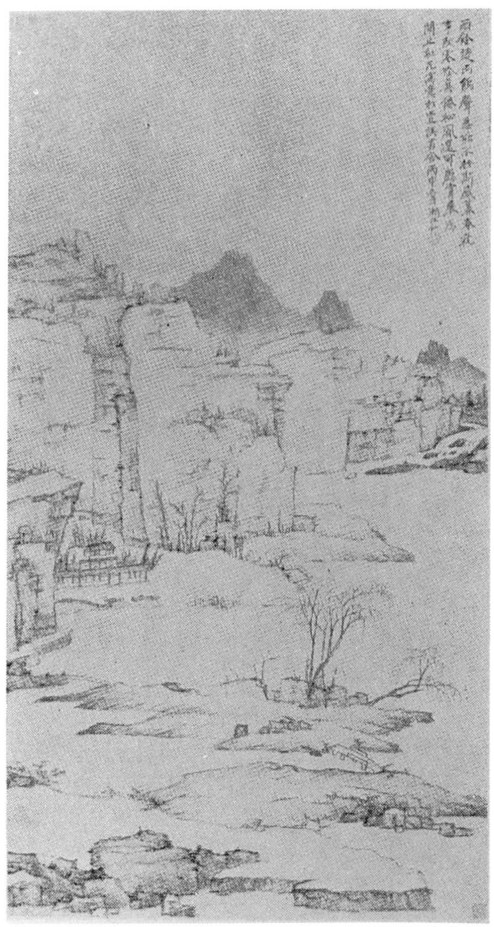

116. Hung-jen: "Landscape," 1656. Hanging scroll. Shanghai Museum.

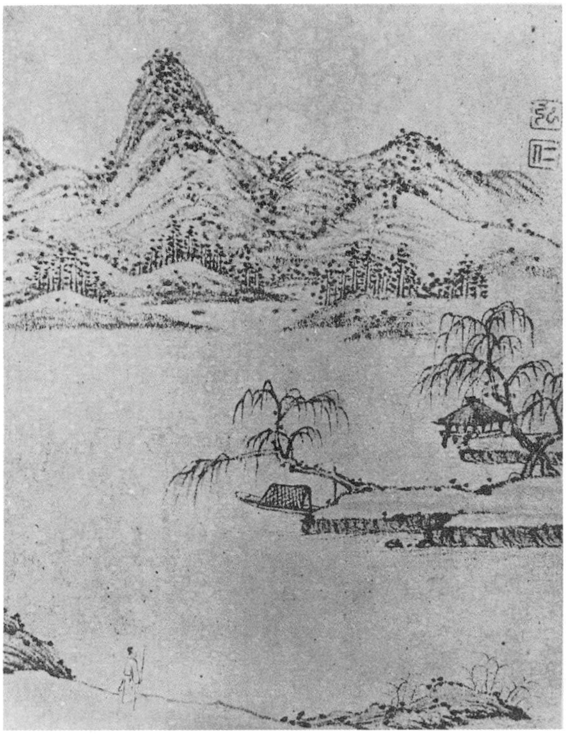

117. Hung-jen: "Album of Landscapes." Undated, 16.1×12.9 cm. each. Palace Museum, Peking. Leaf no. 1.

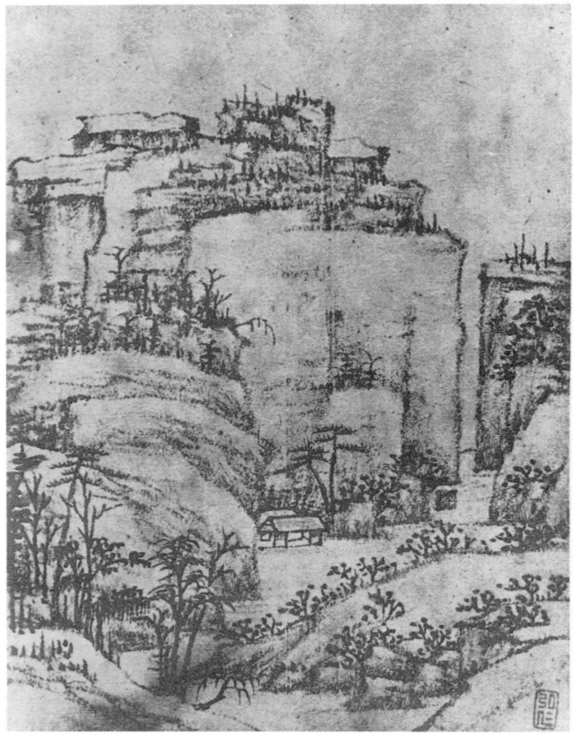

118. Hung-jen: "Album of Landscapes." Leaf no. 2.

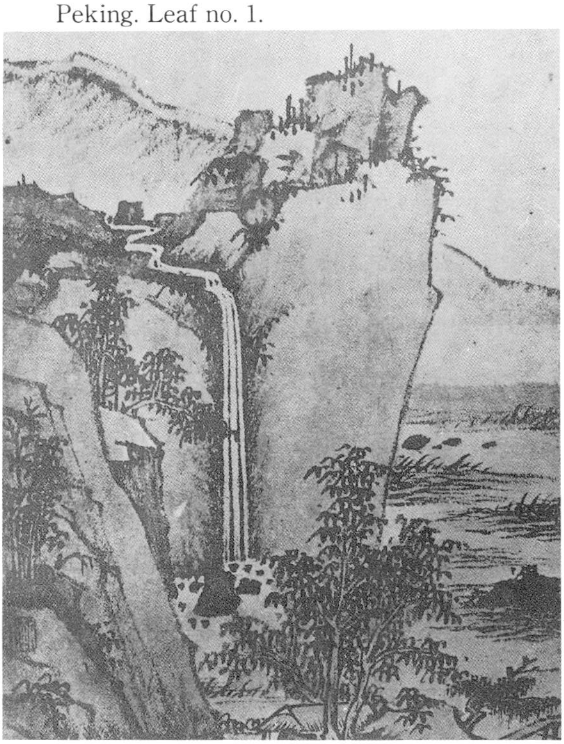

119. Hung-jen: "Album of Landscapes." Leaf no. 3.

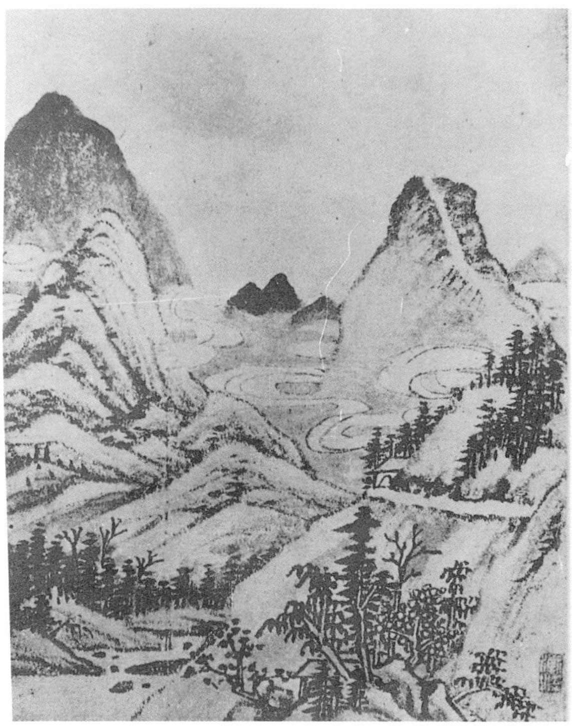

120. Hung-jen: "Album of Landscapes." Leaf no. 4.

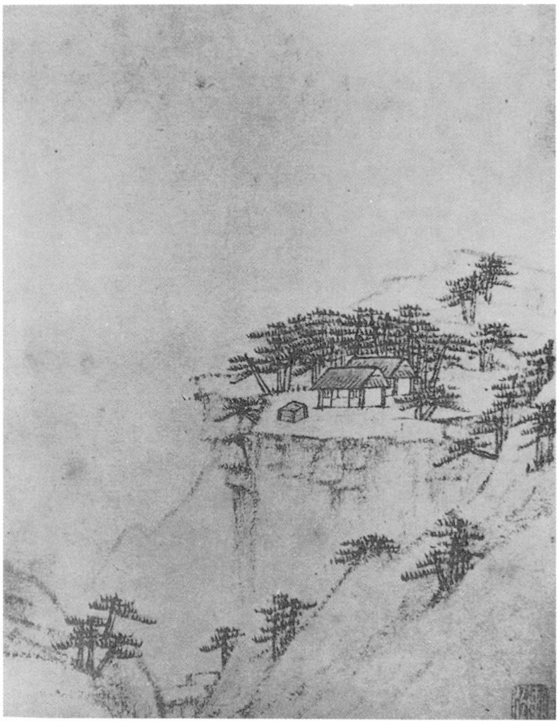

121. Hung-jen: "Album of Landscapes." Leaf no. 5.

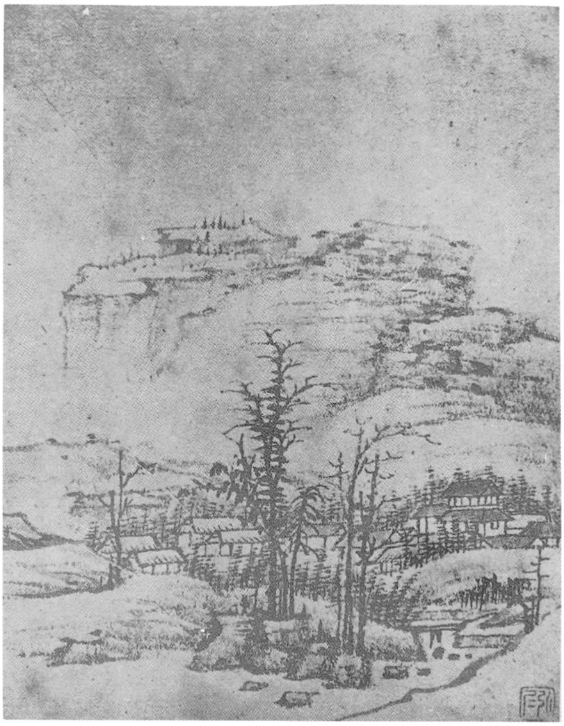

122. Hung-jen: "Album of Landscapes." Leaf no. 6.

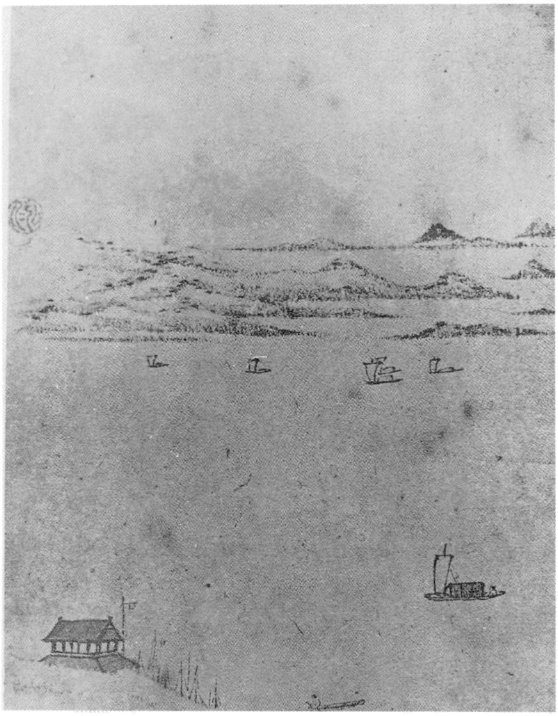

123. Hung-jen: "Album of Landscapes." Leaf no. 7.

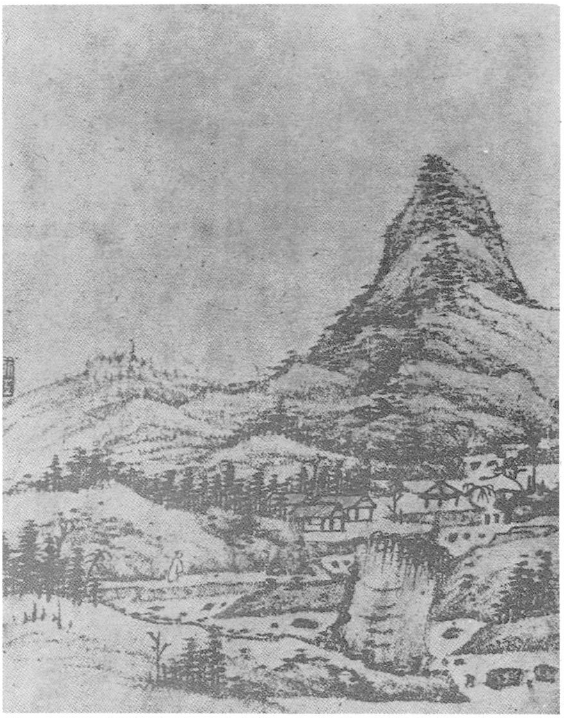

124. Hung-jen: "Album of Landscapes." Leaf no. 8.

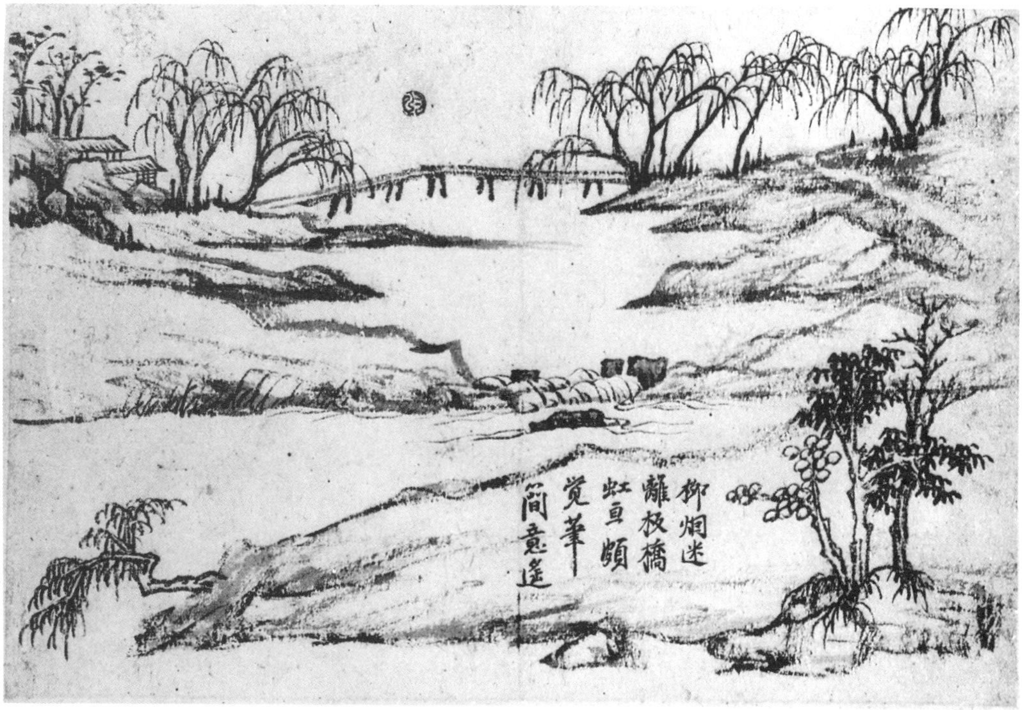

125. Hung-jen: "Album of Landscapes, Plants, and Flowers," 1656. Album, ink on paper, 15×23 cm. each. Drenowatz collection, Zürich. leaf no. 1.

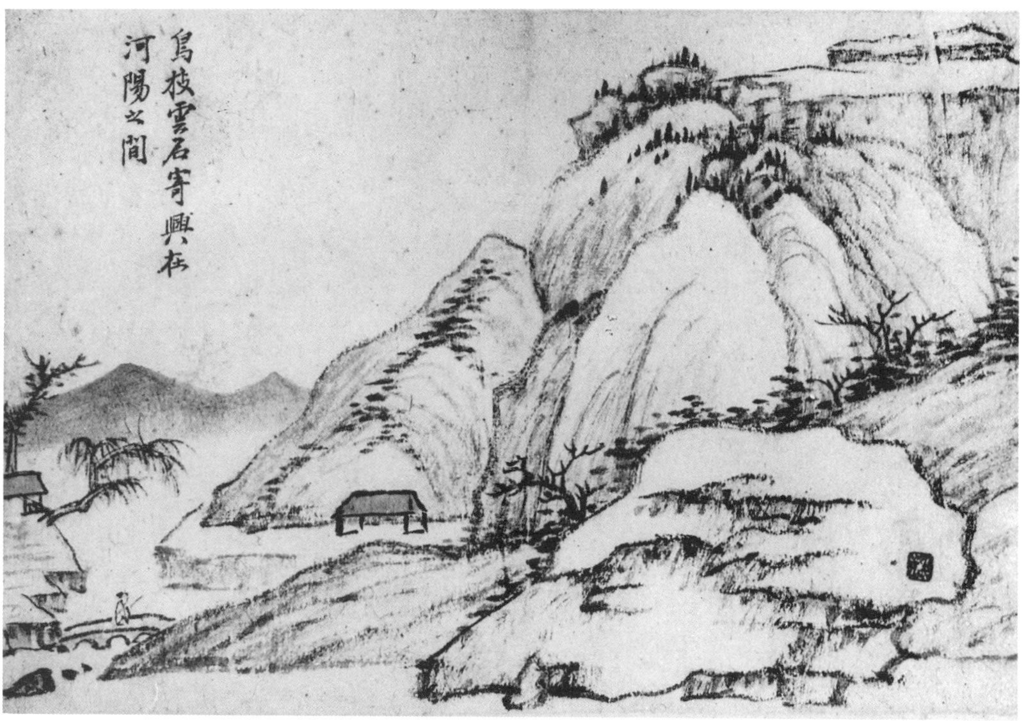

126. Hung-jen: "Album of Landscapes, Plants, and Flowers." Leaf no. 2.

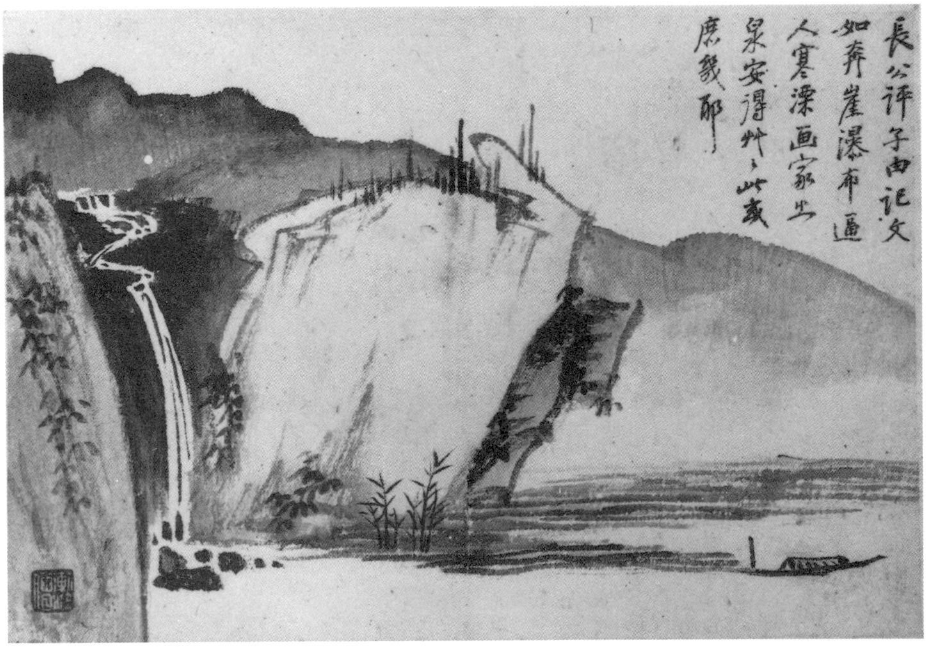

127. Hung-jen: "Album of Landscapes, Plants, and Flowers." Leaf no. 3.

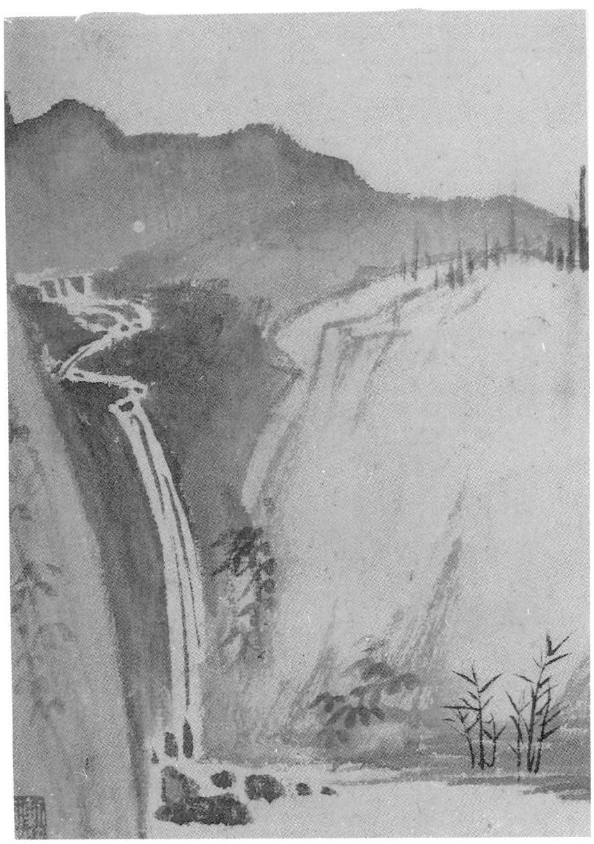

128. Hung-jen: "Album of Landscapes, Plants, and Flowers." Leaf no. 3. Detail.

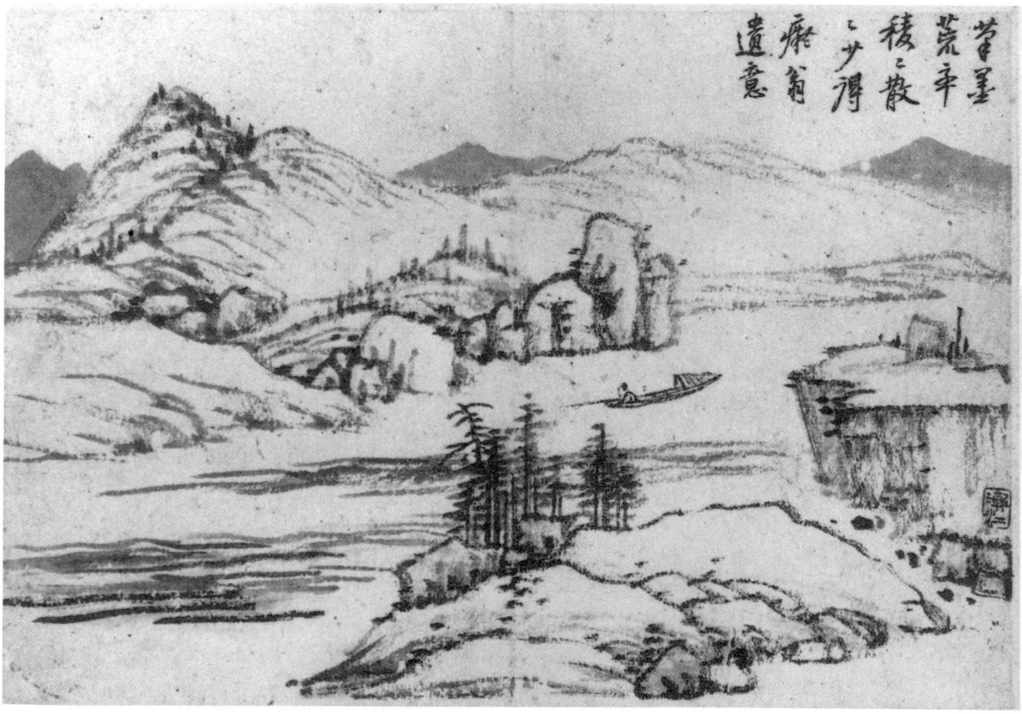

129. Hung-jen: "Album of Landscapes, Plants, and Flowers." Leaf no. 4.

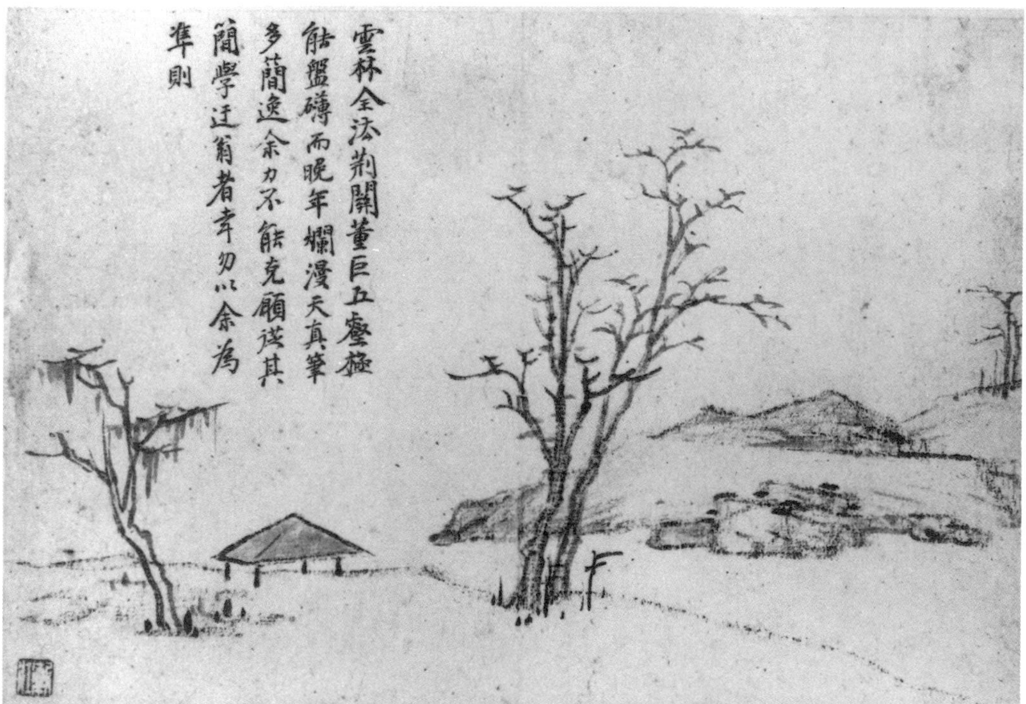

130. Hung-jen: "Album of Landscapes, Plants, and Flowers." Leaf no. 5.

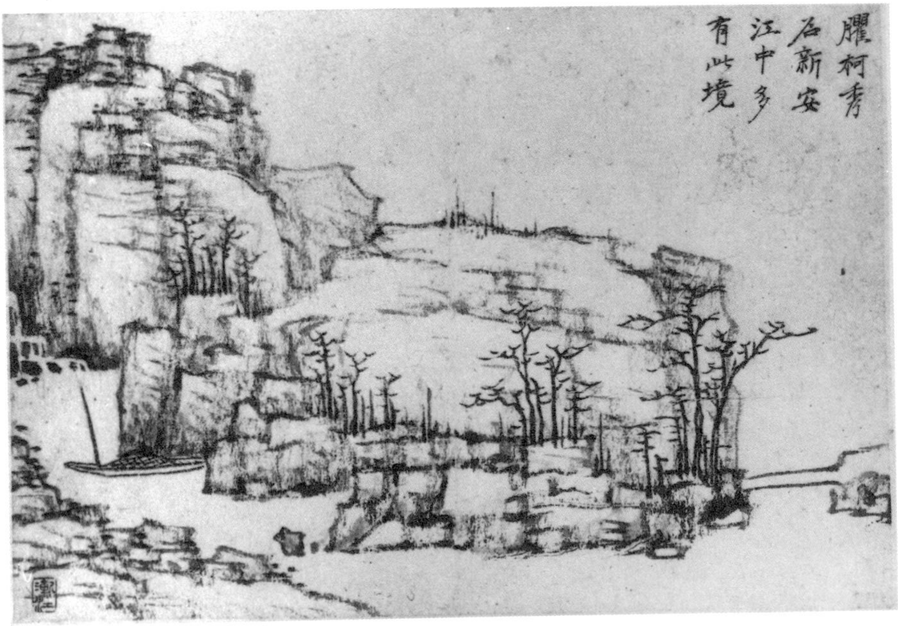

131. Hung-jen: "Album of Landscapes, Plants, and Flowers." Leaf no. 6.

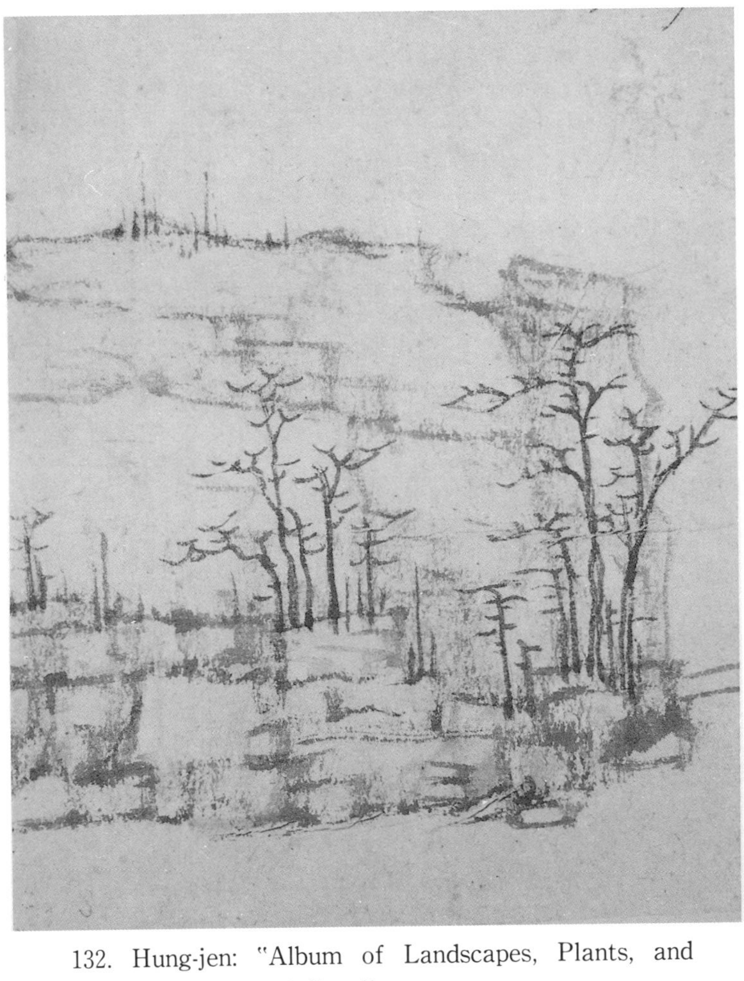

132. Hung-jen: "Album of Landscapes, Plants, and Flowers." Leaf no. 6. Detail.

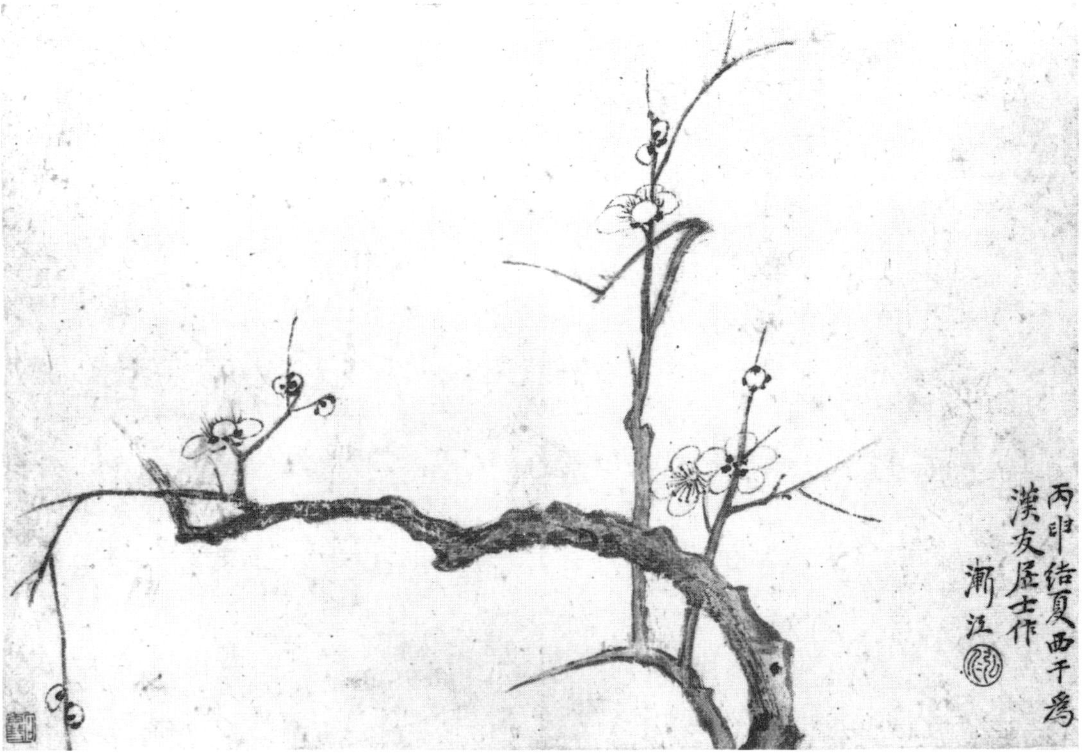

133. Hung-jen: "Album of Landscapes, Plants, and Flowers." Leaf no. 7.

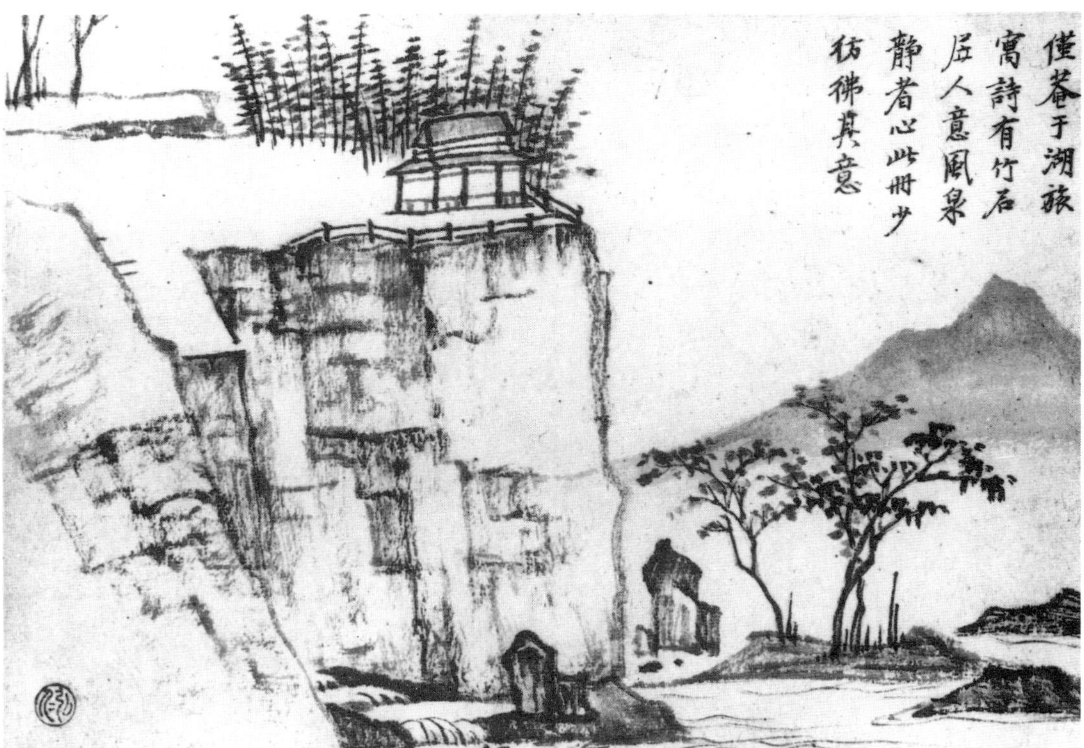

134. Hung-jen: "Album of Landscapes, Plants, and Flowers." Leaf no. 8.

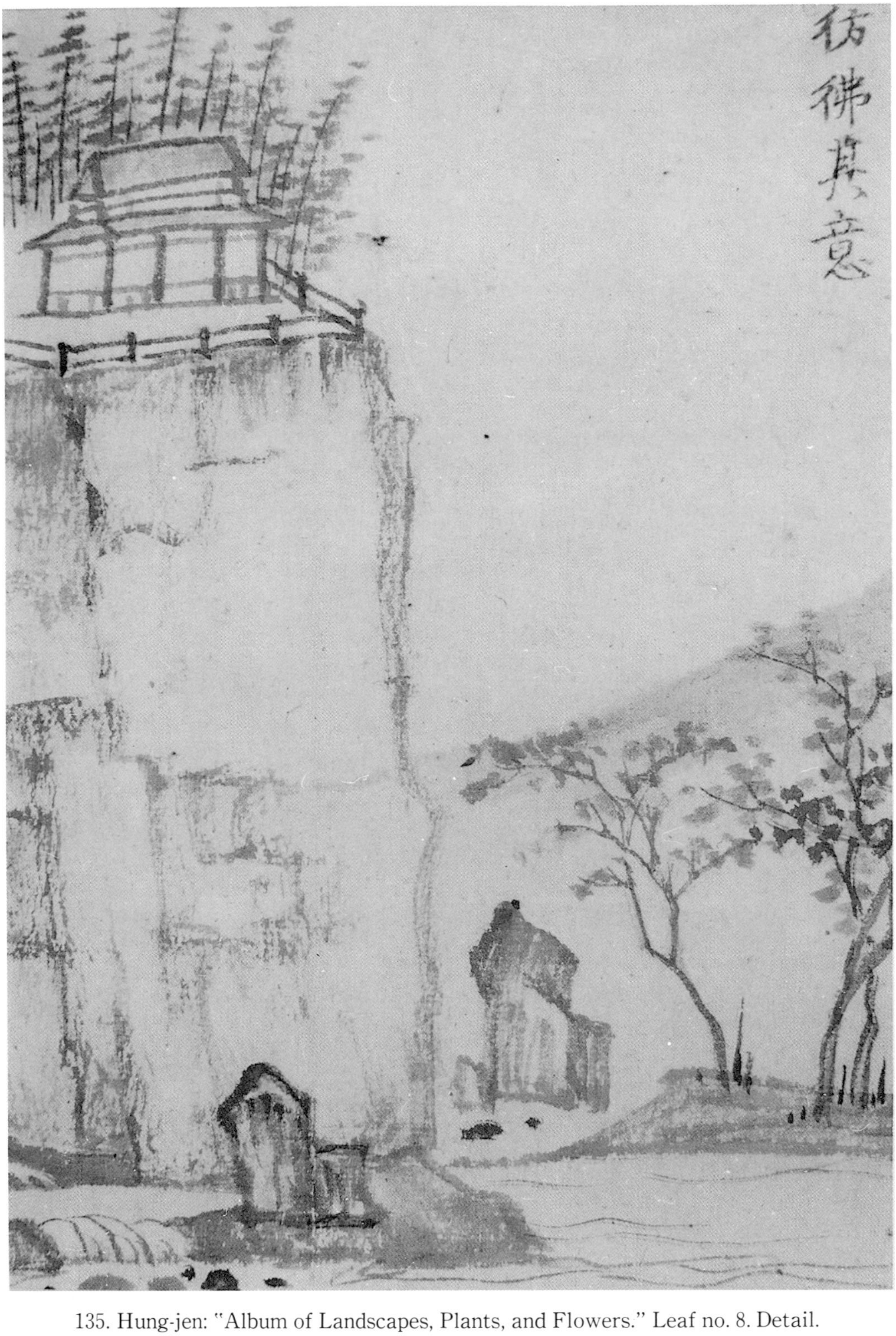

135. Hung-jen: "Album of Landscapes, Plants, and Flowers." Leaf no. 8. Detail.

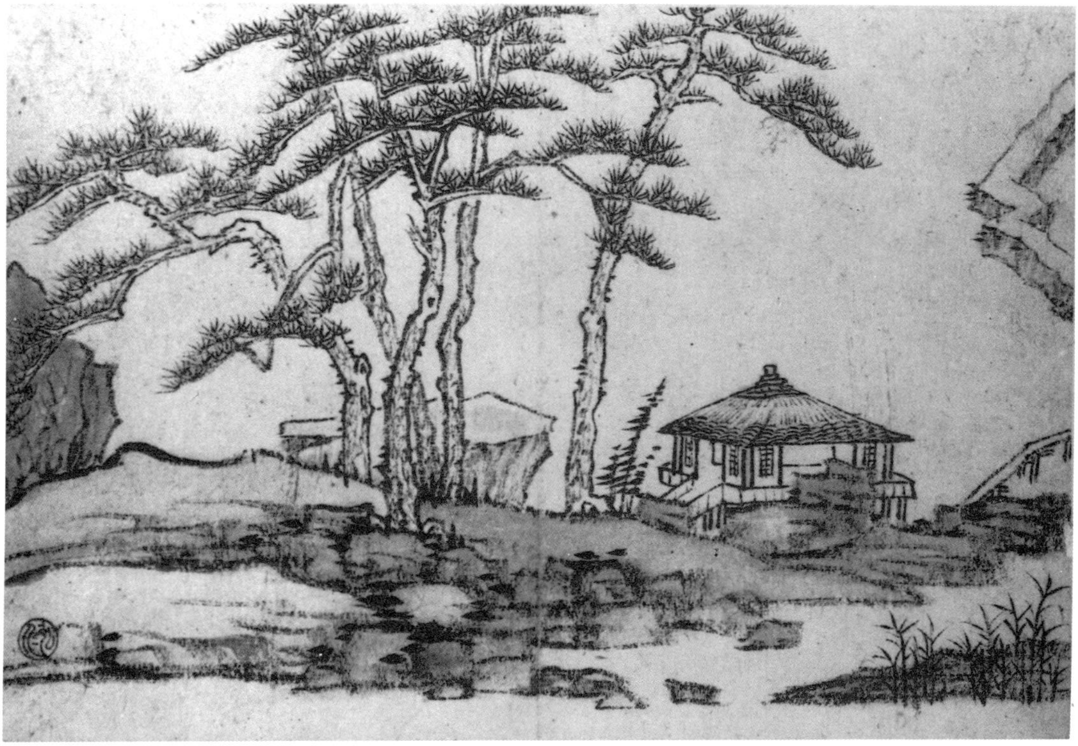

136. Hung-jen: "Album of Landscapes, Plants, and Flowers." Leaf no. 9.

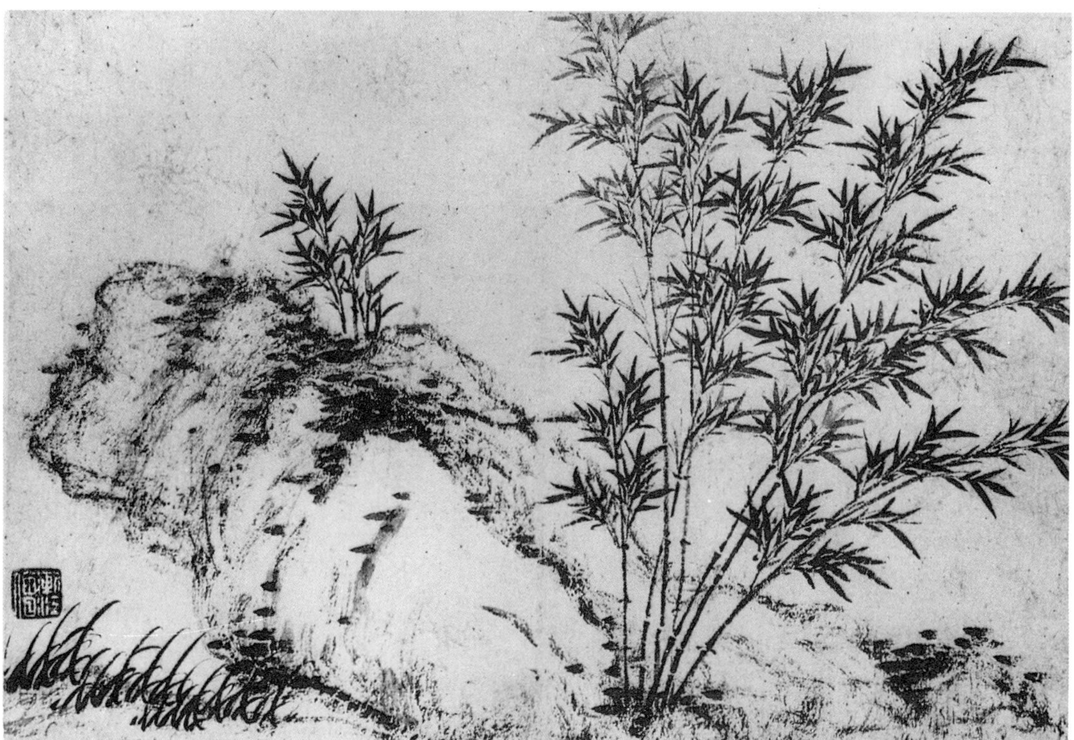

137. Hung-jen: "Album of Landscapes, Plants, and Flowers." Leaf no. 10.

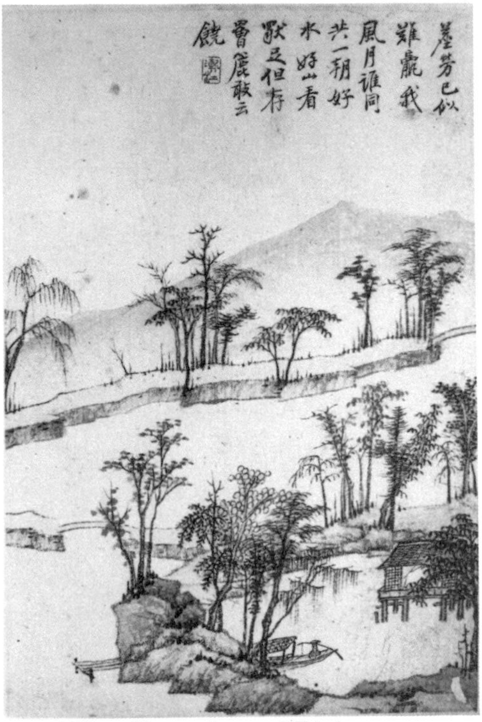

138. Hung-jen: "Fishermen Album," 1656. Album, ink and colors on paper, 23×16 cm. each. Franco Vannotti collection, Lugano. Leaf no. 1.

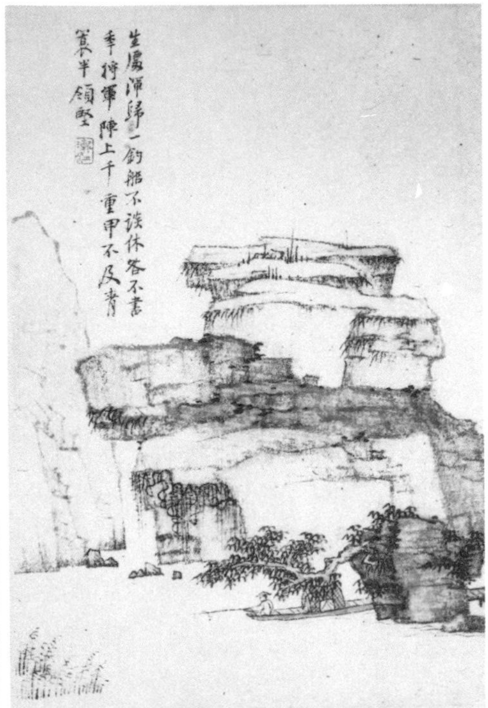

139. Hung-jen: "Fishermen Album." Leaf no. 2.

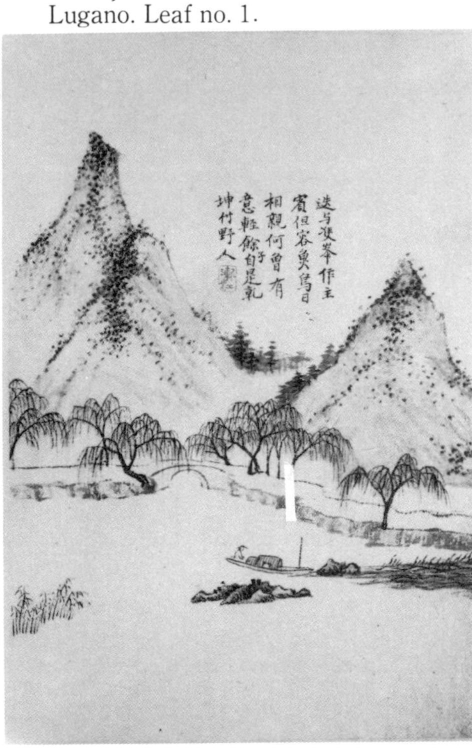

140. Hung-jen: "Fishermen Album." Leaf no. 3.

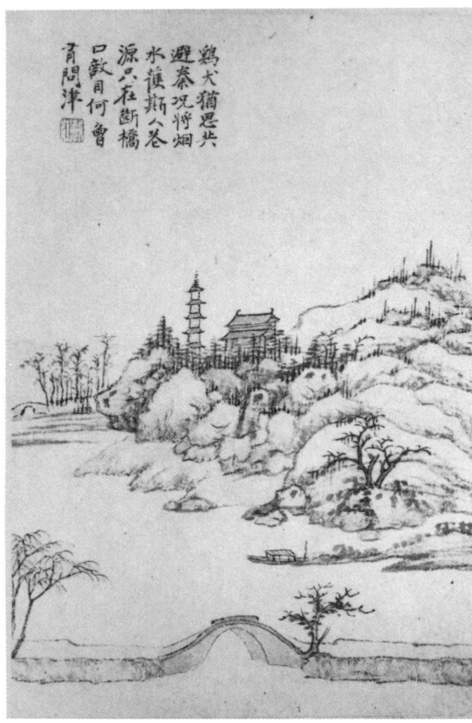

141. Hung-jen: "Fishermen Album." Leaf no. 4.

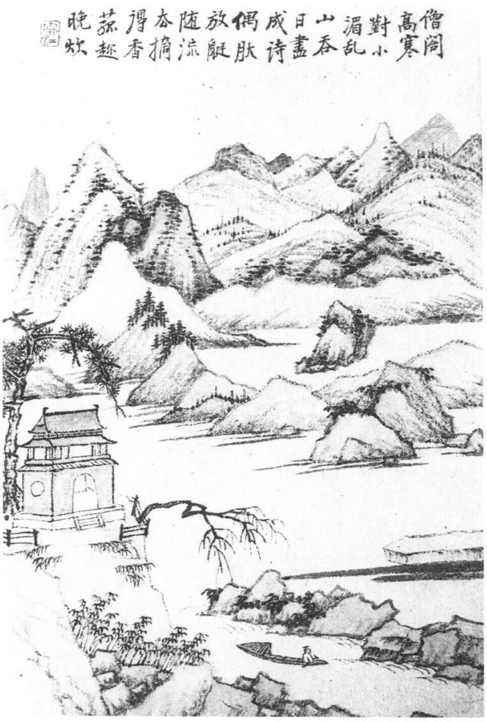

142. Hung-jen: "Fishermen Album." Leaf no. 5.

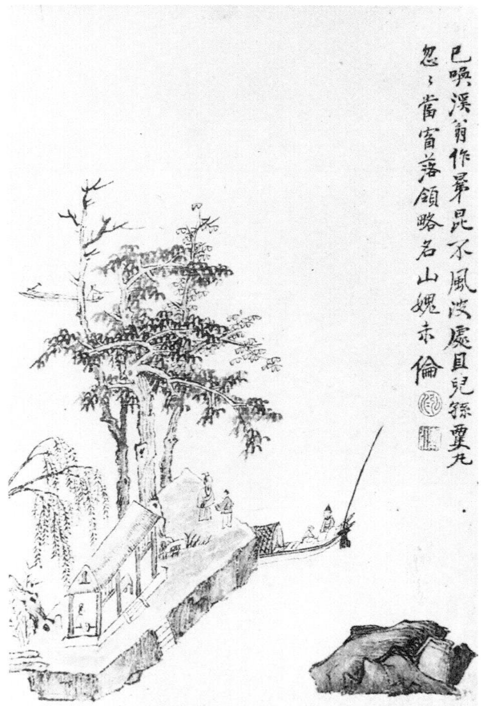

143. Hung-jen: "Fishermen Album." Leaf no. 6.

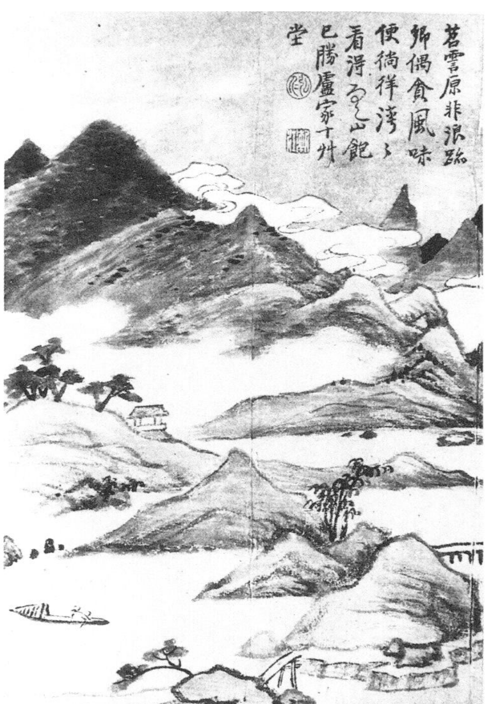

144. Hung-jen: "Fishermen Album." Leaf no. 7.

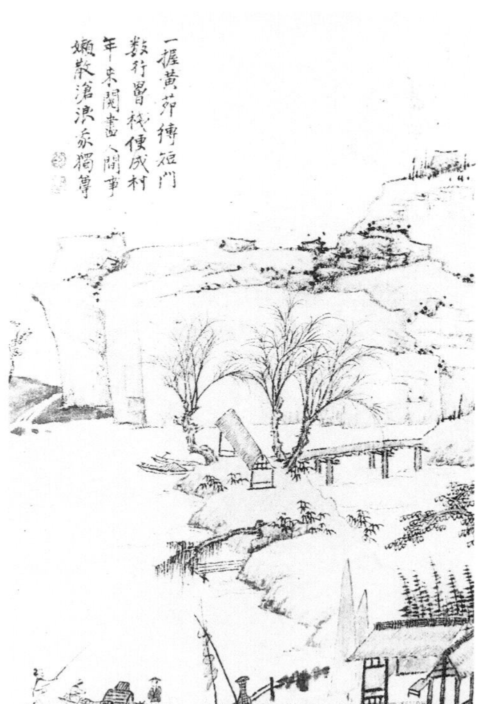

145. Hung-jen: "Fishermen Album." Leaf no. 8.

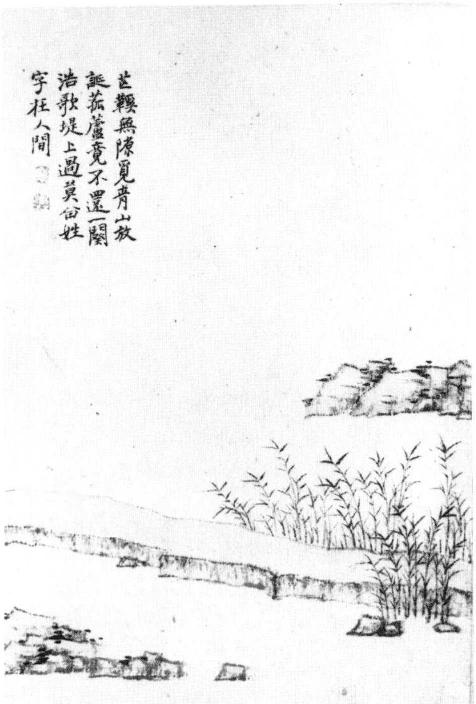

146. Hung-jen: "Fishermen Album." Leaf no. 9.

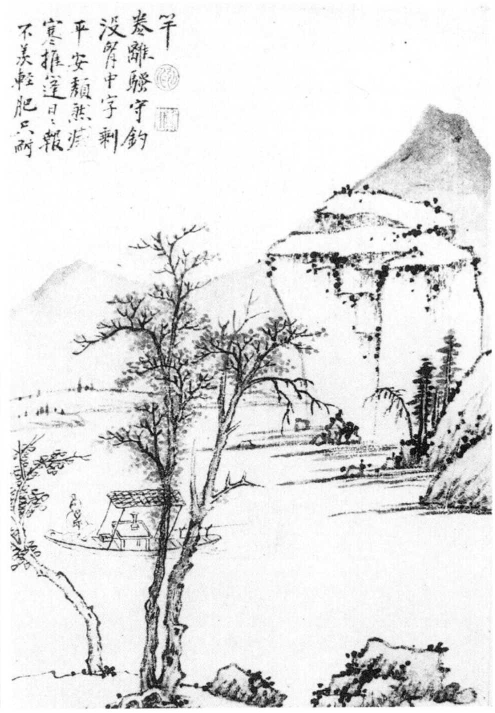

147. Hung-jen: "Fishermen Album." Leaf no. 10.

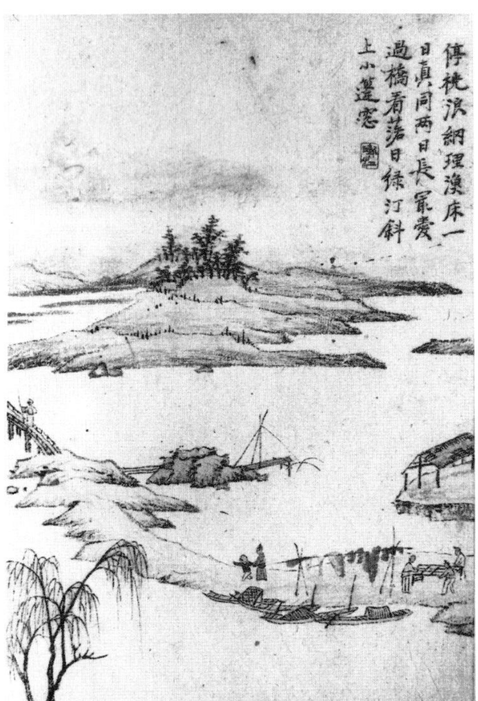

148. Hung-jen: "Fishermen Album." Leaf no. 11.

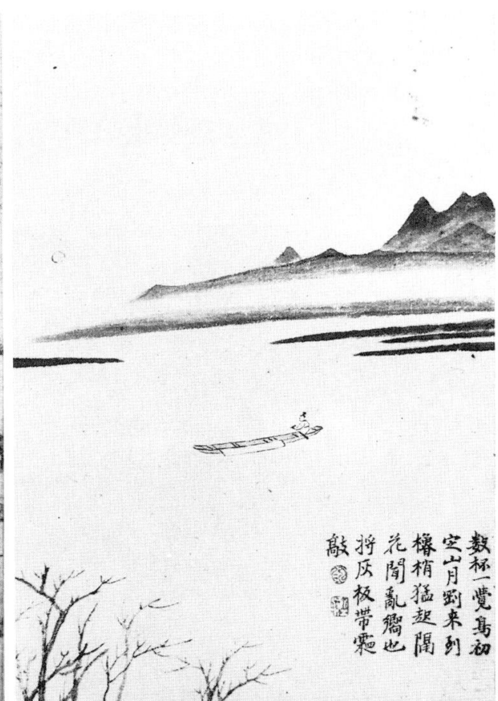

149. Hung-jen: "Fishermen Album." Leaf no. 12.

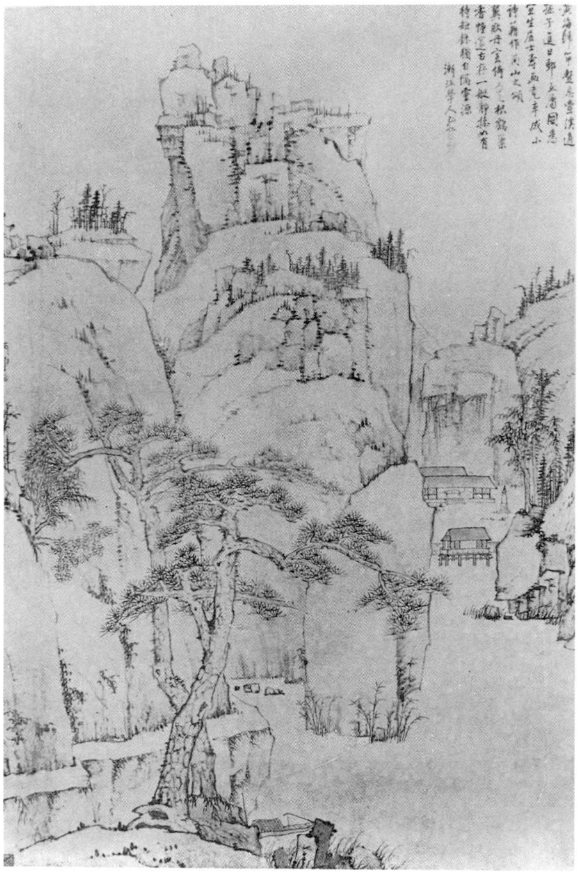

150. Hung-jen: "Landscape." 1656. Hanging scroll, ink and colors on paper, 147×102 cm. Chih-lo-lou collection, Hong Kong.

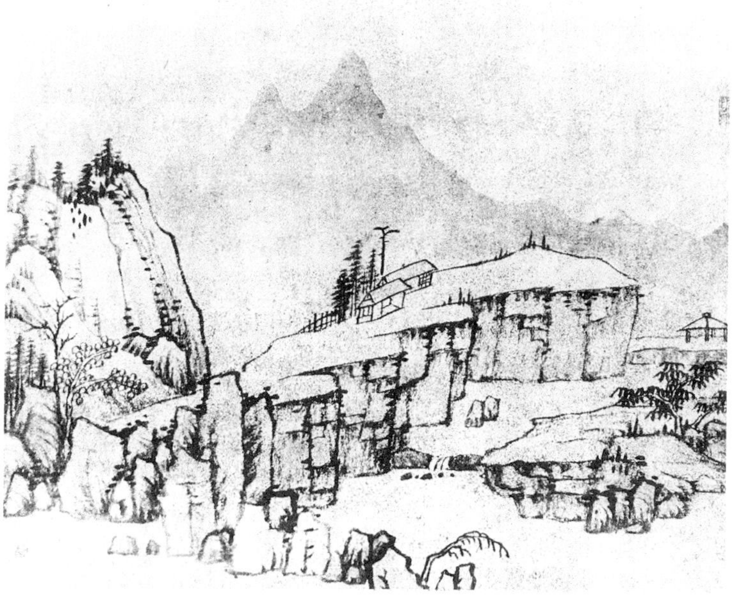

151. Hung-jen: "Landscape Album." Album, ink on gold paper. Palace Museum, Peking. Leaf no. 1.

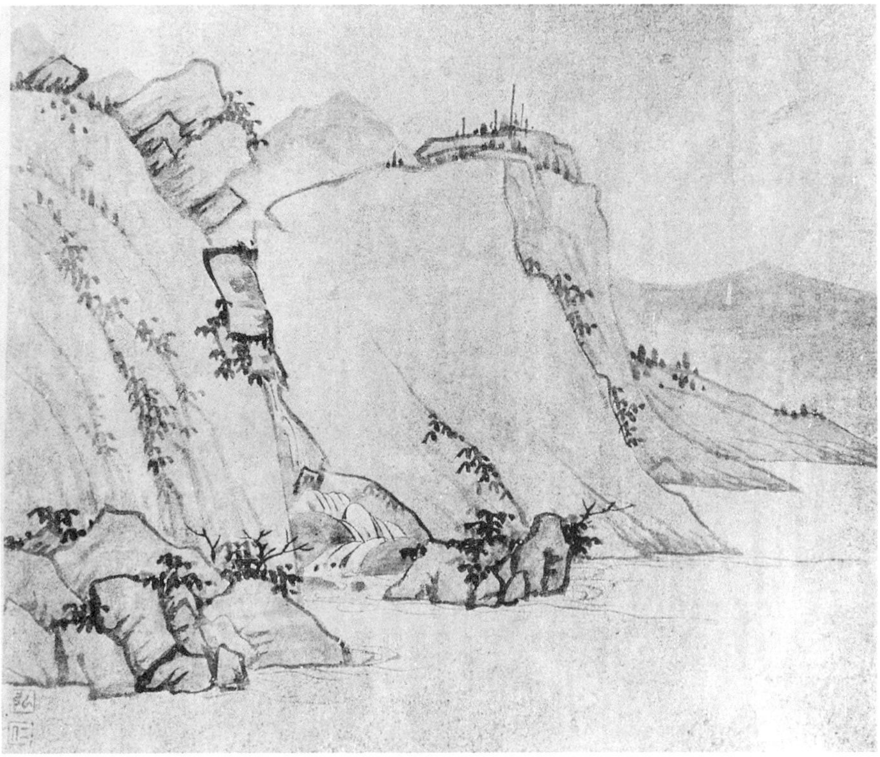

152. Hung-jen: "Landscape Album." Leaf no. 2.

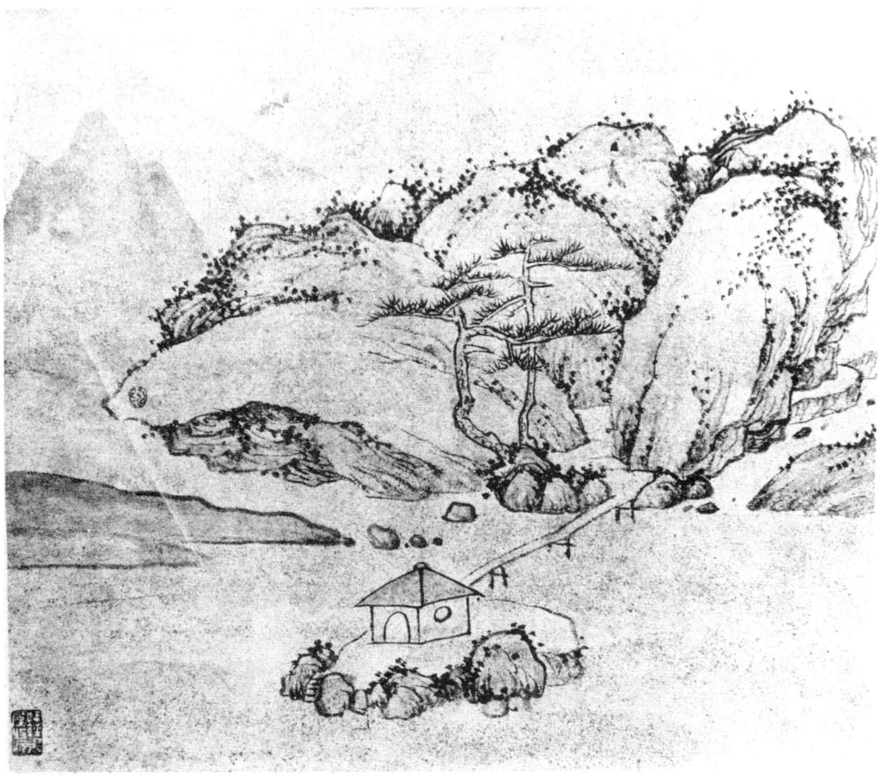

153. Hung-jen: "Landscape Album." Leaf no. 3.

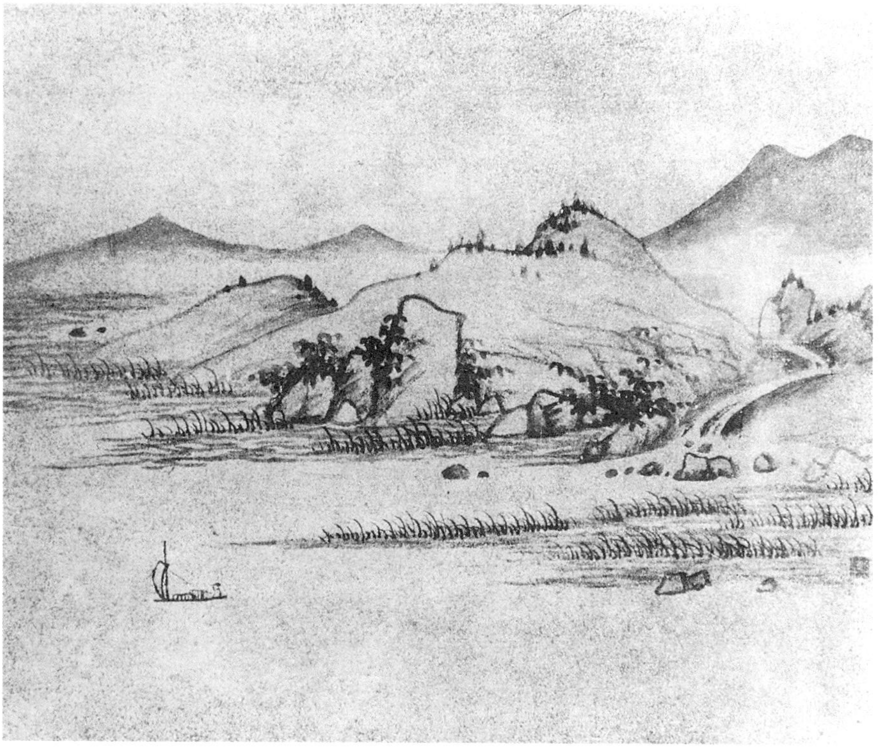

154. Hung-jen: "Landscape Album." Leaf no. 4.

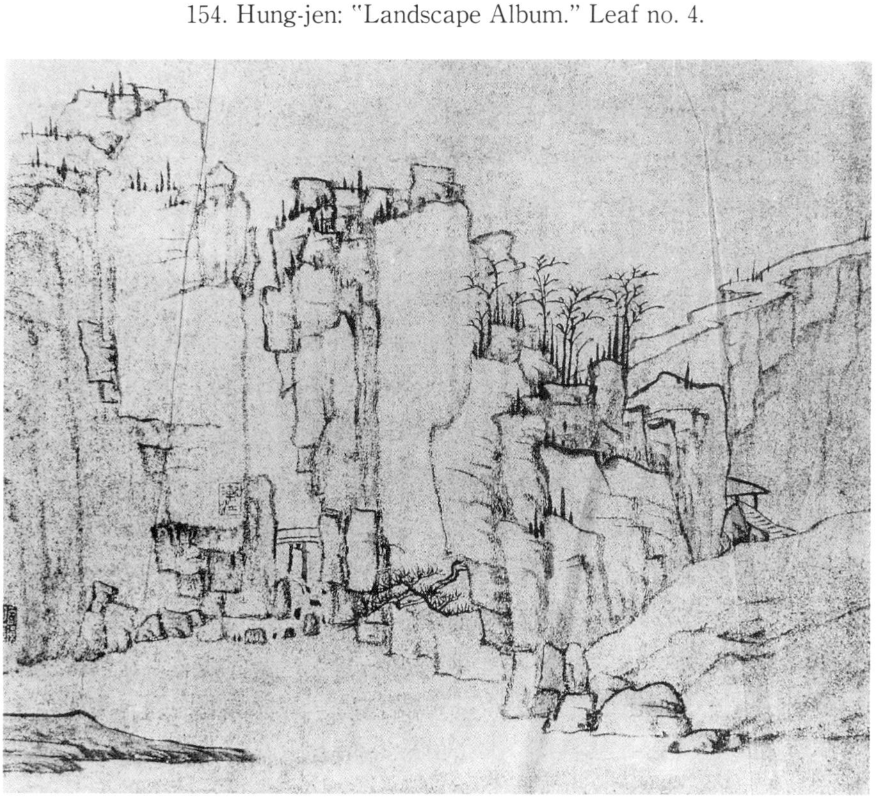

155. Hung-jen: "Landscape Album." Leaf no. 5.

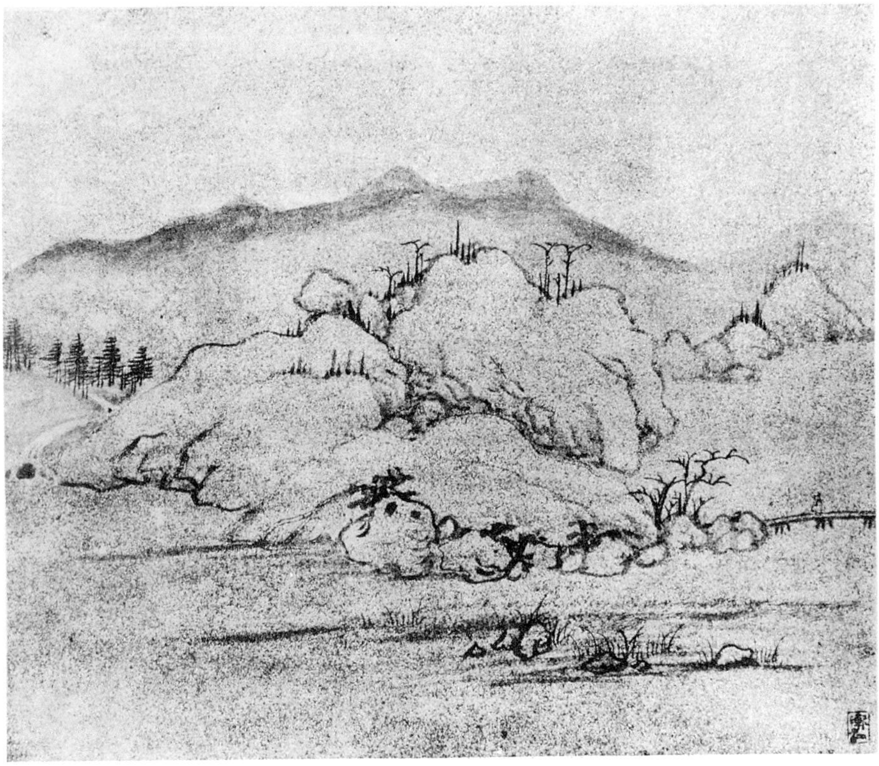

156. Hung-jen: "Landscape Album." Leaf no. 6.

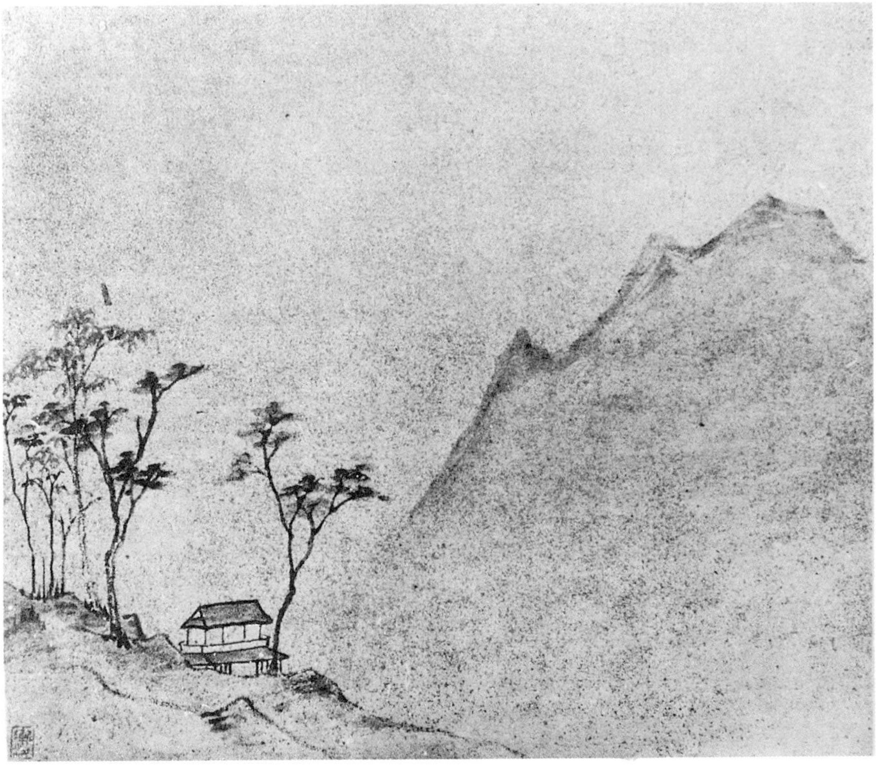

157. Hung-jen: "Landscape Album." Leaf no. 7.

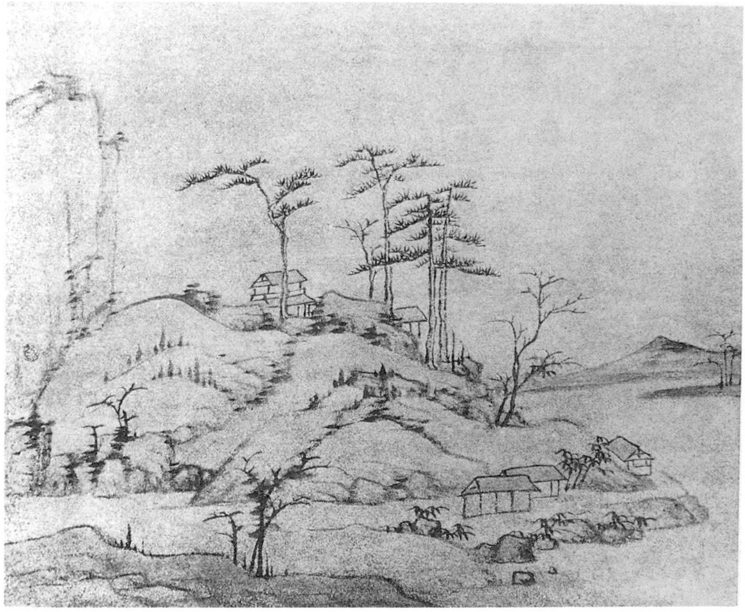

158. Hung-jen: "Landscape Album." Leaf no. 8.

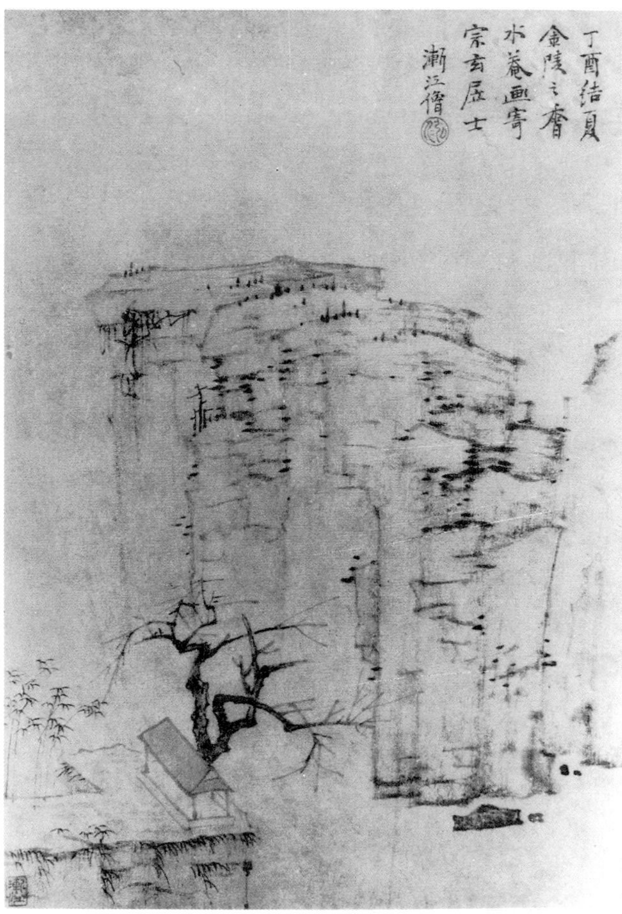

159. Hung-jen: "Landscape Album." 1657. Album. Whereabouts unknown. Leaf no. 1.

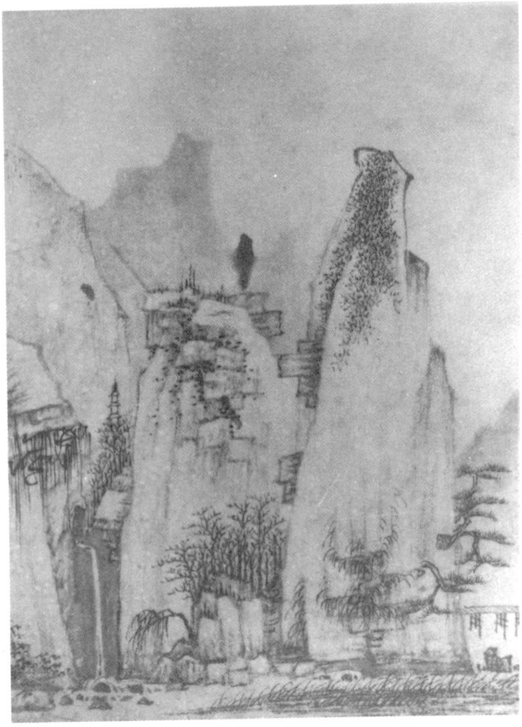

160. Hung-jen: "Landscape Album." Leaf no. 2.

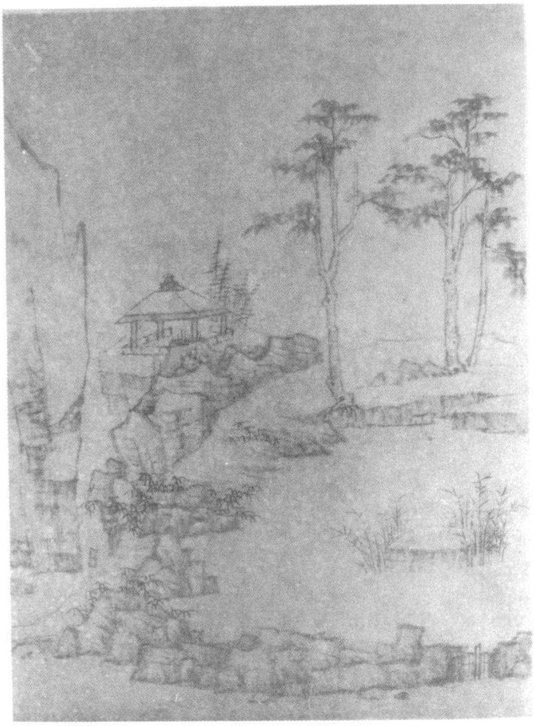

161. Hung-jen: "Landscape Album." Leaf no. 3.

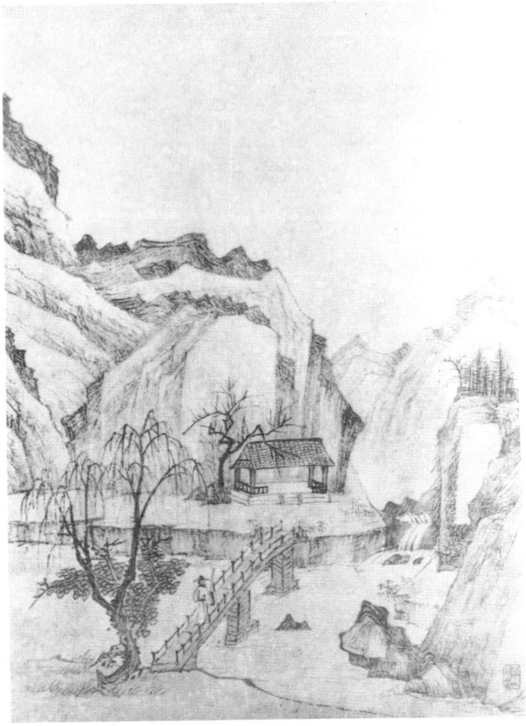

162. Hung-jen: "Landscape Album." Leaf no. 4.

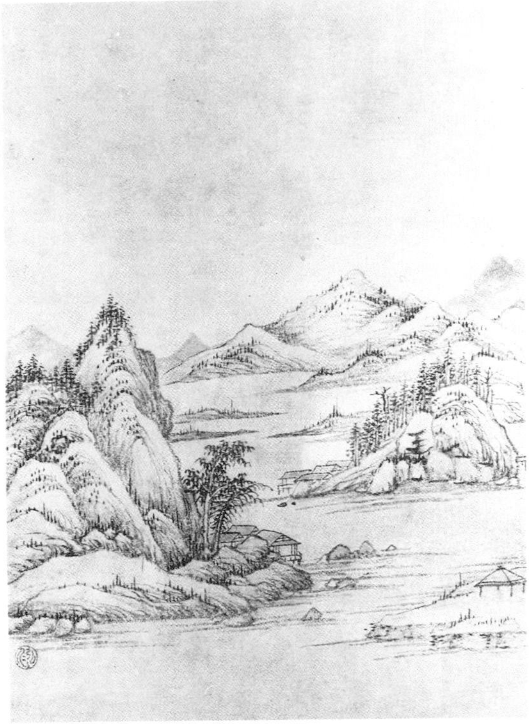

163. Hung-jen: "Landscape Album." Leaf no. 5.

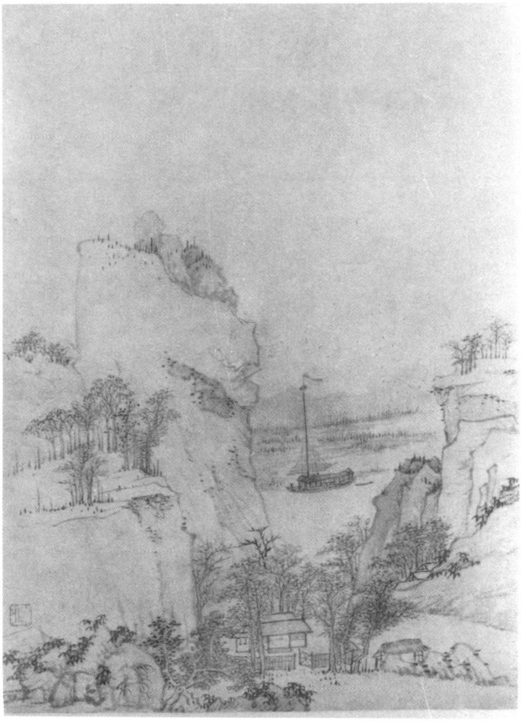

164. Hung-jen: "Landscape Album." Leaf no. 6.

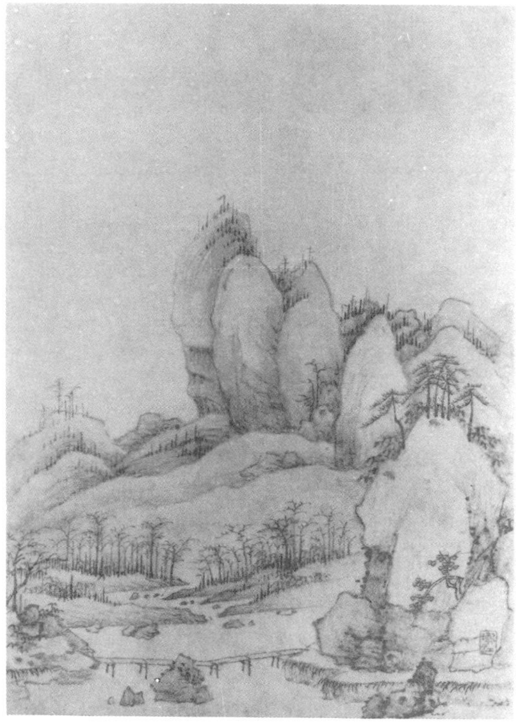

165. Hung-jen: "Landscape Album." Leaf no. 7.

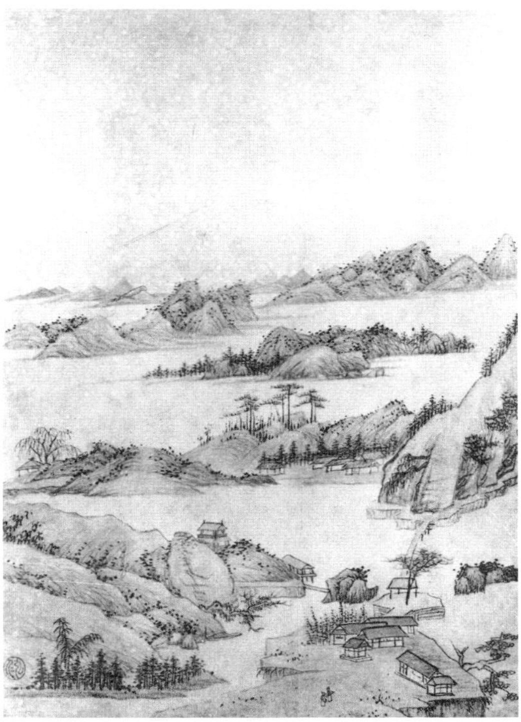

166. Hung-jen: "Landscape Album." Leaf no. 8.

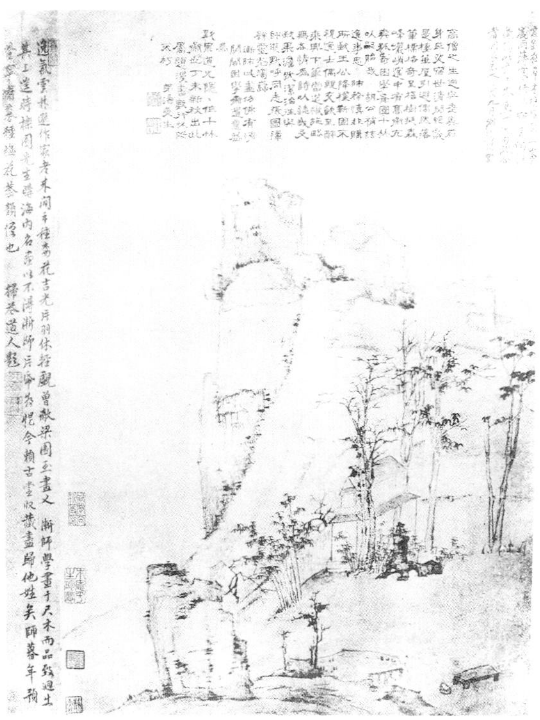

167. Hung-jen: "The Ten-Bamboo Studio," 1657. Hanging scroll , ink on paper. Whereabouts unknown (formerly Ch'eng Mu-an collection).

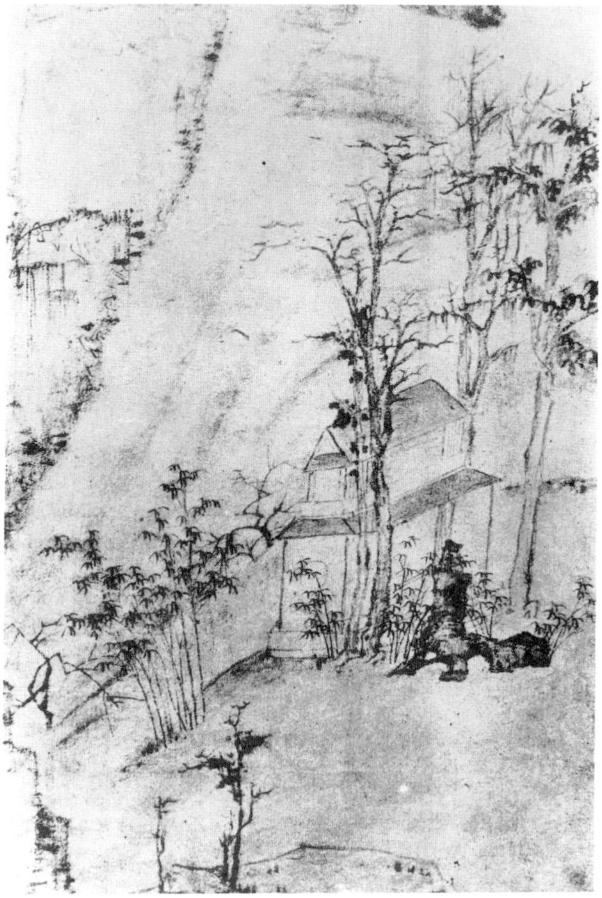

168. Hung-jen: "The Ten-Bamboo Studio." Detail.

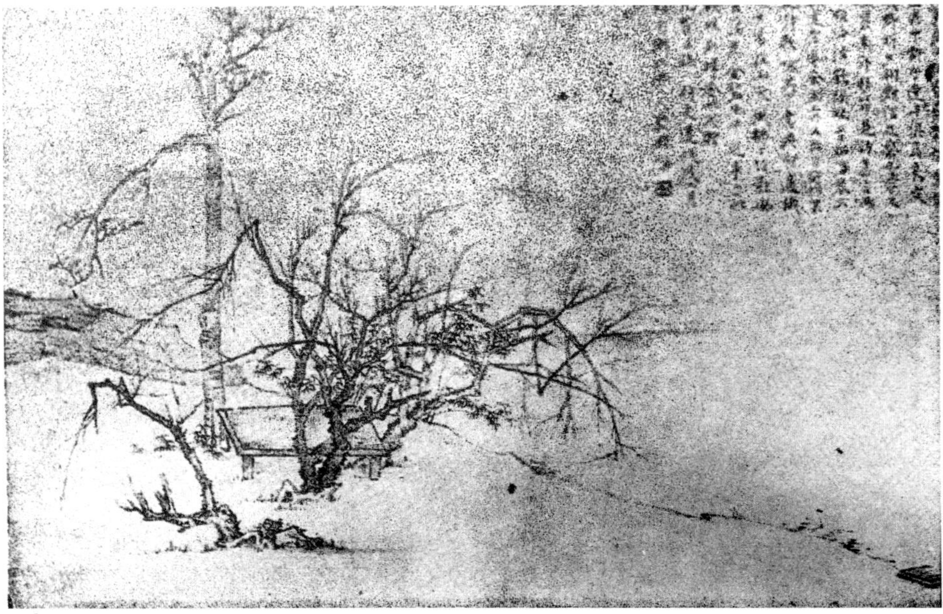

169. Hung-jen: "Plum-Blossom Pavilion," 1658. Short handscroll, ink on paper, 52.5×84.5 cm. Whereabouts unknown (formerly Shen Chien-fu collection).

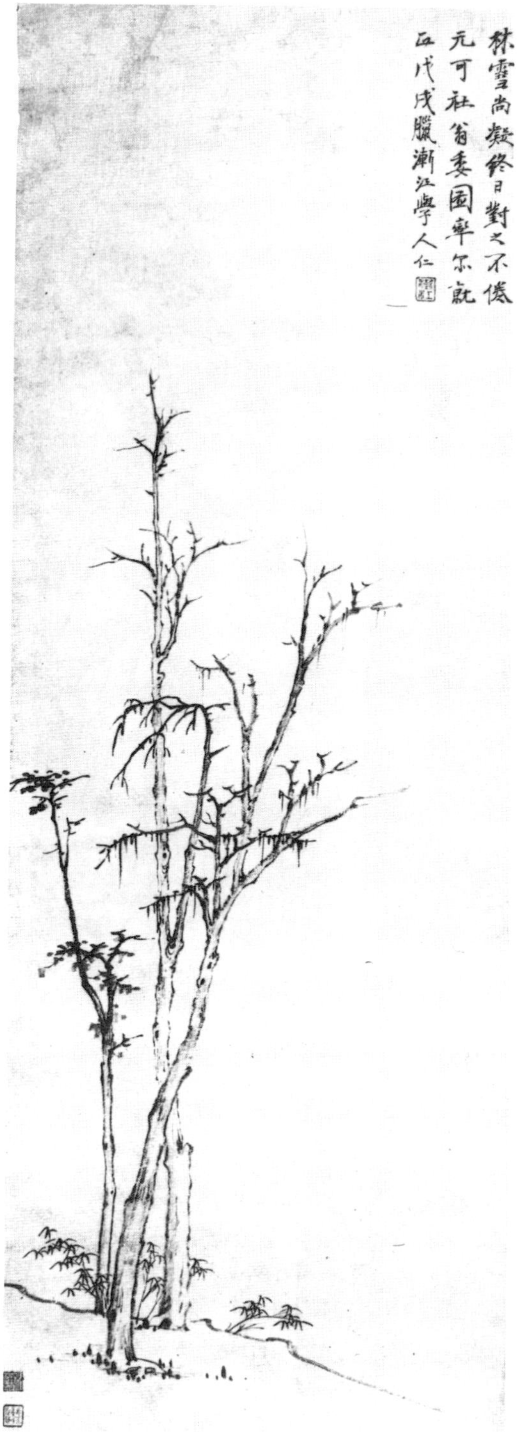

170. Hung-jen: "Frosted Trees," 1658. Hanging scroll, ink on paper, 83.5×30.2 cm. Musée Guimet, Paris.

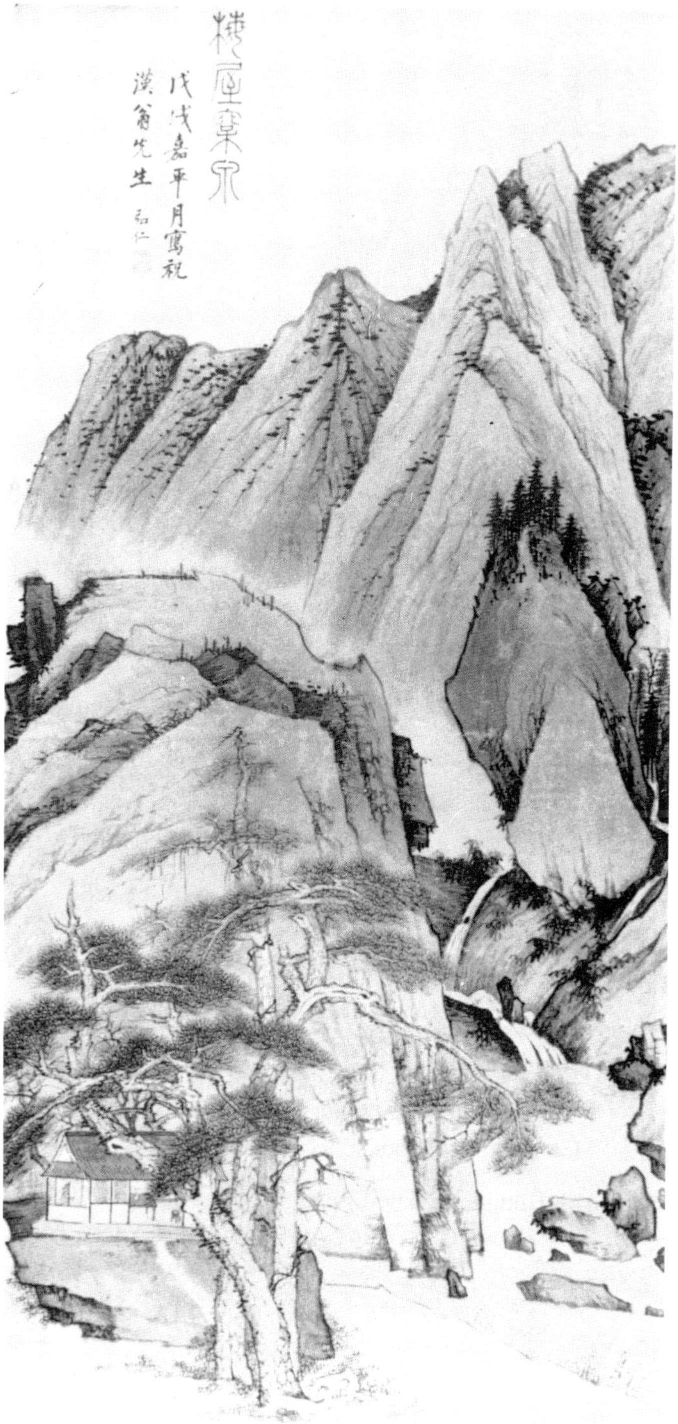

171. Hung-jen: "Plum House and Pine Spring," 1658. Hanging scroll, ink and colors on paper, 185.5×86.5 cm. Shanghai Museum.

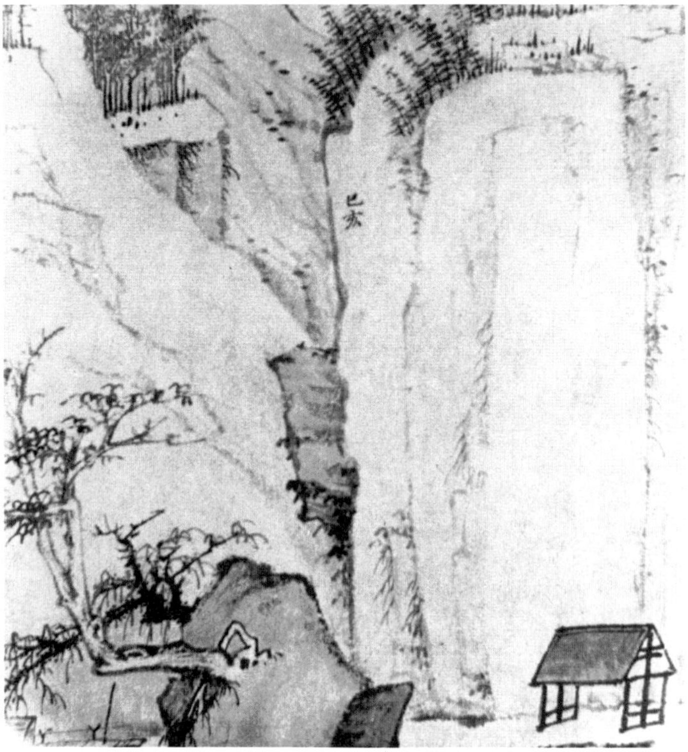

176. Hung-jen: "Landscape." Detail.

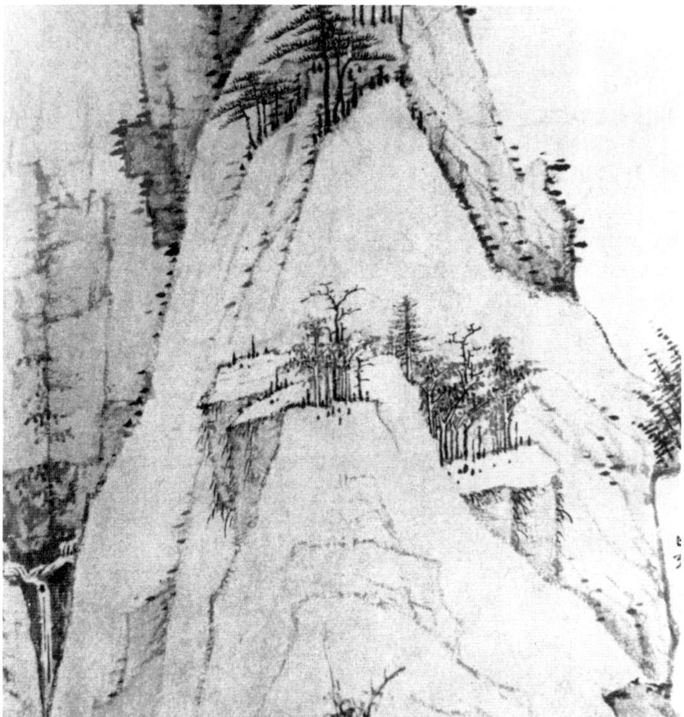

177. Hung-jen: "Landscape." Detail.

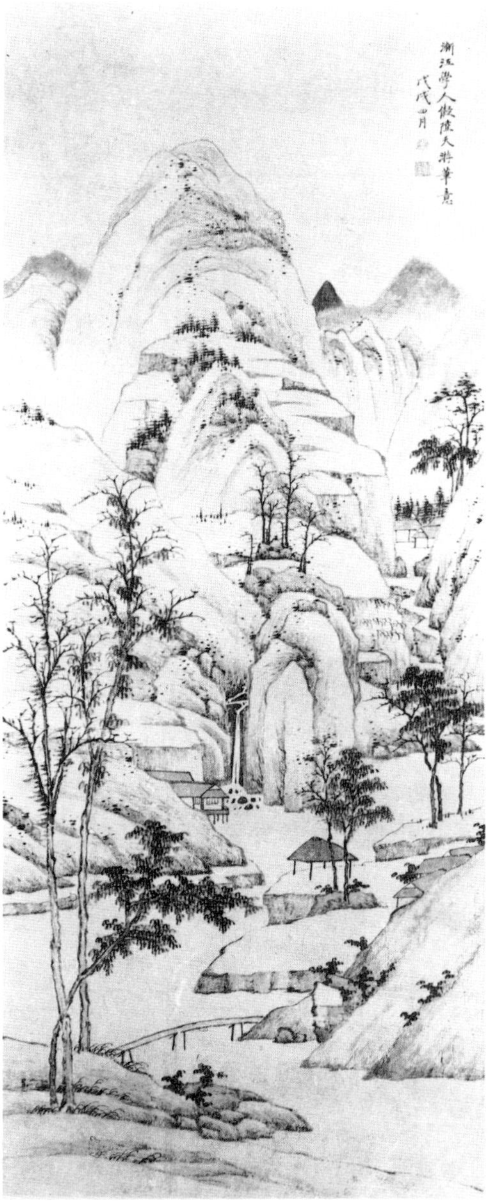

178. Hung-jen: "Landscape after Lu Kuang," 1658. Hanging scroll. Ching Yüan Chai Collection, Berkeley.

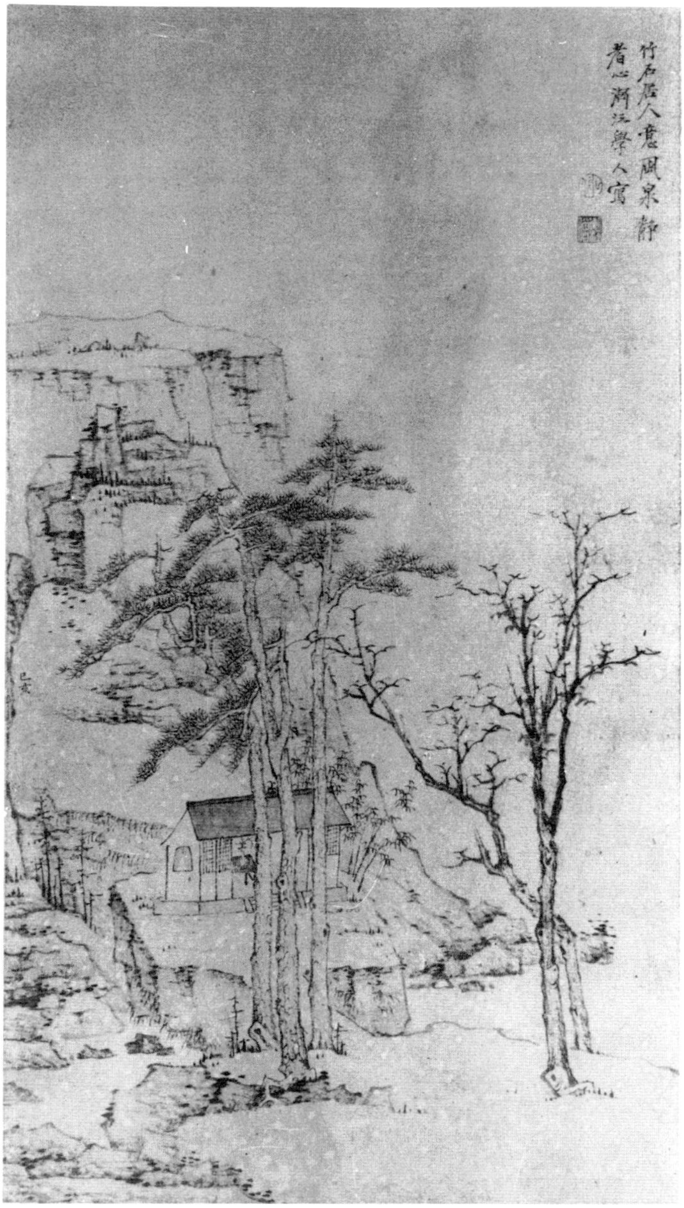

179. Hung-jen: "Quiet House amid Bamboo and Rocks," 1659. Hanging scroll, ink on paper, 58.5×34 cm. T'ien-tsin Museum.

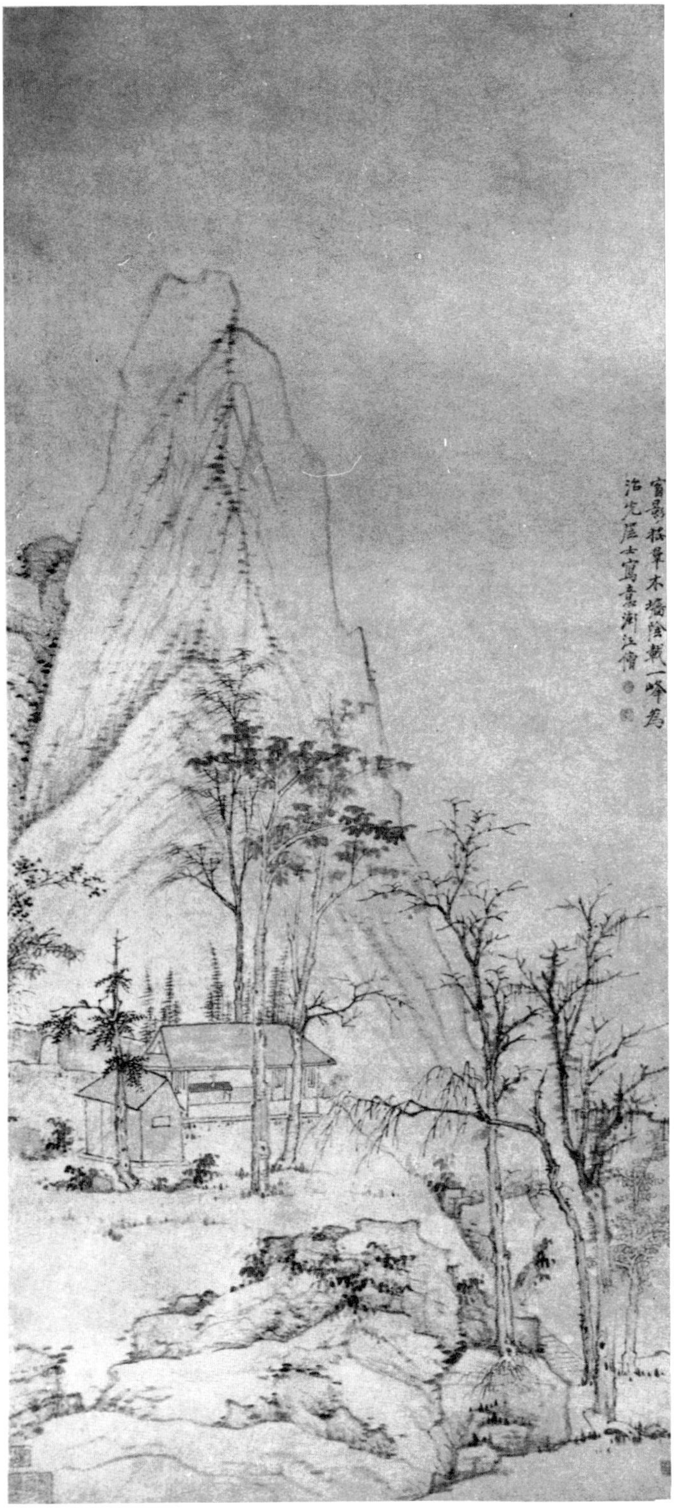

180. Hung-jen: "Landscape." Hanging scroll, 103×46.5 cm. Li Ch'u-li collection.

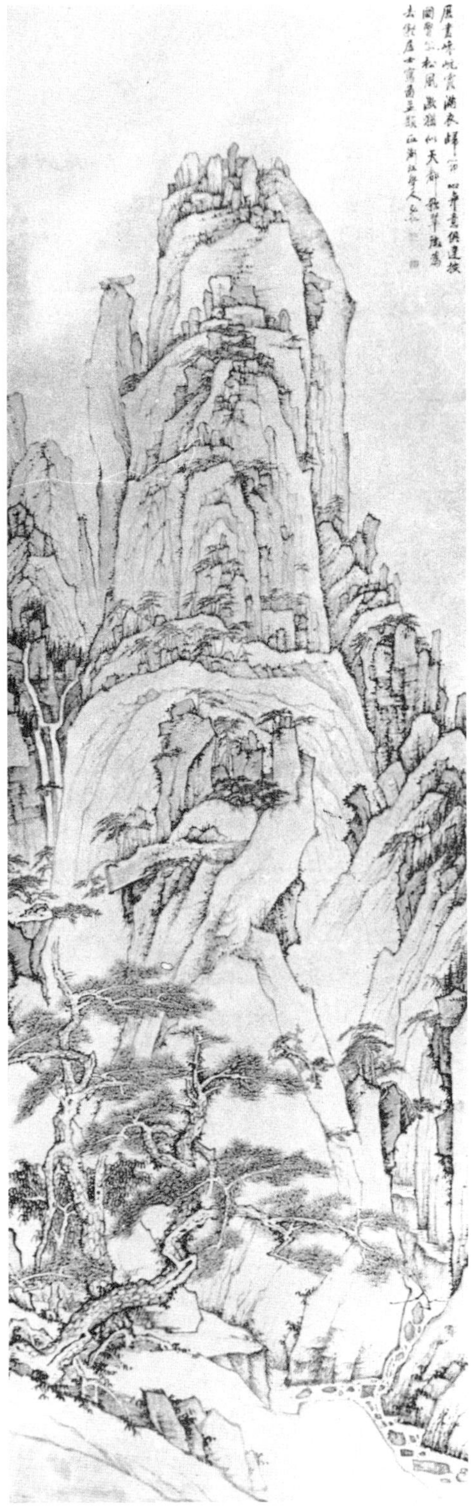

181. Hung-jen: "Heavenly Capital Peak," 1660. Hanging scroll, height about 280 cm. Nanking Museum.

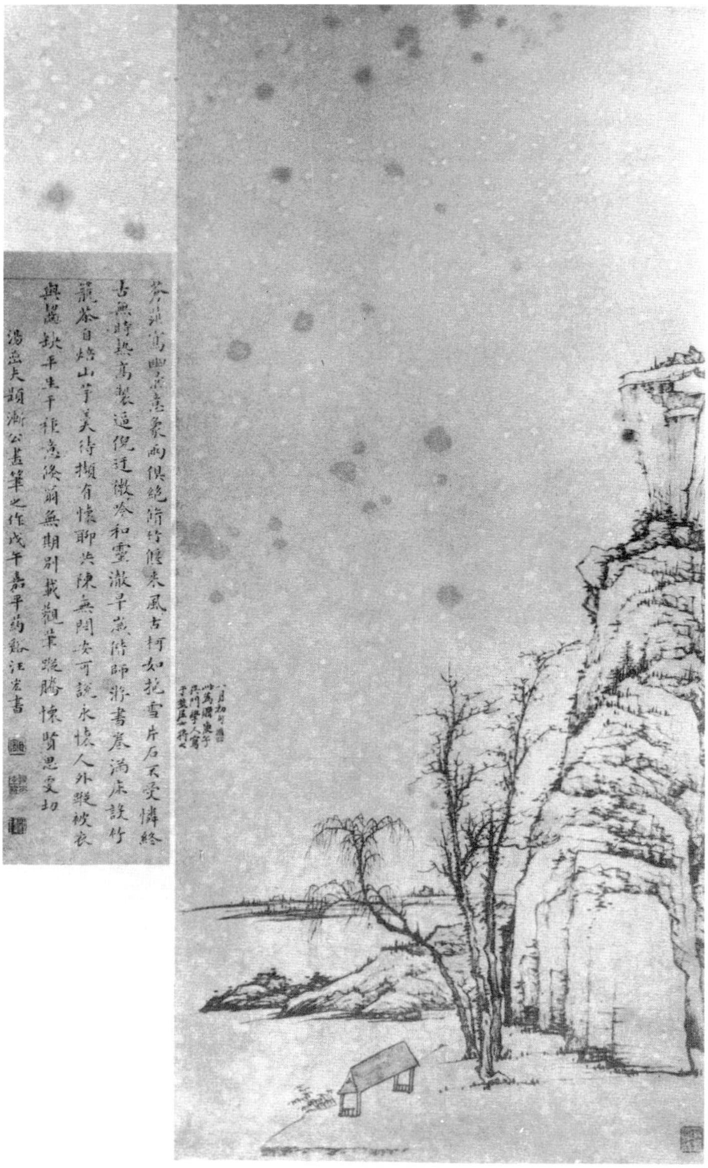

182. Hung-jen: "Landscape in the Style of Ni Yün-lin," 1660. Hanging scroll. Whereabouts unknown (formerly Meng-shui shan-fang collection).

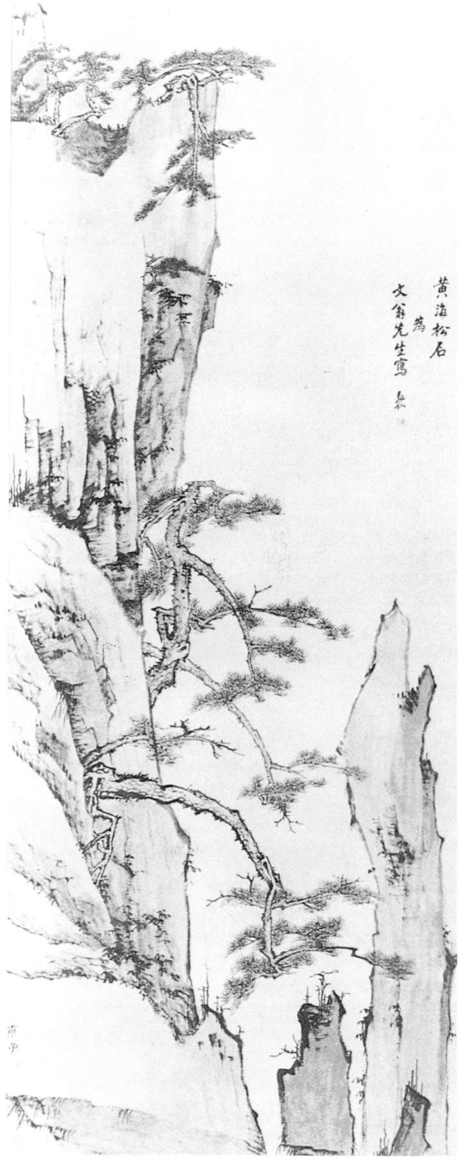

183. Hung-jen: "Pines and Rocks at Mt. Huang," 1660. Hanging scroll, ink and colors on paper, 198.7×81 cm. Shanghai Museum.

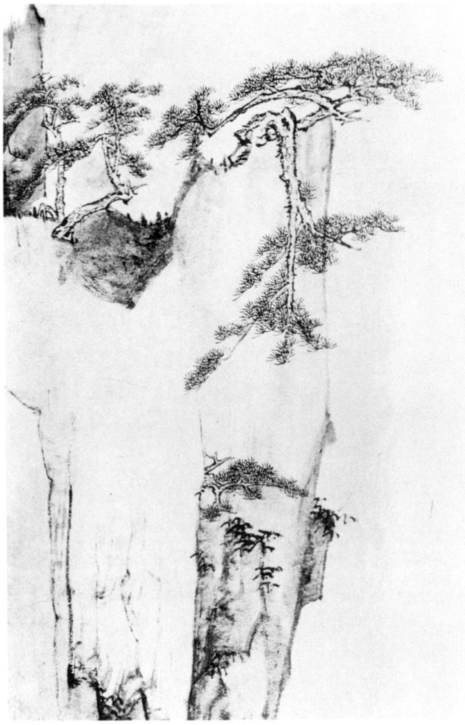

184. Hung-jen: "Pines and Rocks at Mt. Huang." Detail.

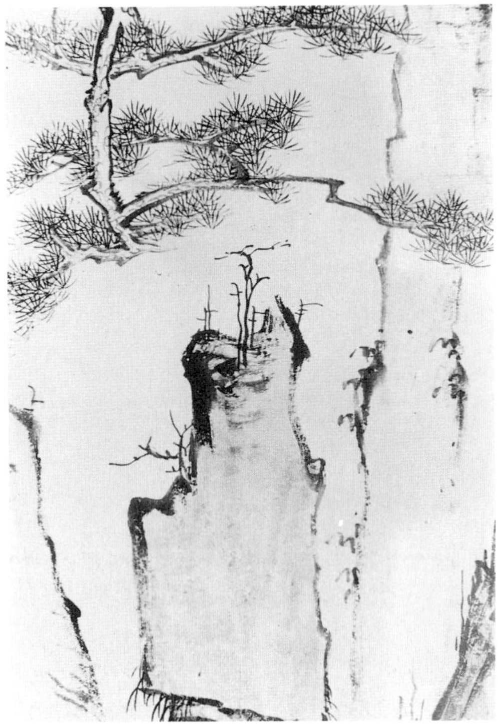

185. Hung-jen: "Pines and Rocks at Mt. Huang." Detail.

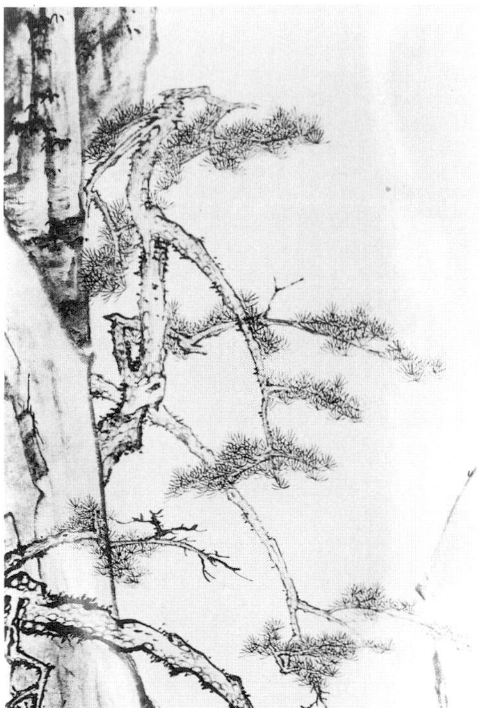

186. Hung-jen: "Pines and Rocks at Mt. Huang." Detail.

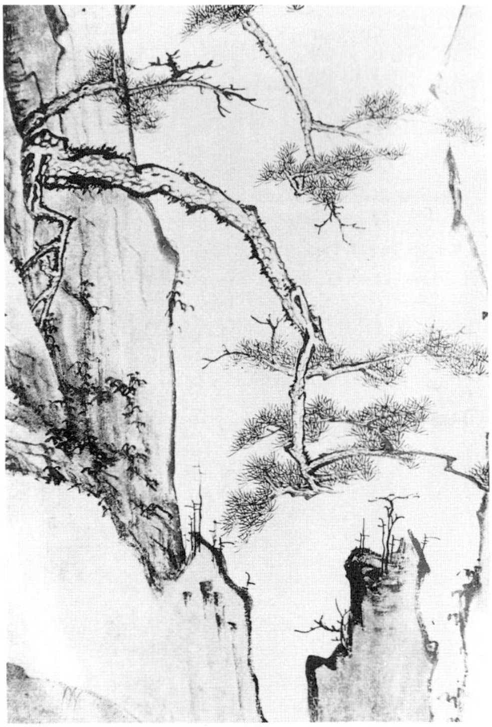

187. Hung-jen: "Pines and Rocks at Mt. Huang." Detail.

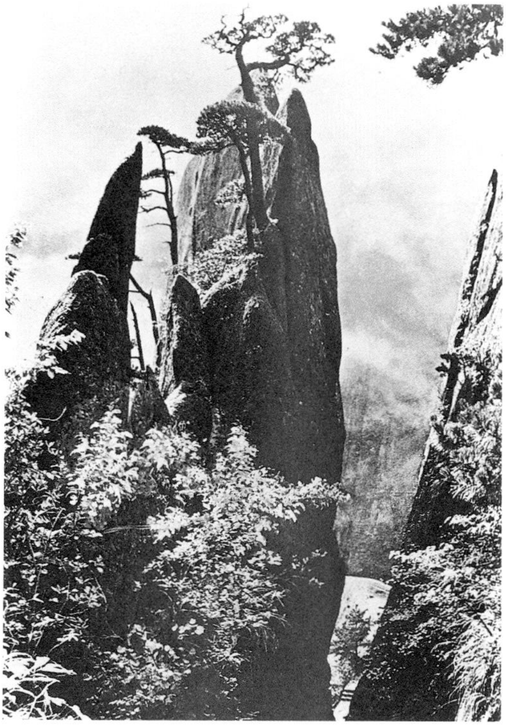

188. Pines and rocks at Mt. Huang. From Huang Yen-p'ei and Lü I-shou, *Huang-shan*.

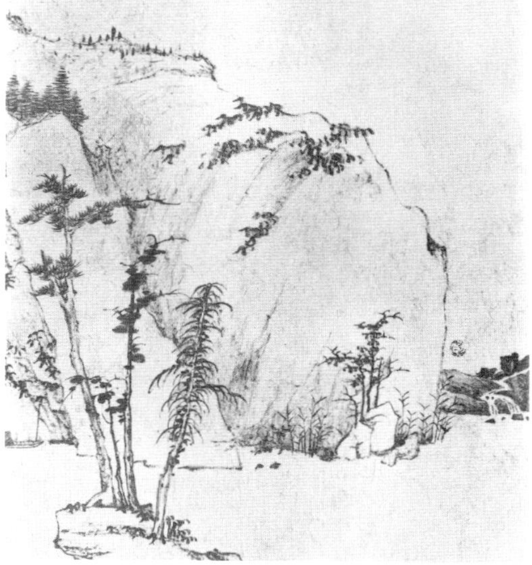

189. Hung-jen: "Landscape." Album, ink and colors on paper, 29.5×22 cm. T'ien-tsin Museum.

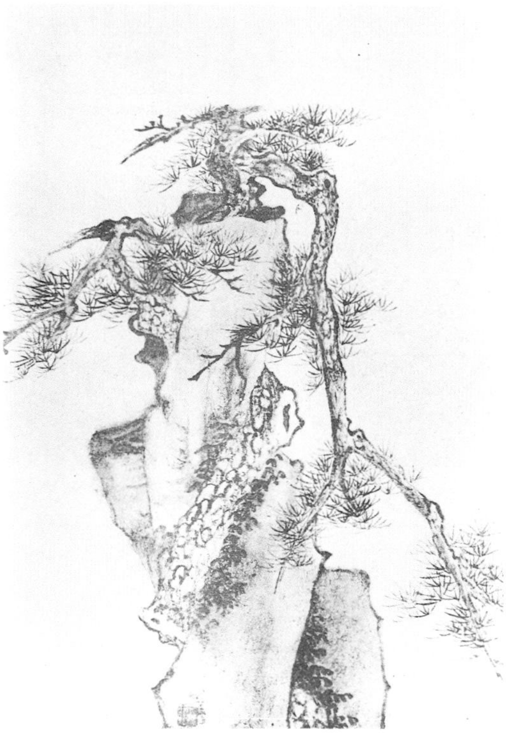

190. Hung-jen: "Landscape." Leaf no. 2.

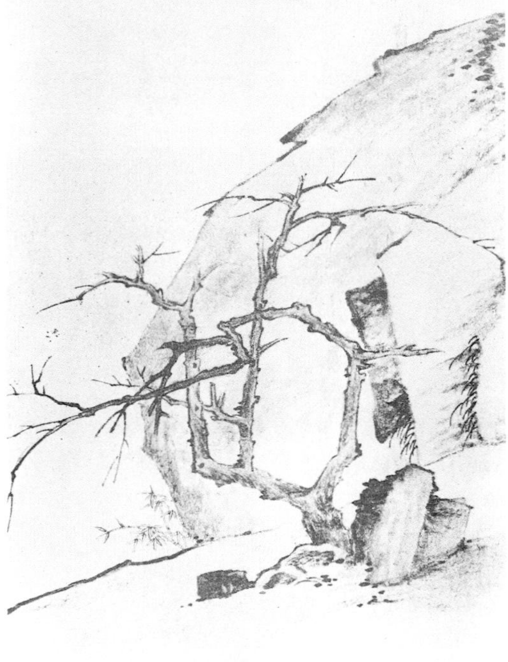

191. Hung-jen: "Landscape." Leaf no. 3.

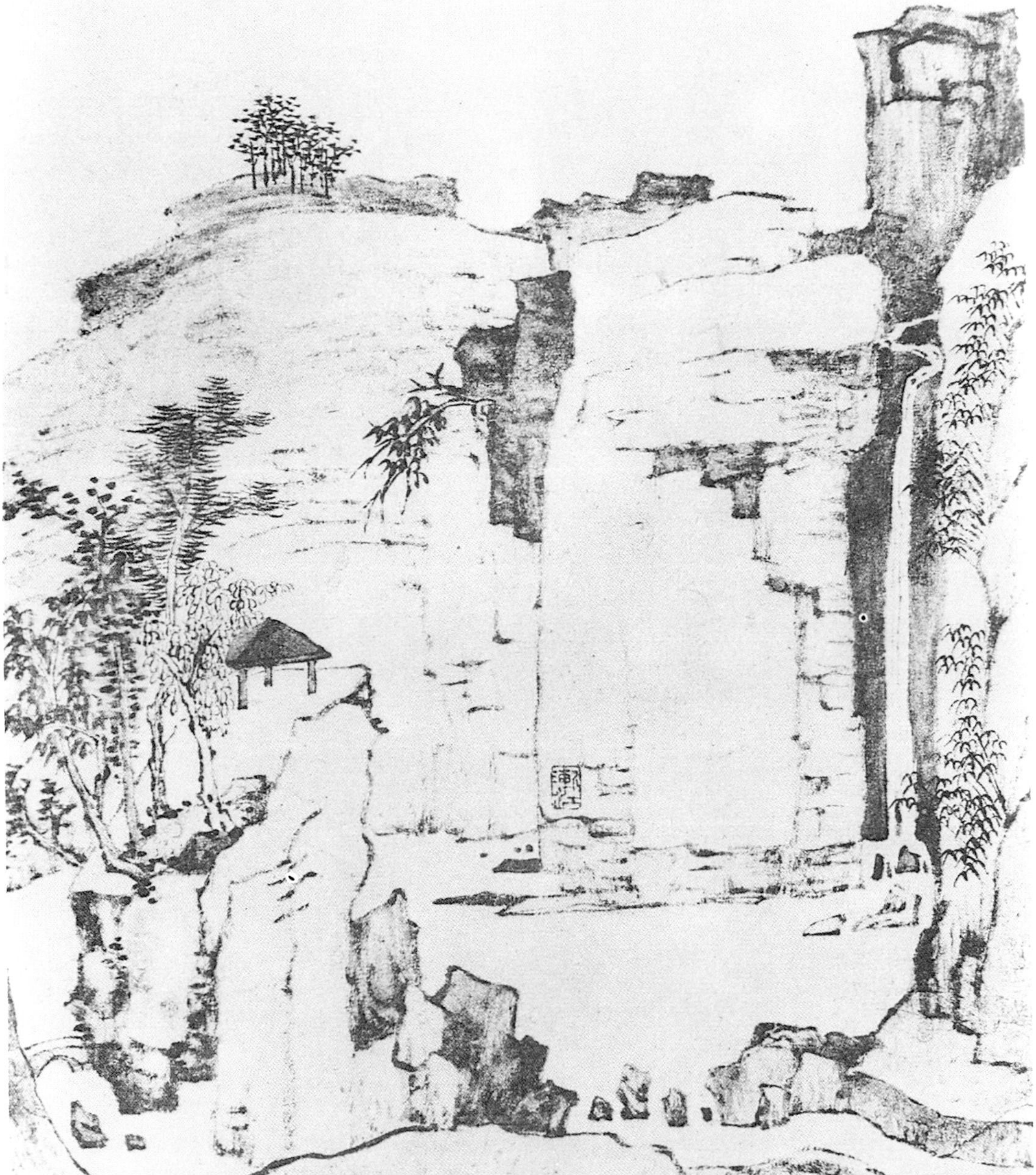

192. Hung-jen: "Landscape." Leaf no. 4.

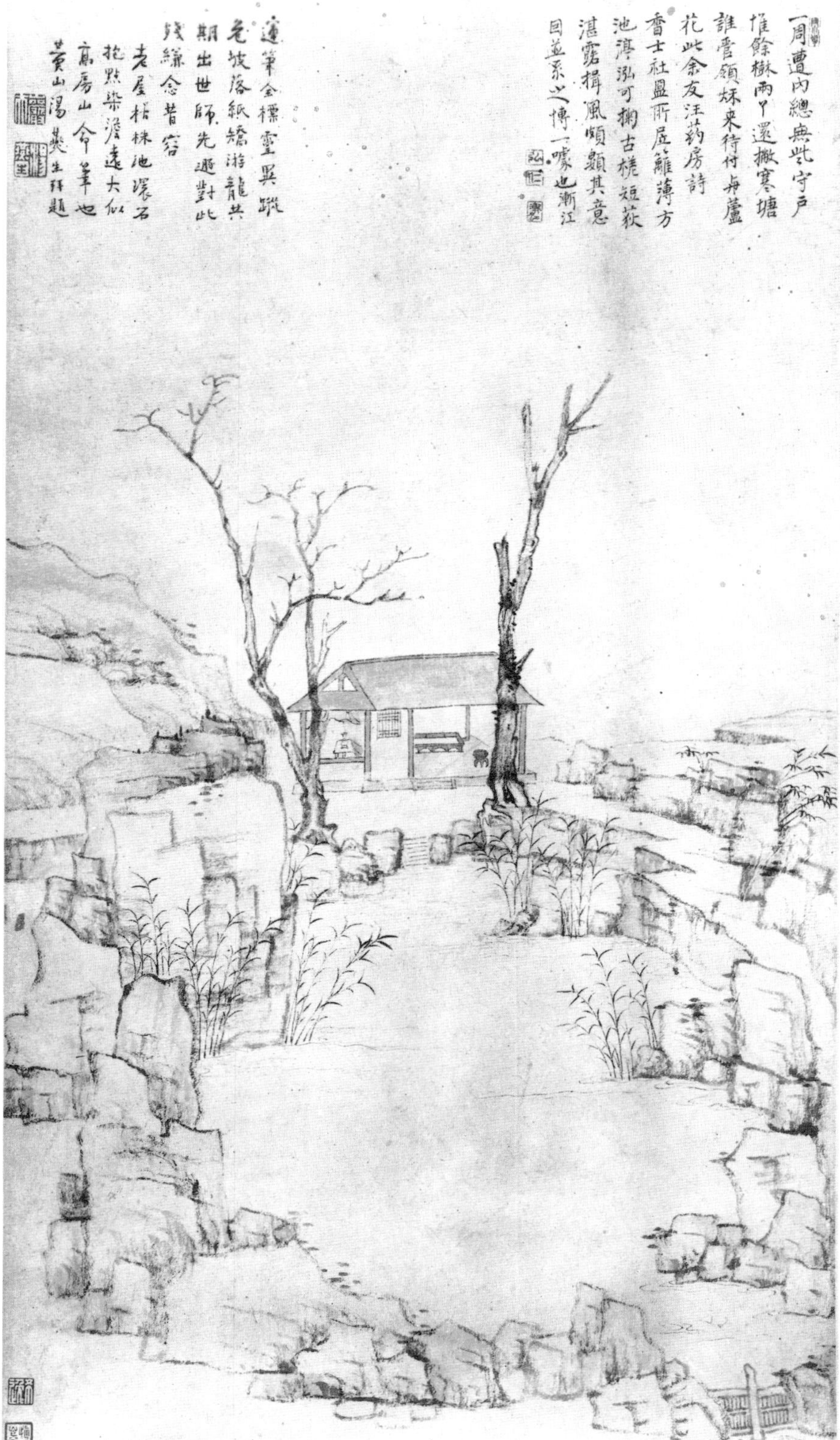

193. Hung-jen: "Old Trees and Short Reeds." Hanging scroll, ink on paper, 61.9×35.5 cm. Palace Museum, Peking.

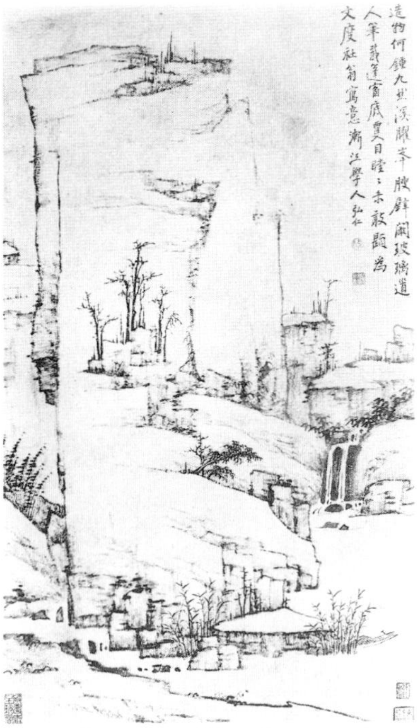

194. Hung-jen: "Landscape at Mt. Wu-i." Hanging scroll. Whereabouts unknown.

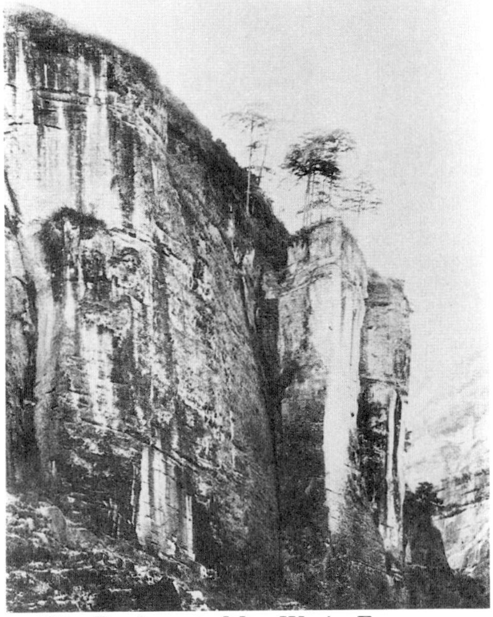

195. Rocks at Mt. Wu-i. From Chiang Hsi-chao, *Wu-i-shan.*

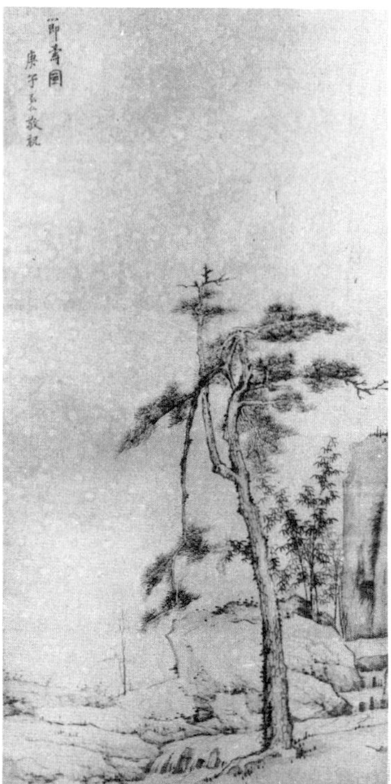

196. Hung-jen: "Pine, Bamboo, and Rocks for Birthday," 1660. Hanging scroll on satin. Palace Museum, Peking.

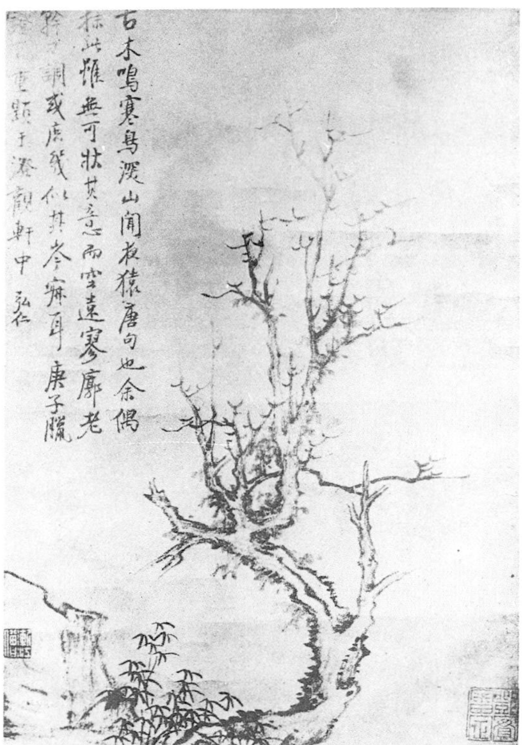

197. Hung-jen: "Old Trees, Bamboo, and Rocks," ca. 1660. Palace Museum, Peking.

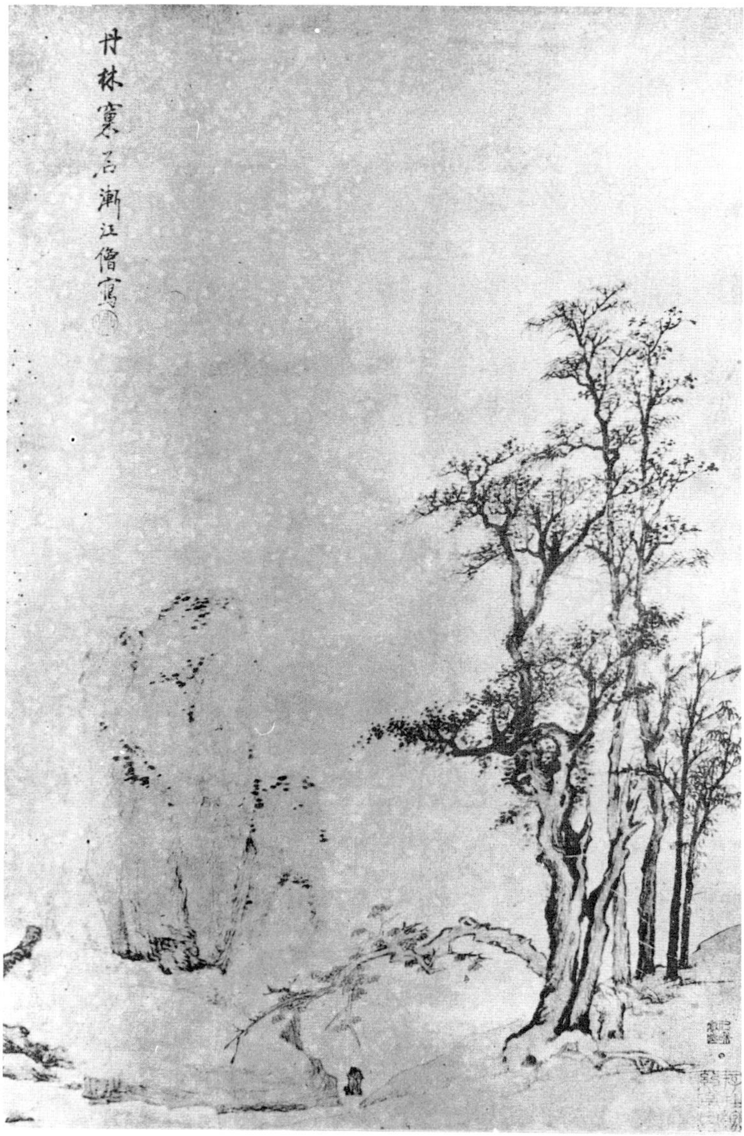

198. Hung-jen: "Red Forest and Burrowed Rocks." Hanging scroll, ink and colors on paper, 54.1×36 cm. Whereabouts unknown (formerly Fujii Seido collection).

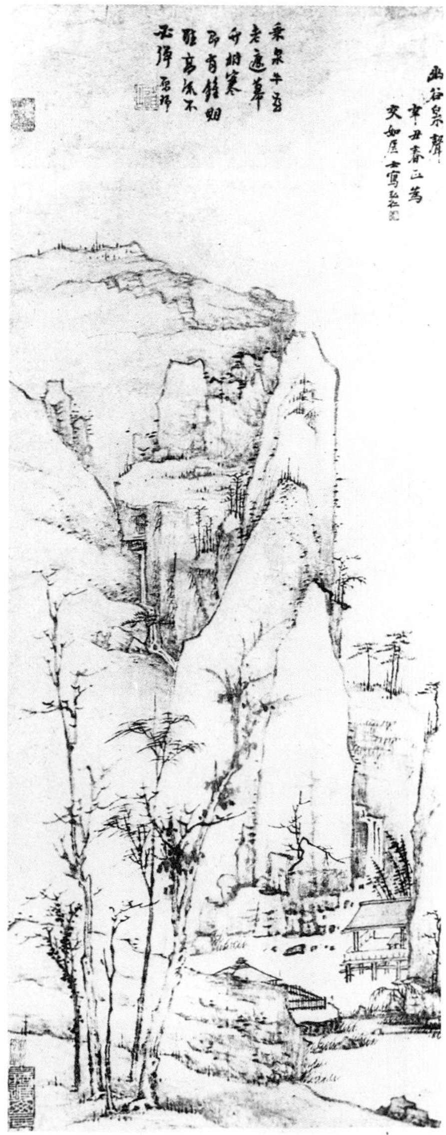

199. Hung-jen: "The Sound of Spring in a Quiet Valley," 1661. Hanging scroll, ink on paper, 101×41 cm. John Crawford, Jr. collection, New York.

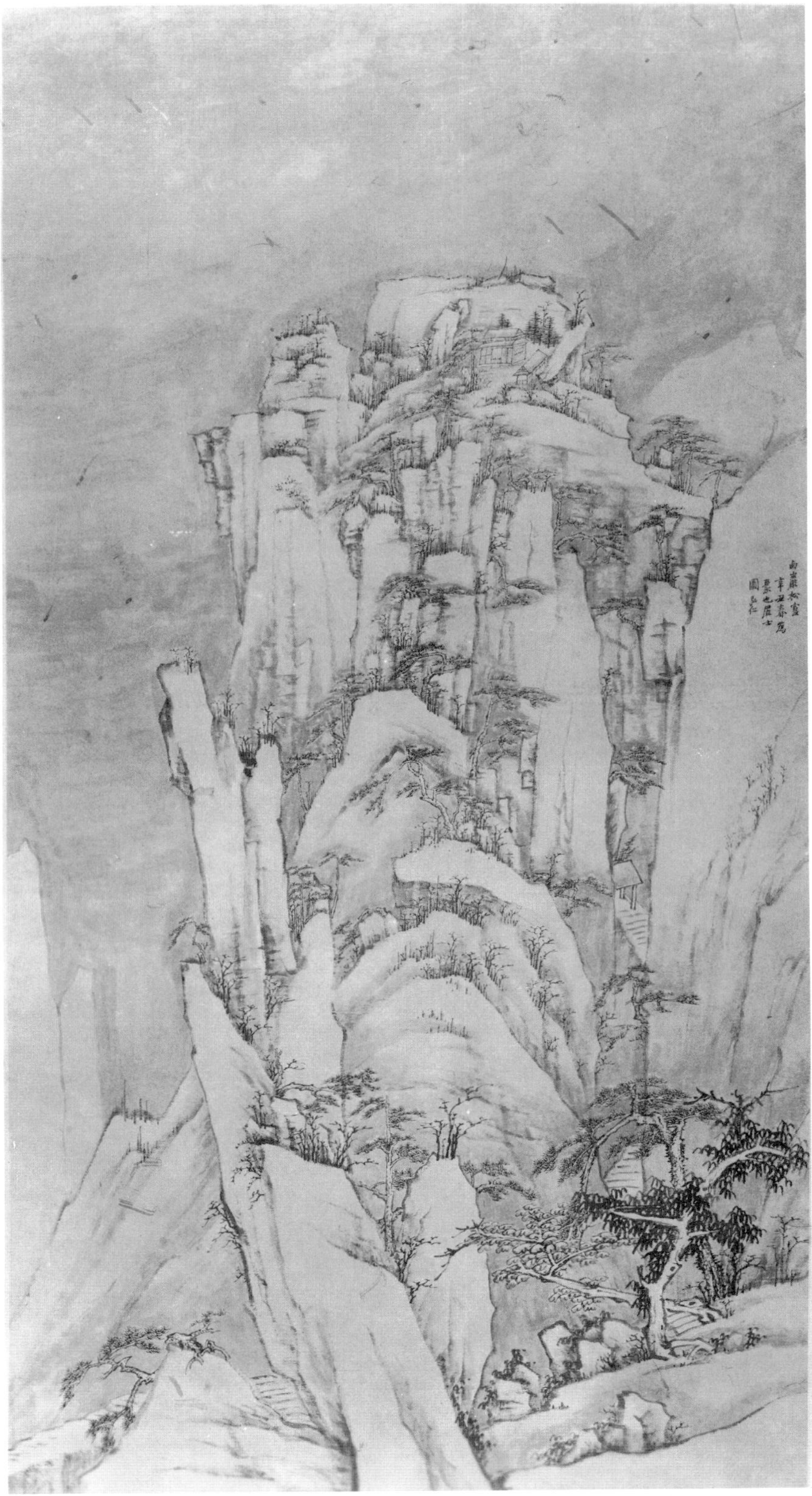

200. Hung-jen: "Snow on the Pines at Western Cliffs," 1661. Hanging scroll. Palace Museum, Peking.

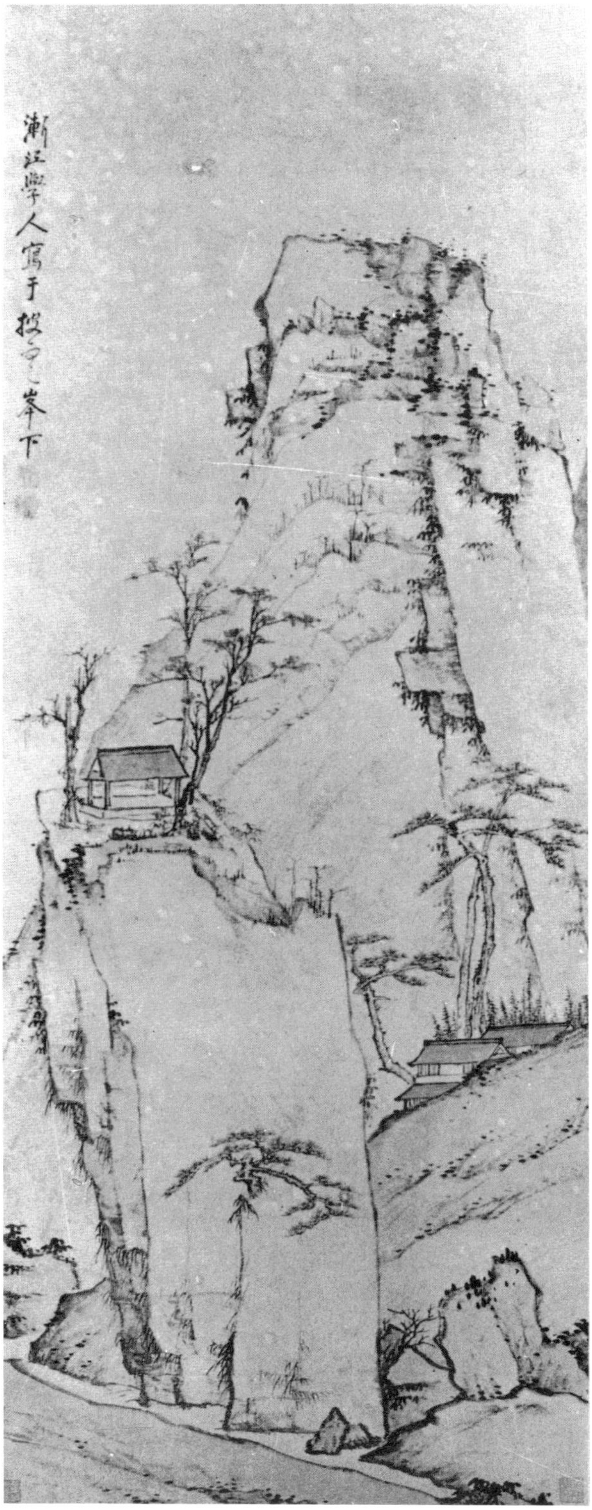

201. Hung-jen: "Pines, Rocks, and Distant Temple." Hanging scroll, ink on paper, 107×43.5 cm. Chih-lo-lou collection, Hong Kong.

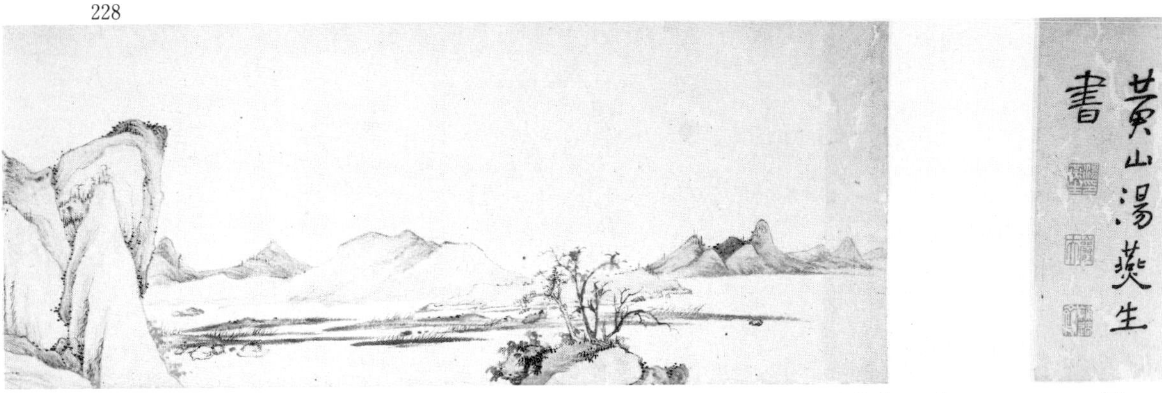

202. Hung-jen: "Landscape of Feng River," 1661. Handscroll, ink and light colors on paper, 28. 5× 292.8 cm. Sumitomo collection, Japan. Frontispiece by T'ang Yen-sheng and Section 1.

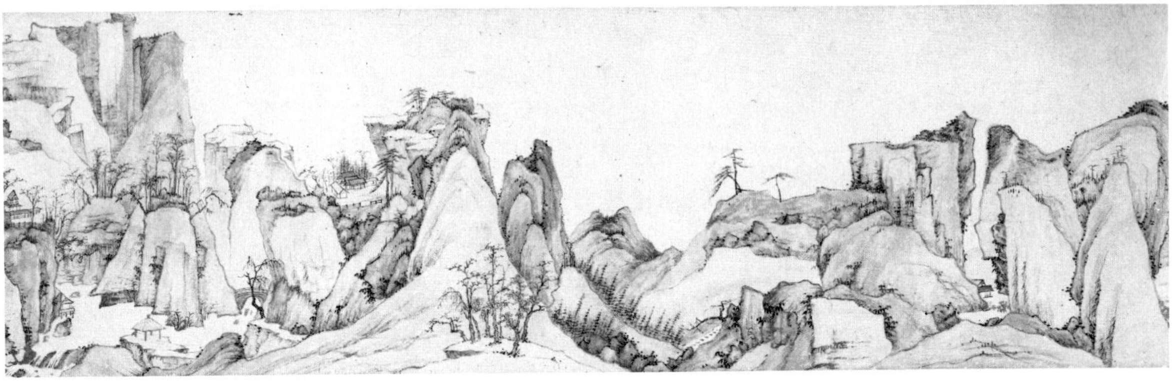

203. Hung-jen: "Landscape of Feng River." Section 2.

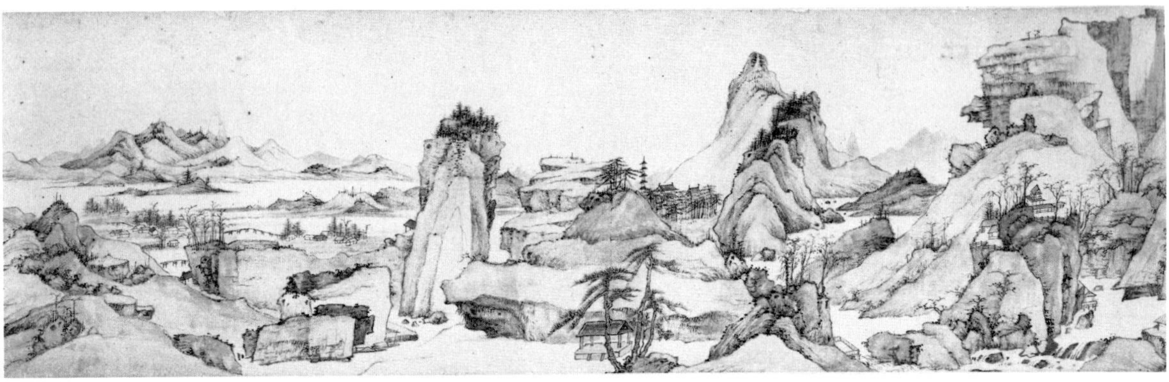

204. Hung-jen: "Landscape of Feng River." Section 3.

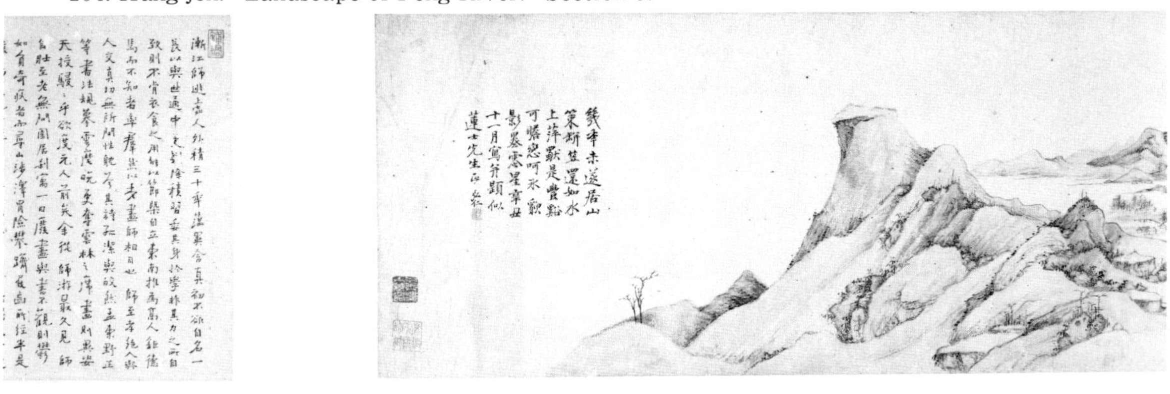

205. Hung-jen: "Landscape of Feng River." Section 4.

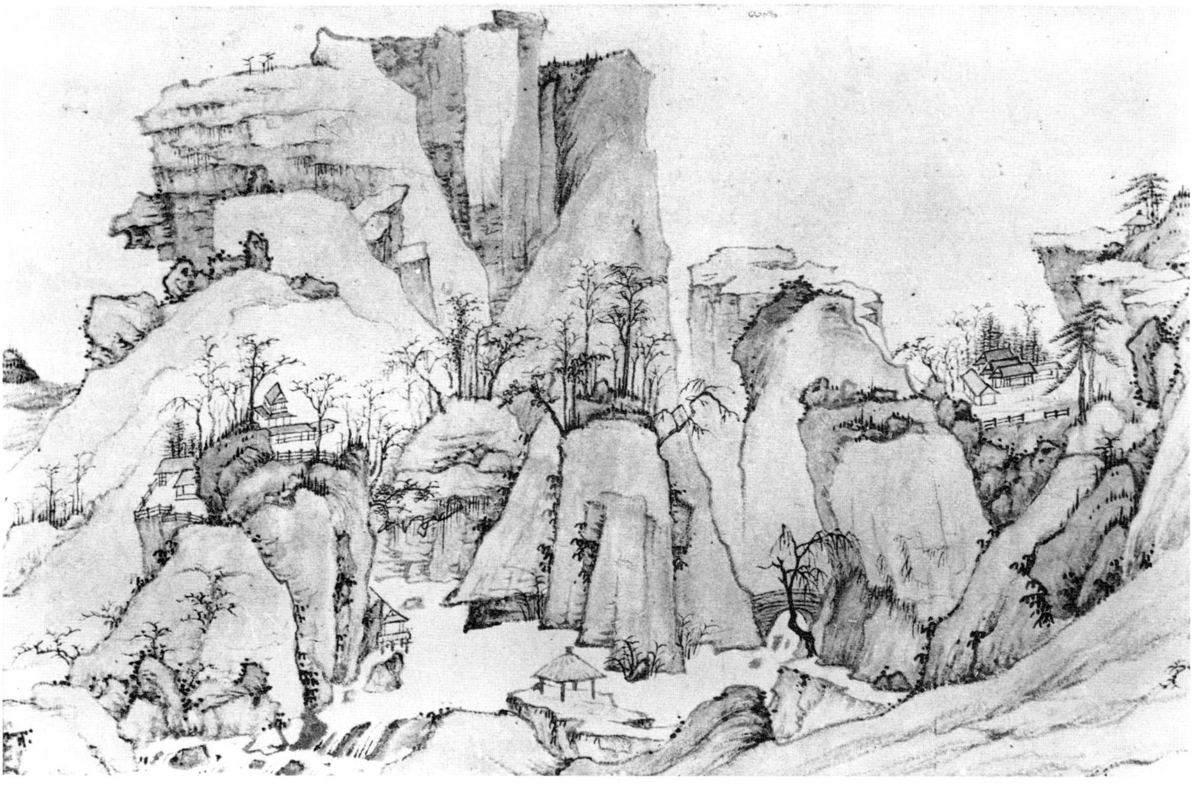

206. Hung-jen: "Landscape of Feng River." Detail.

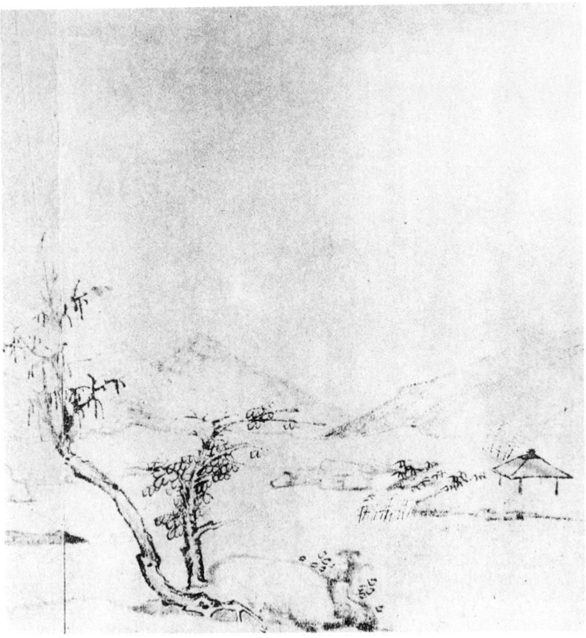

208. Hung-jen: "Dawn on the River." Section 2.

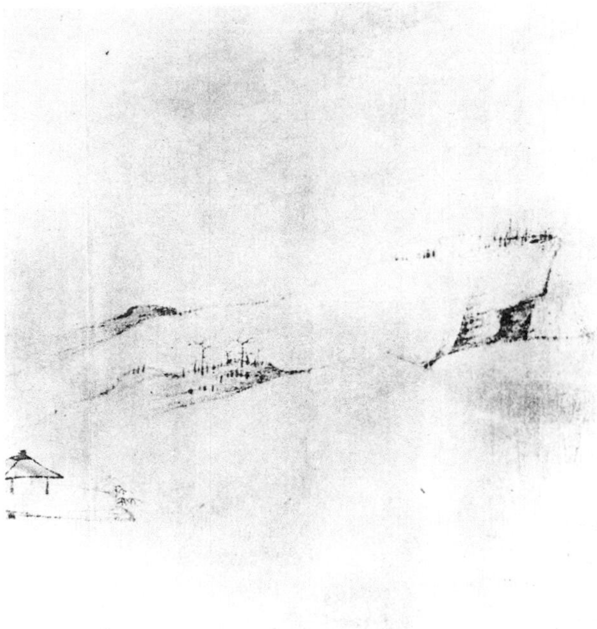

207. Hung-jen: "Dawn on the River," 1661. Handscroll. Anhui Provincial Museum. Section 1.

210. Hung-jen: "Dawn on the River." Section 4.

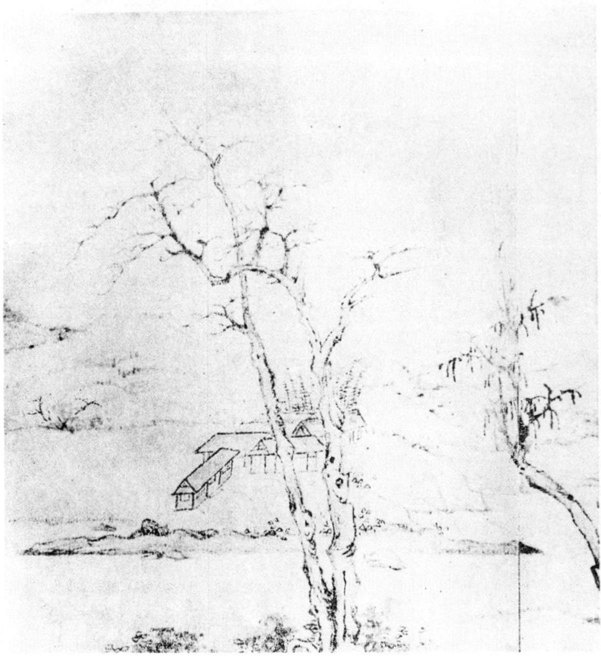

209. Hung-jen: "Dawn on the River." Section 3.

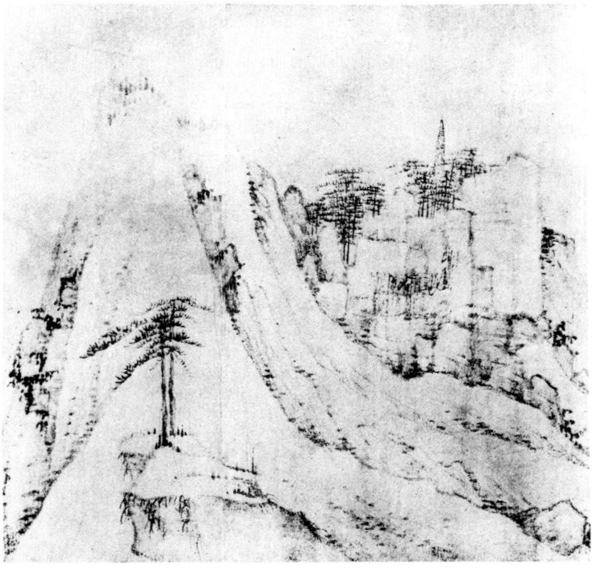

212. Hung-jen: "Dawn on the River." Section 6.

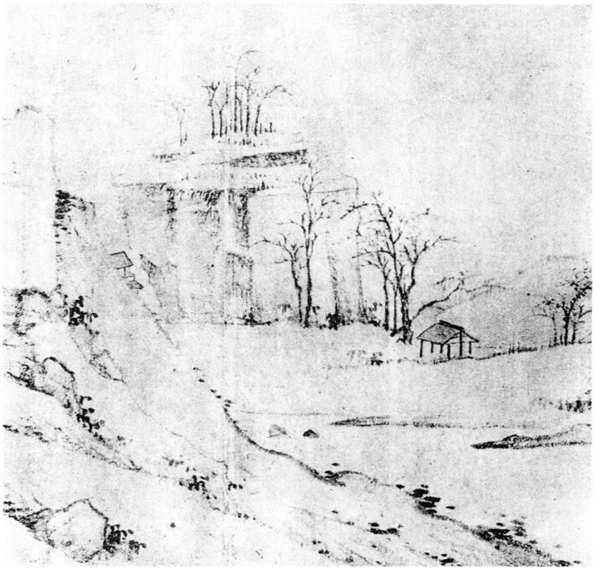

211. Hung-jen: "Dawn on the River." Section 5.

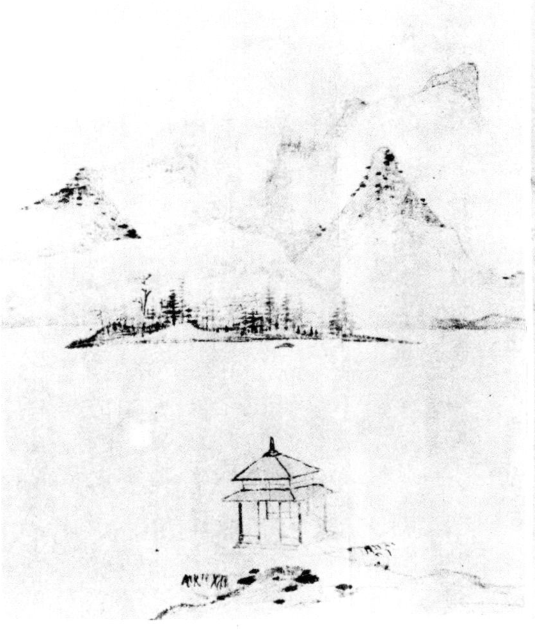

214. Hung-jen: "Dawn on the River." Section 8.

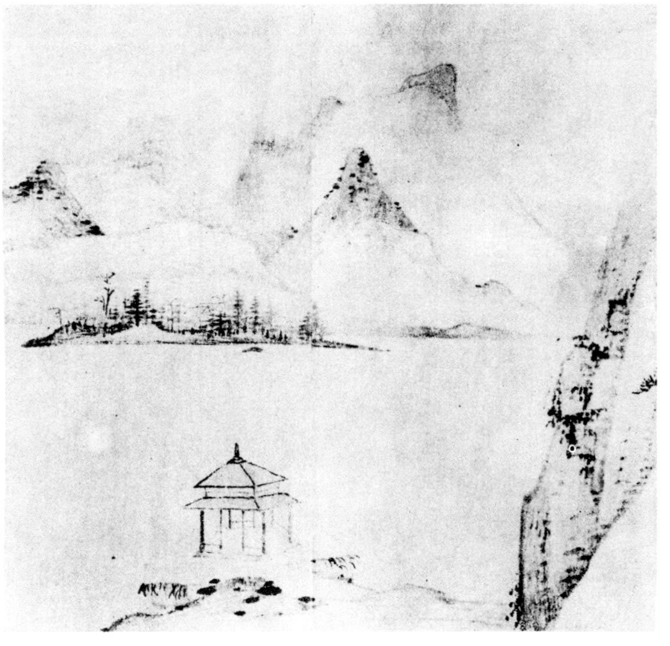

213. Hung-jen: "Dawn on the River." Section 7.

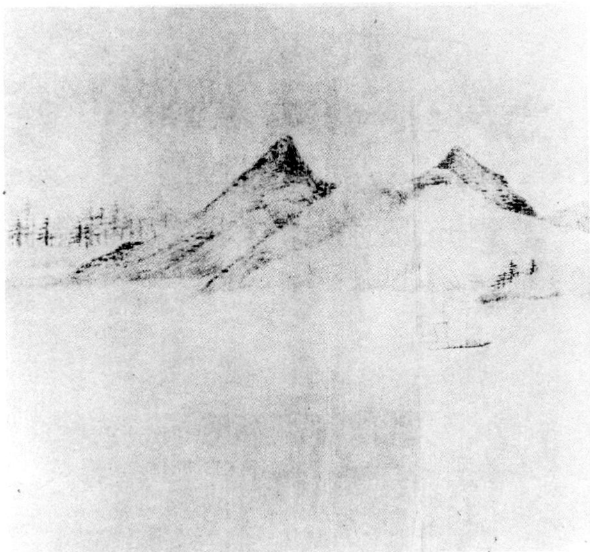

216. Hung-jen: "Dawn on the River." Section 10.

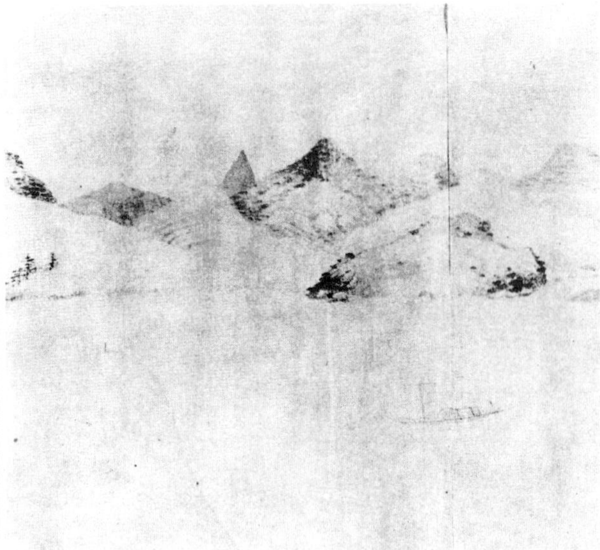

215. Hung-jen: "Dawn on the River." Section 9.

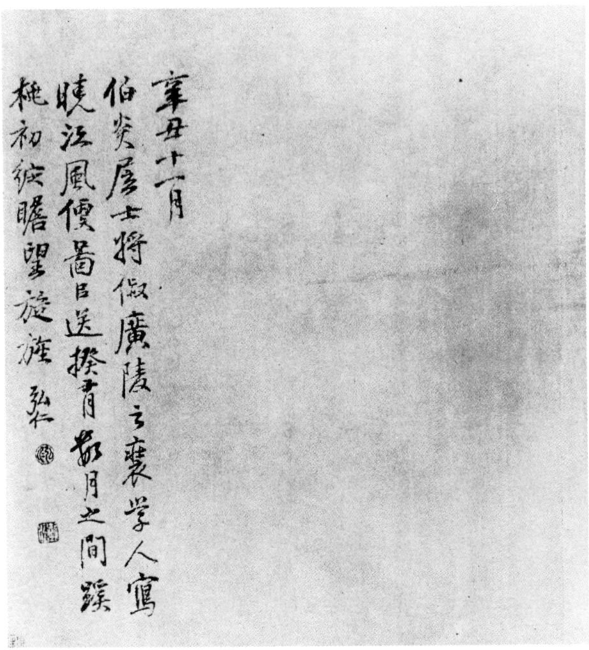

218. Hung-jen: "Dawn on the River." Section 12.

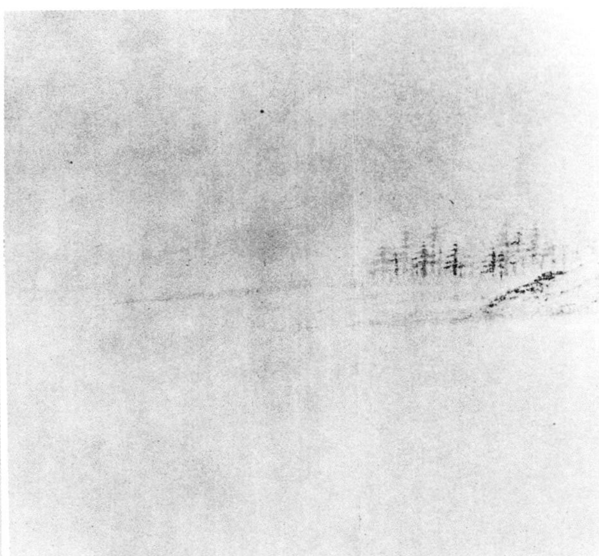

217. Hung-jen: "Dawn on the River." Section 11.

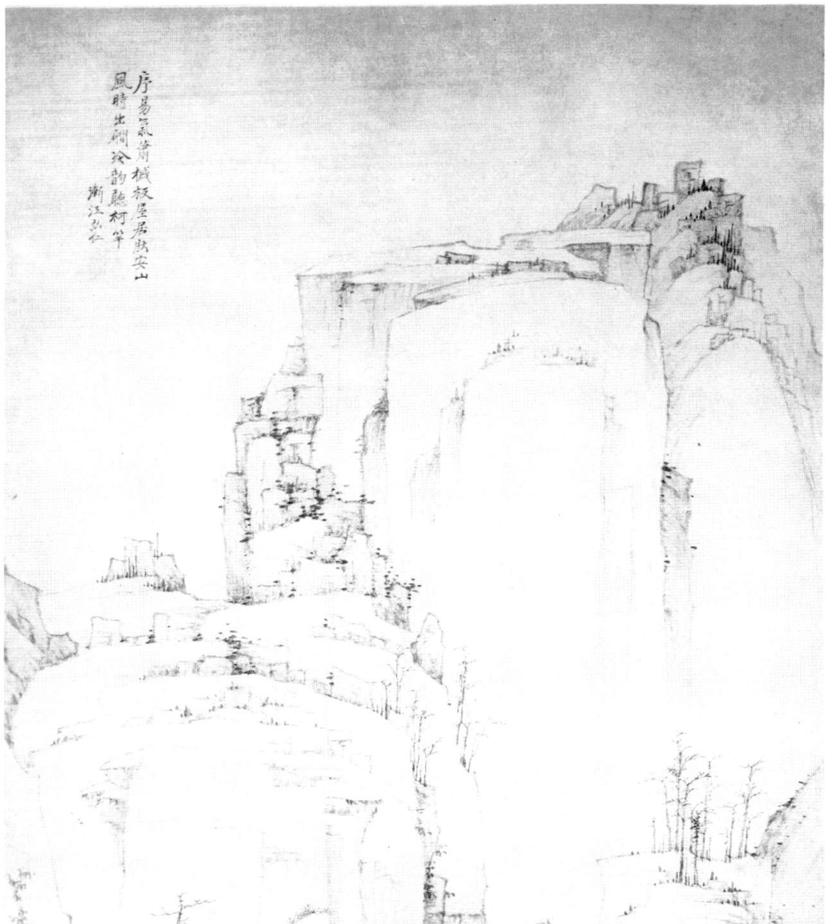

219. Hung-jen: "The Coming of Autumn." Hanging scroll, ink on paper, 122×63 cm. Honolulu Academy of Arts.

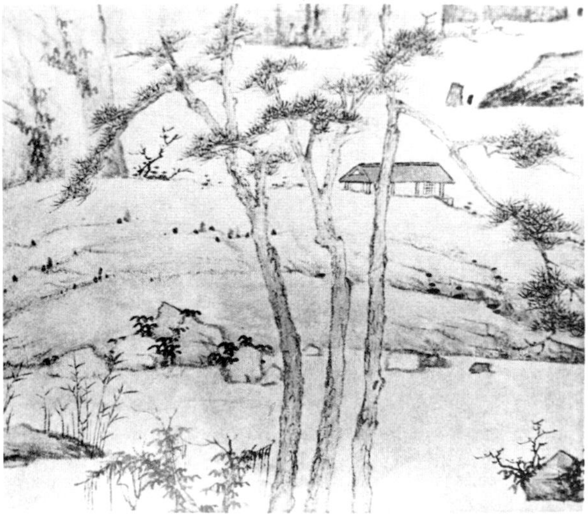

220. Hung-jen: "The Coming of Autumn." Detail.

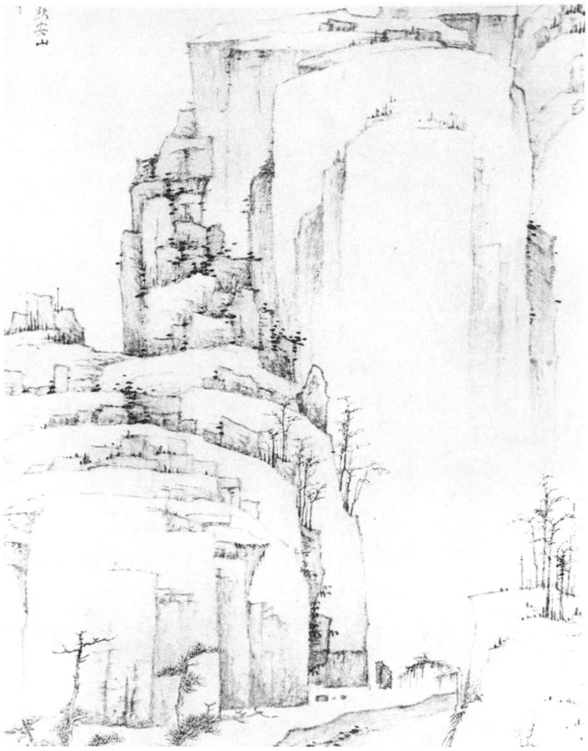

221. Hung-jen: "The Coming of Autumn." Detail.

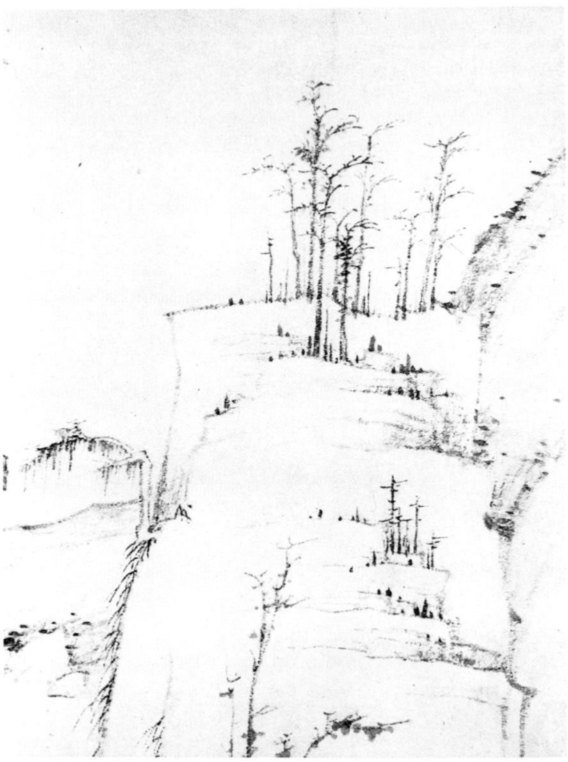

222. Hung-jen: "The Coming of Autumn." Detail.

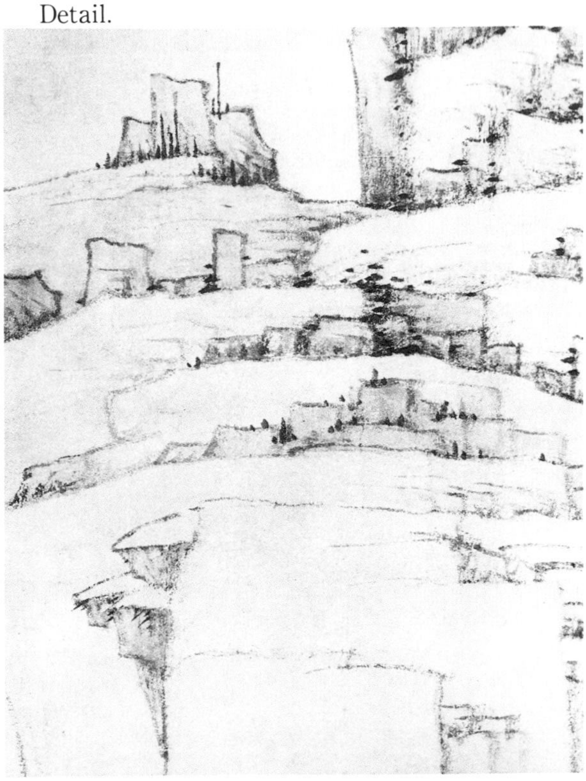

223. Hung-jen: "The Coming of Autumn." Detail.

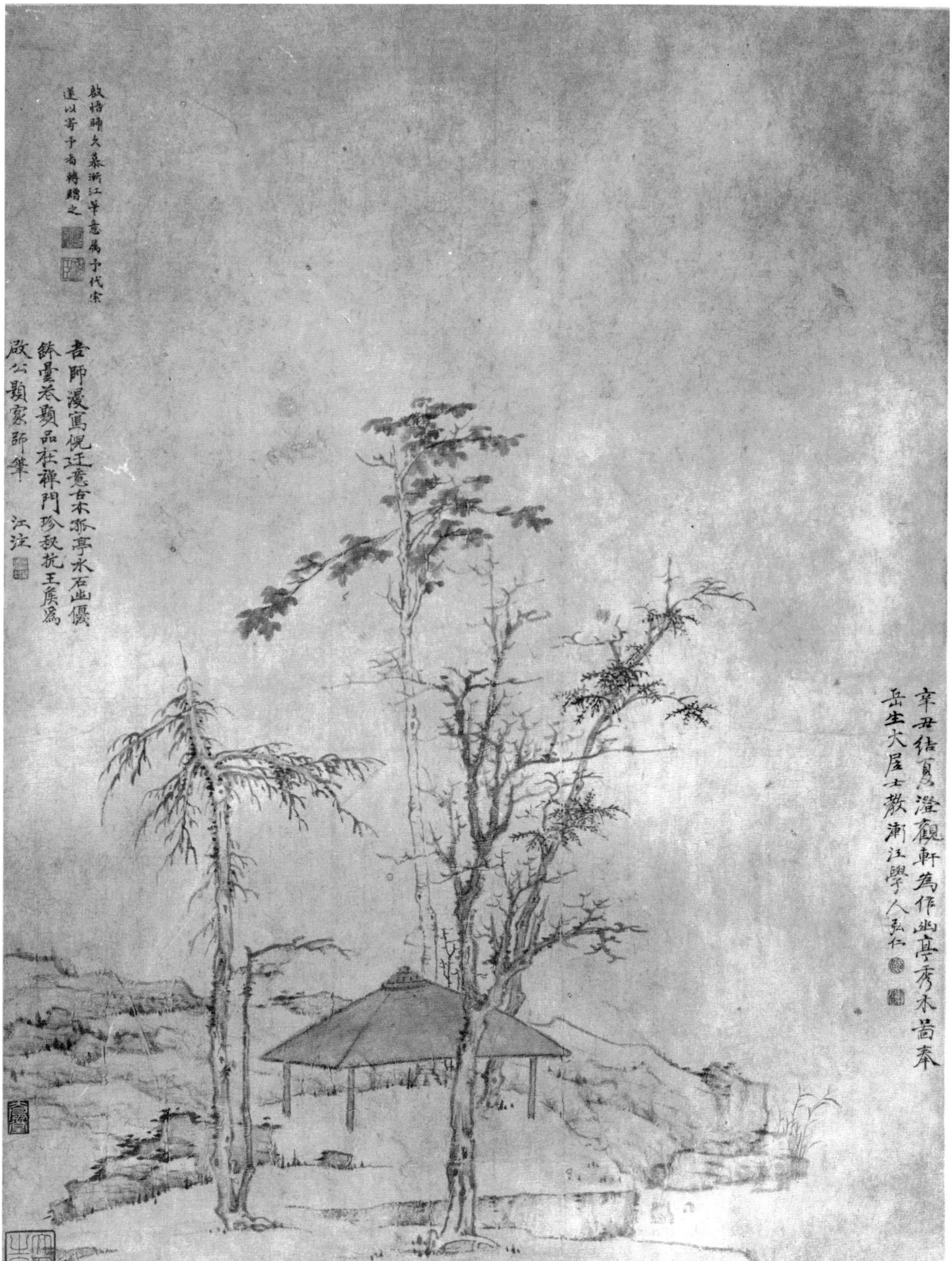

224. Hung-jen: "Quiet Pavilion and Elegant Trees," 1661. Hanging scroll. Palace Museum, Peking.

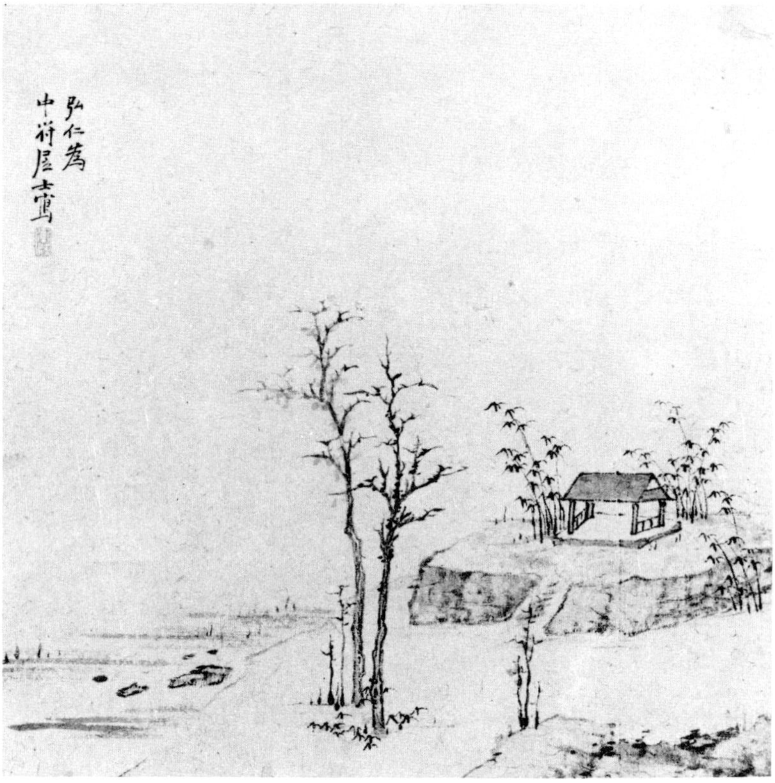

225. Hung-jen: "Landscape," 1663. Album leaf, ink and color on paper. Ching Yüan Chai collection, Berkeley.

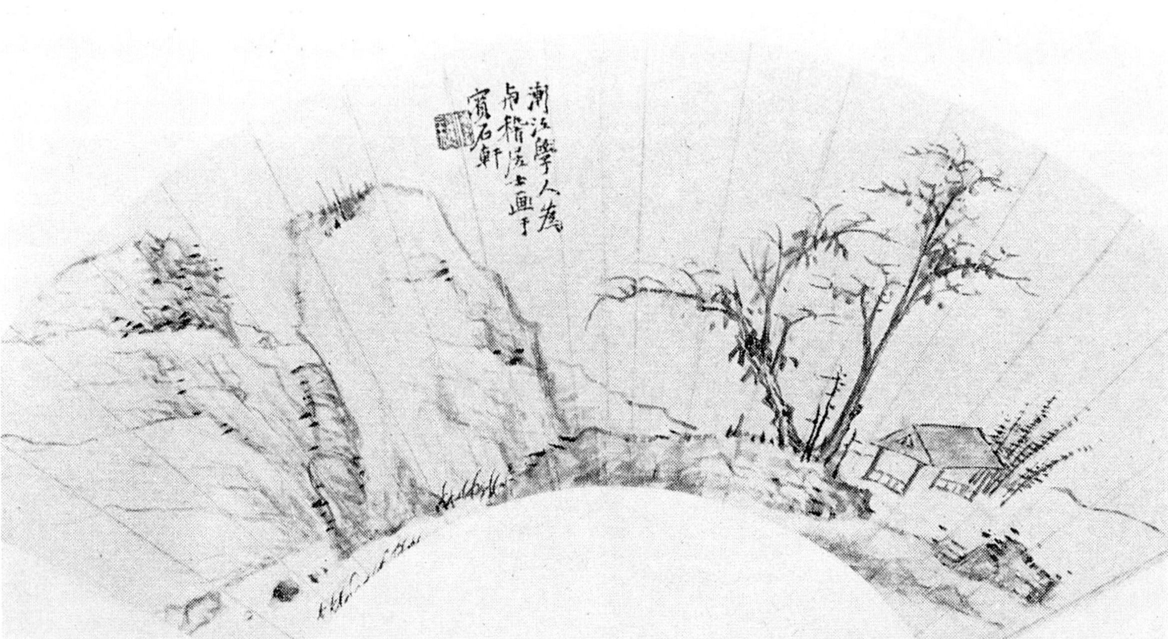

226. Hung-jen: "Landscape in the Style of Ni Tsan." Fan, ink on paper, $23\frac{7}{8} \times 12\frac{3}{4}$ in. Franco Vannotti collection, Lugano.

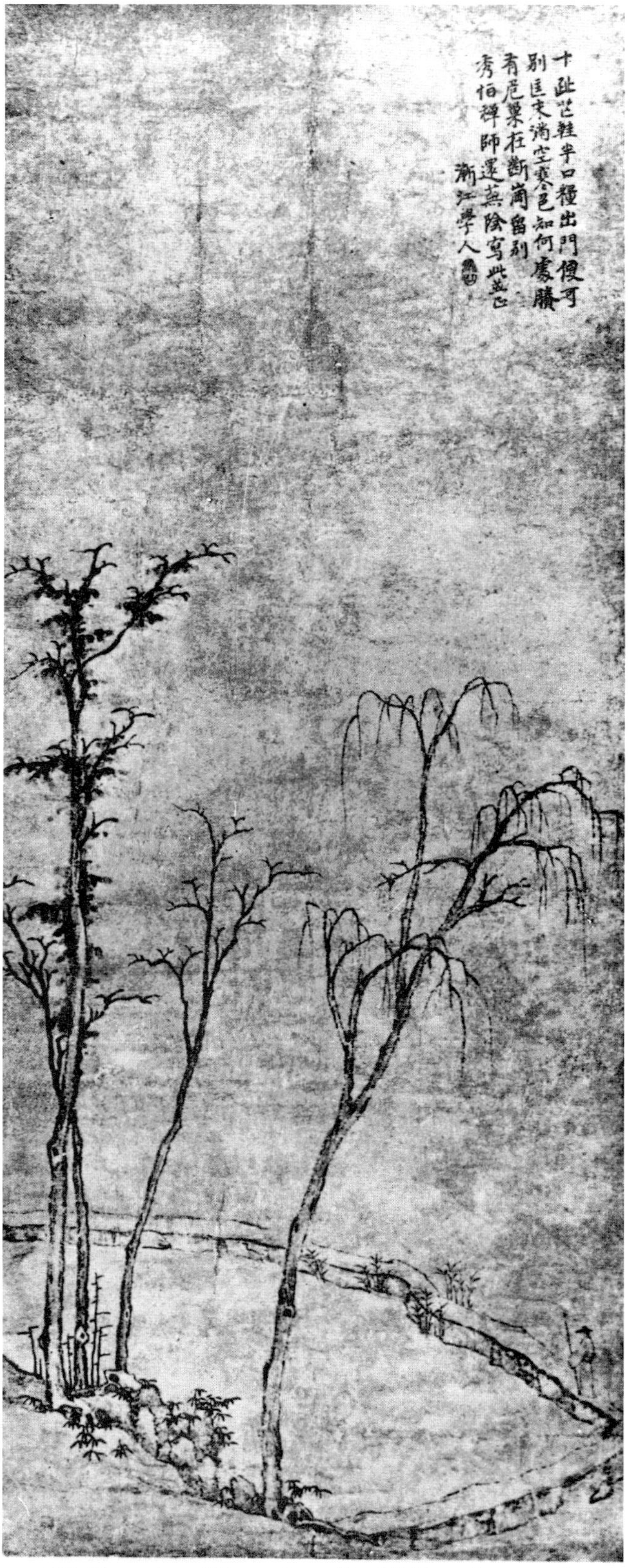

227. Hung-jen: "Trees by a Pool." Hanging scroll, ink on paper, 69.9×29.3 cm. Museum für Ostasiatische Kunst der Stadt Köln.

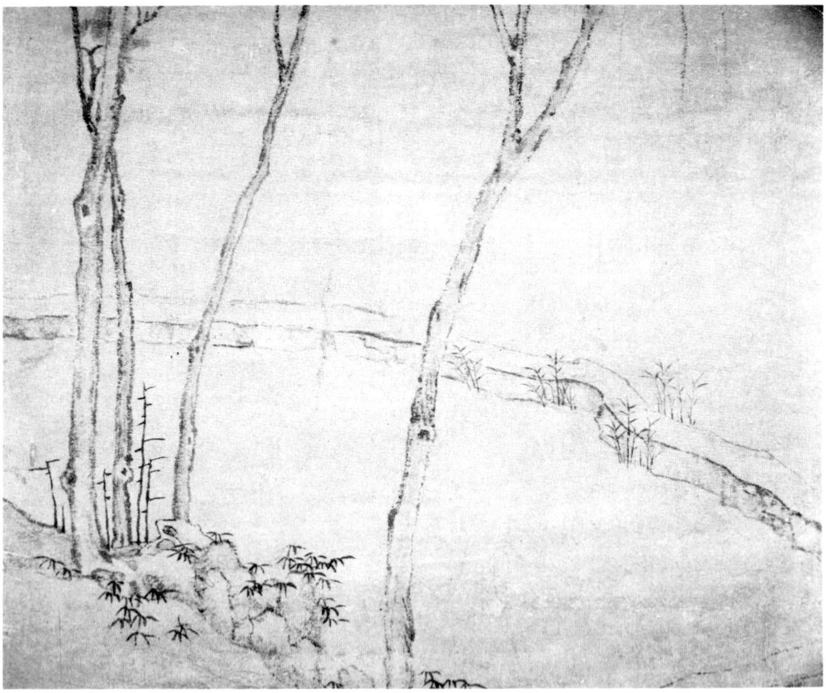

228. Hung-jen: "Trees by a Pool." Detail.

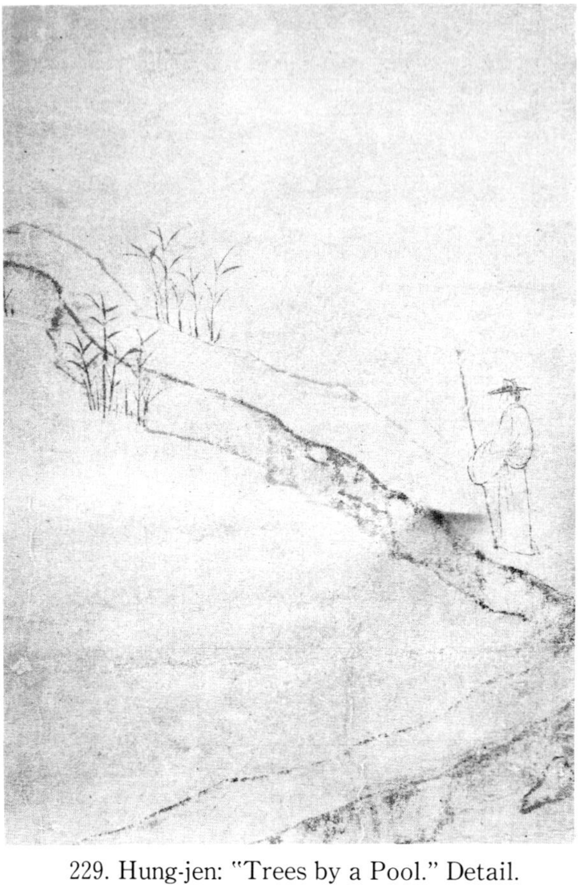

229. Hung-jen: "Trees by a Pool." Detail.

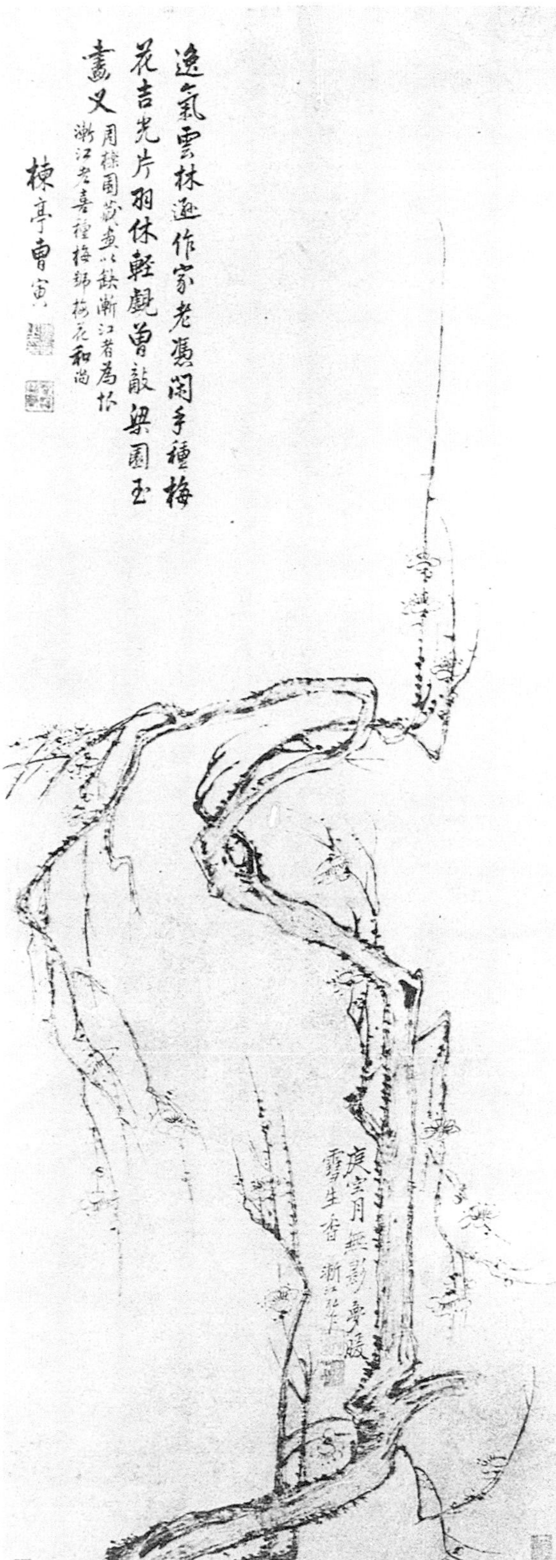

230. Hung-jen: "Plum Blossoms." Hanging scroll. Whereabout unknown (formerly Lu Shu-t'ung collection).

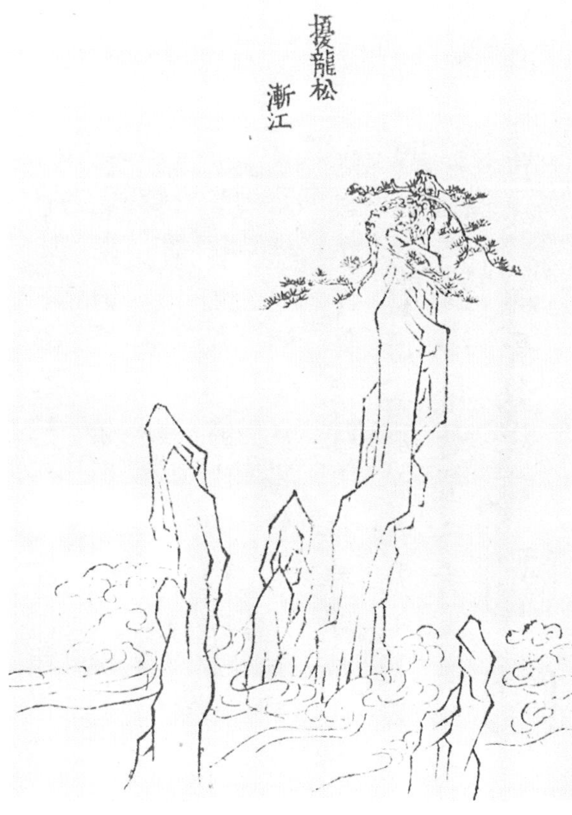

231. "Unsettling Dragon Pine." Woodcut after Hung-jen. From *Huang-shan-chih*, 1: 20a.

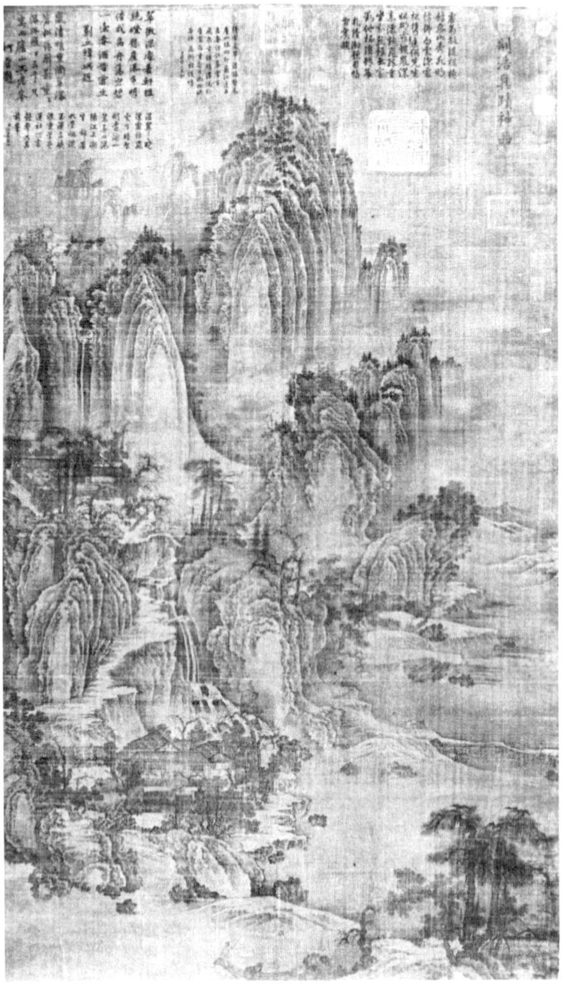

232. Attributed to Ching Hao: "The K'uang Lu Mountains." Hanging scroll, ink on silk, 185.8×106.8 cm. National Palace Museum, Taipei.

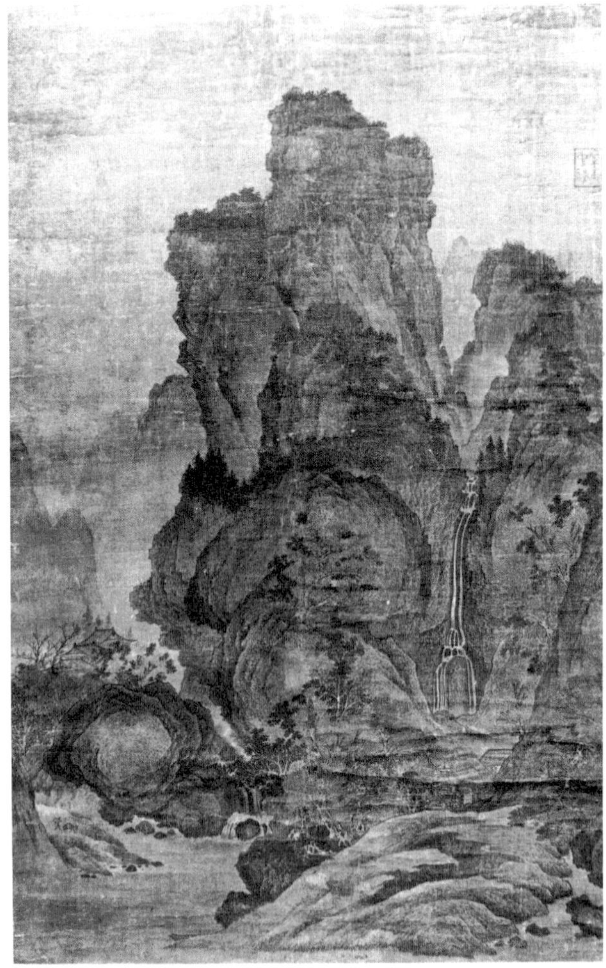

233. Attributed to Kuan T'ung: "Waiting for the Ferry at a Mountain-stream." Hanging scroll, ink and light colors on silk, 156.6×99.6 cm. National Palace Museum, Taipei.

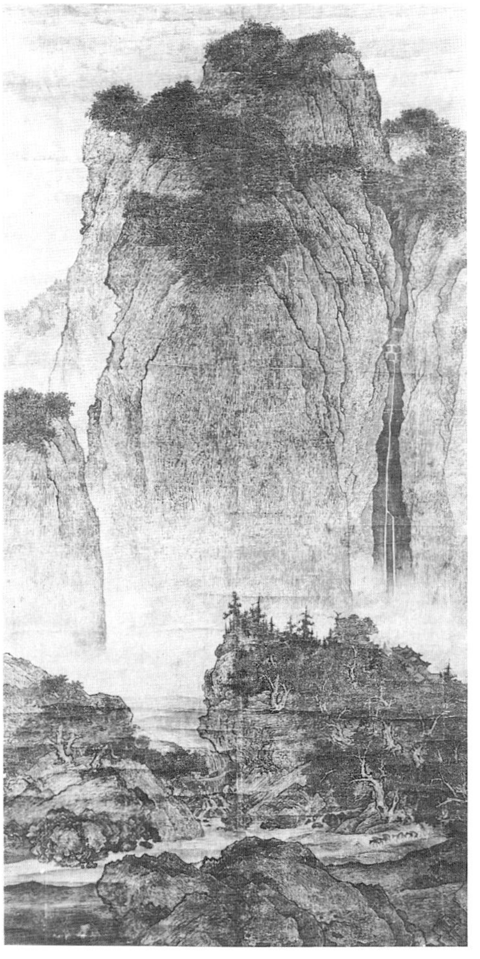

234. Fan K'uan: "Travellers among Mountains and Streams." Hanging scroll, ink on silk, 206.3×103.3 cm. National Palace Museum, Taipei.

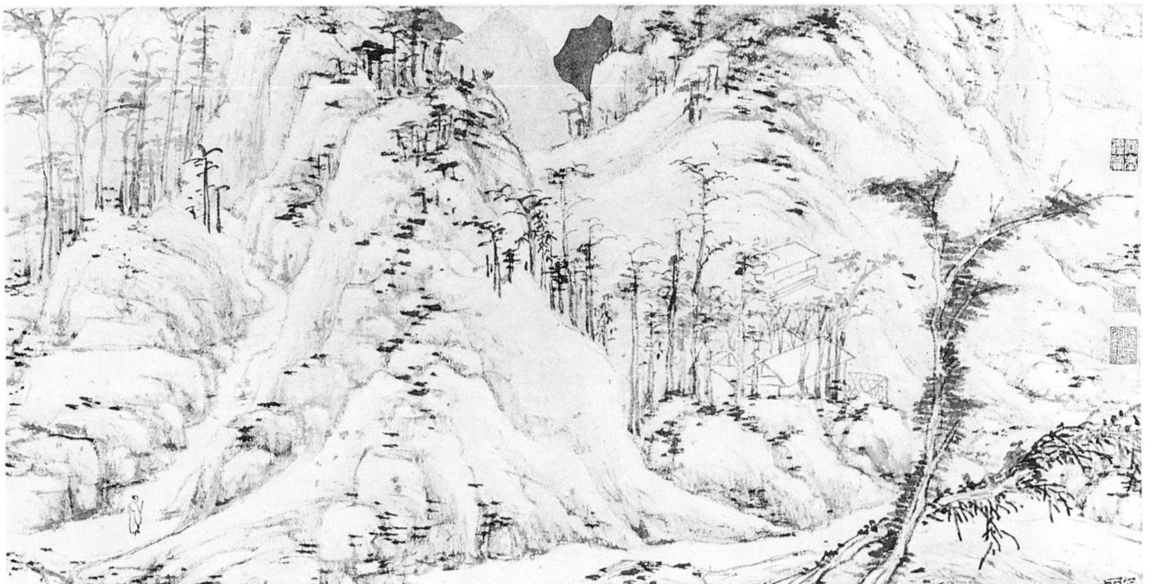

235. Wen Cheng-ming: "Autumn in the Mountains." Handscroll, ink on paper, 31.8×120.8 cm. The Art Institute of Chicago, Section 1.

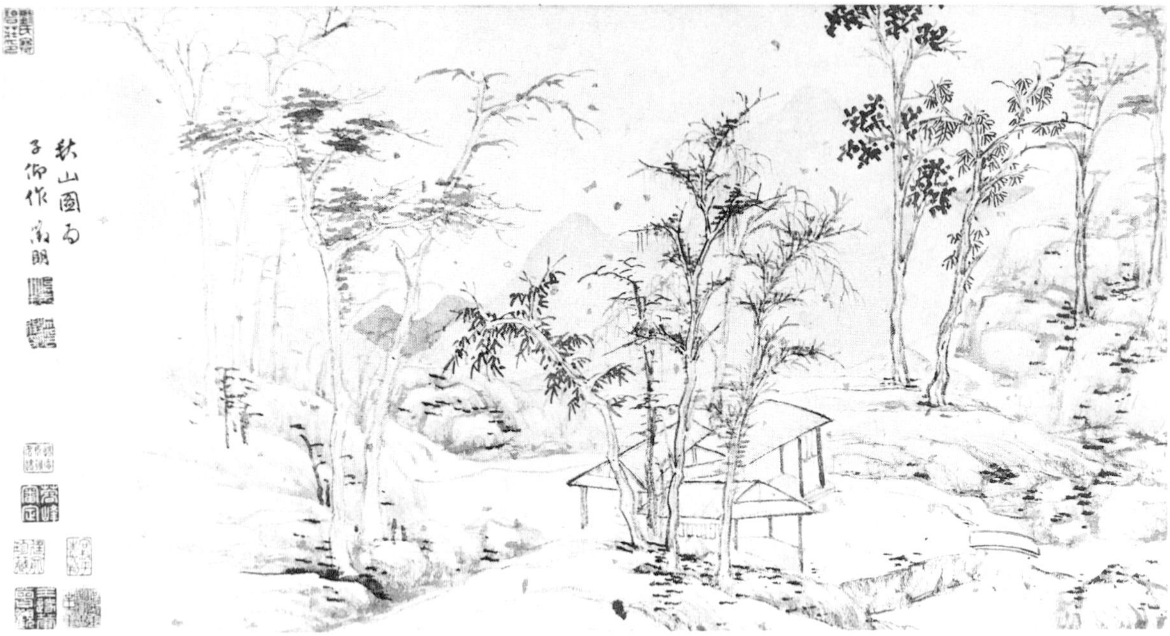

236. Wen Cheng-ming: "Autumn in the Mountains." Section 2.

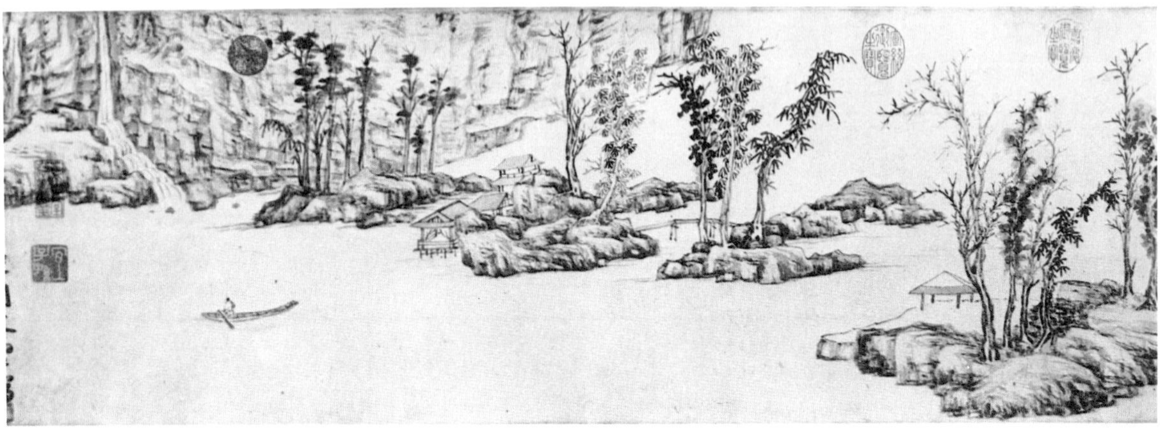

237. Wen Cheng-ming: "Summer Retreat in the Eastern Grove," 1515. Handscroll, ink on paper, 31.8×107.8 cm. John Crawford , Jr. collection, New York.

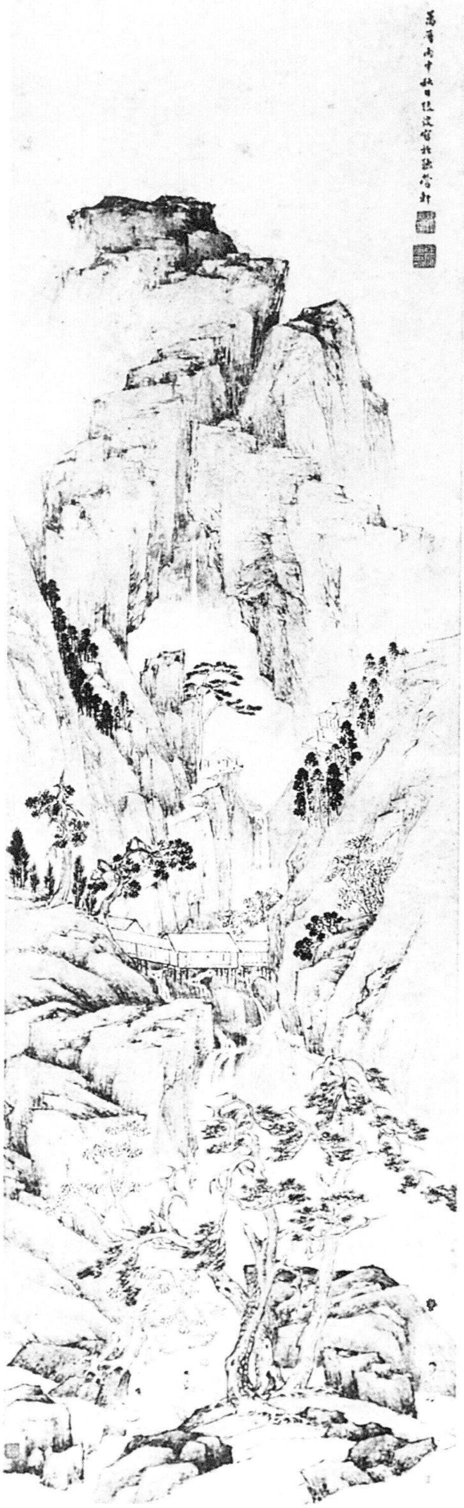

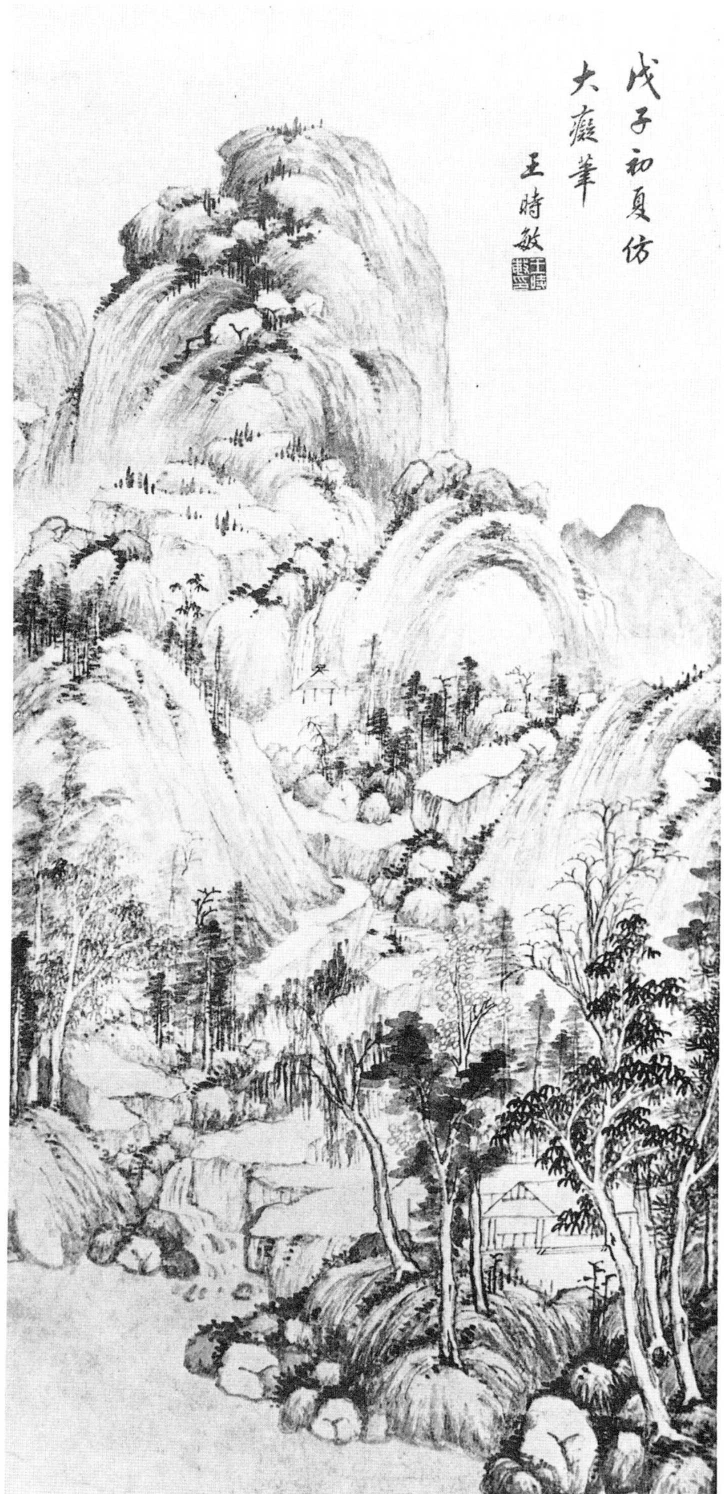

238. Chang Fu: "Landscape," 1596. Hanging scroll, ink and light colors on paper, $40\frac{1}{4} \times 12\frac{1}{4}$ in. Ching Yüan Chai collection, Berkeley.

239. Wang Shih-min: "Landscape in the Style of Huang Kung-wang," 1648. Hanging scroll. Whereabouts unknown.

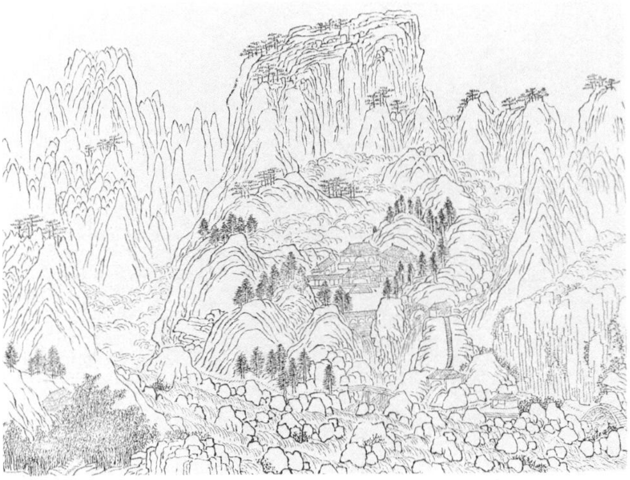

240. "Mt. Huang." Woodcut after Cheng Chung.
From *Ming-k'an ming-shan-t'u pan-hua-chi*, P1. 7.

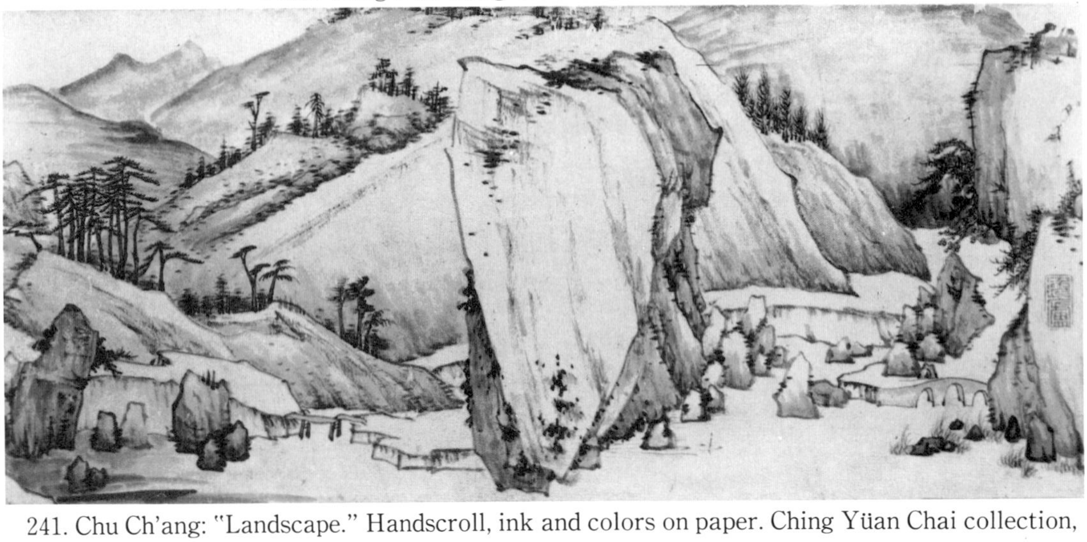

241. Chu Ch'ang: "Landscape." Handscroll, ink and colors on paper. Ching Yüan Chai collection, Berkeley. Section 1.

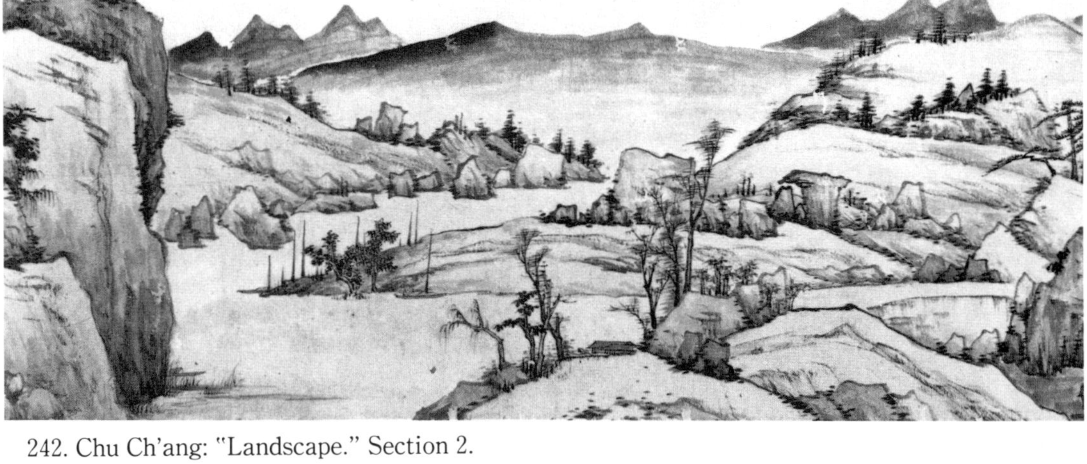

242. Chu Ch'ang: "Landscape." Section 2.

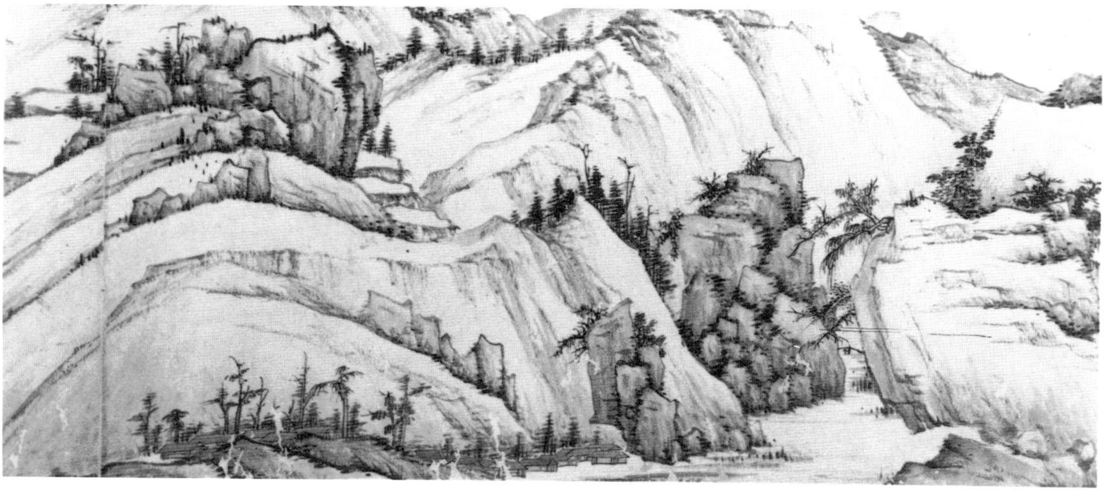

243. Chu Ch'ang: "Landscape." Section 3.

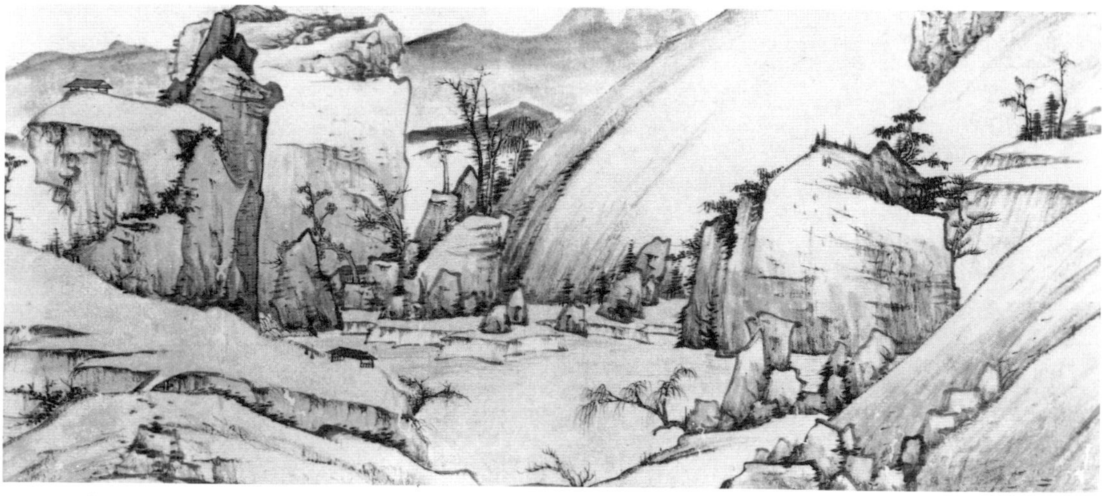

244. Chu Ch'ang: "Landscape." Section 4.

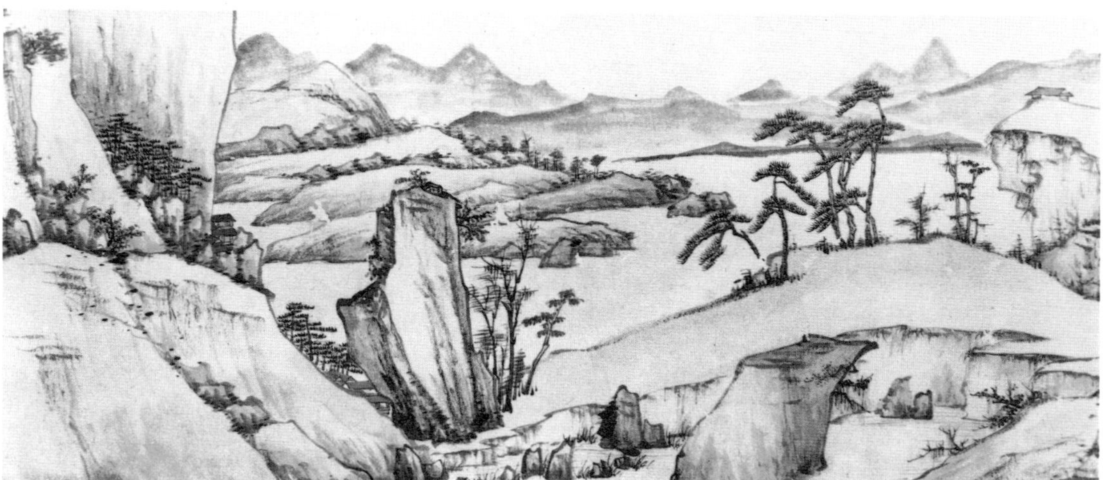

245. Chu Ch'ang: "Landscape." Section 5.

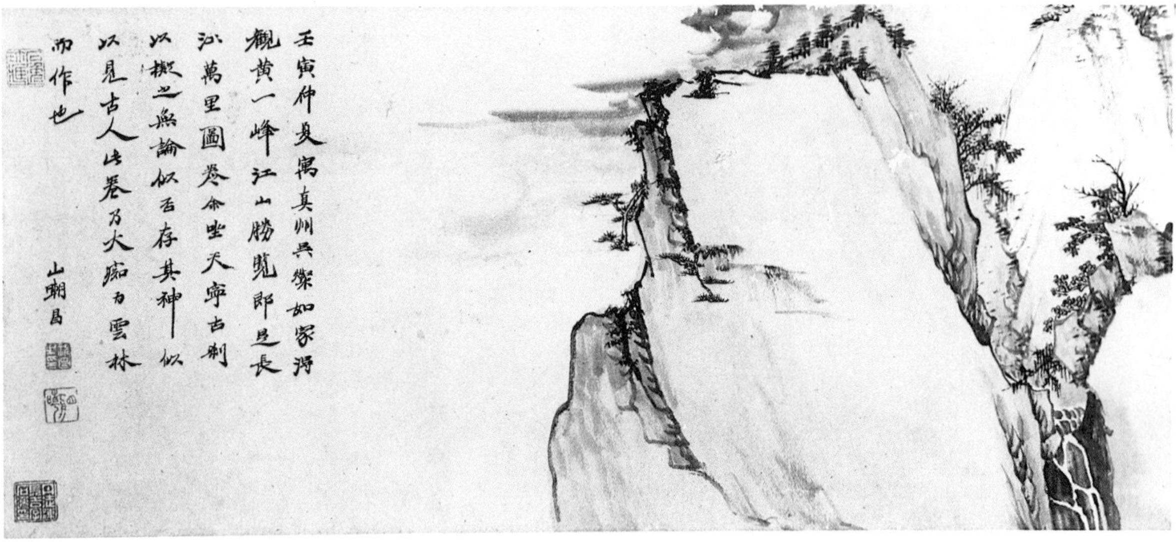

246. Chu Ch'ang: "Landscape." Section 6.

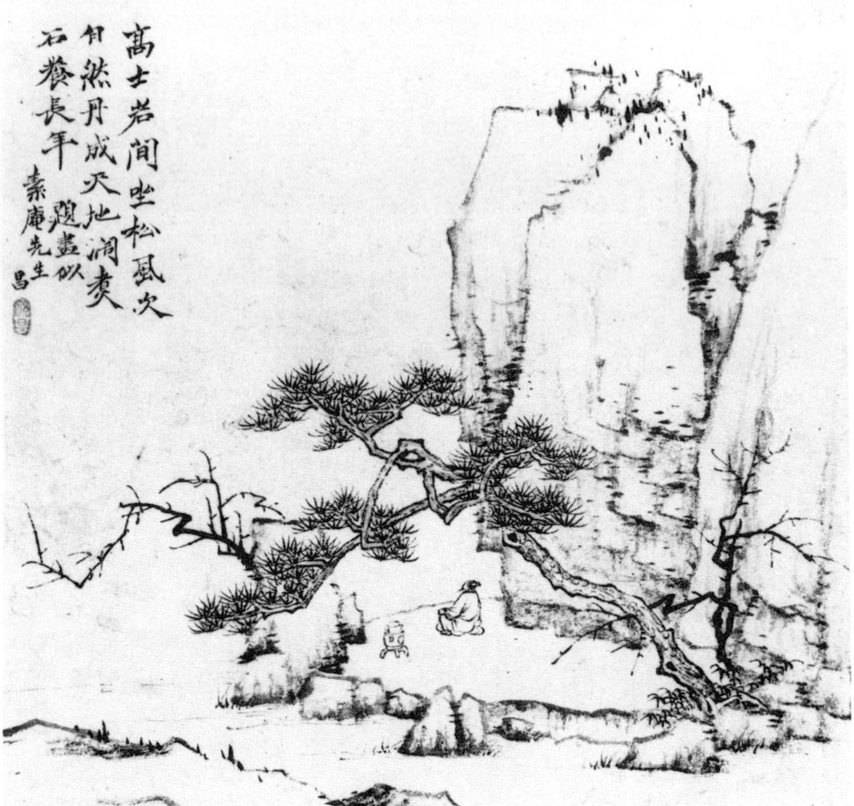

247. Chu Ch'ang: "Figure in Landscape," 1663. Album leaf, ink on paper. Ching Yüan Chai collection, Berkeley.

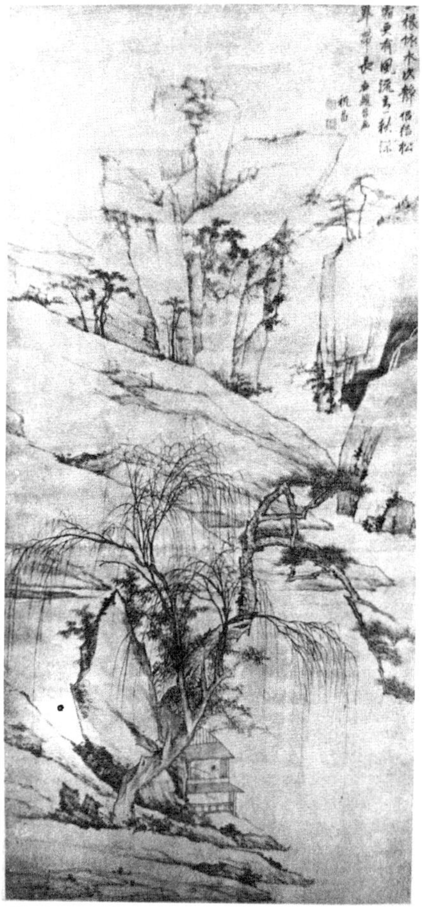

248. Chu Ch'ang: "Landscape." Hanging scroll. Whereabouts unknown.

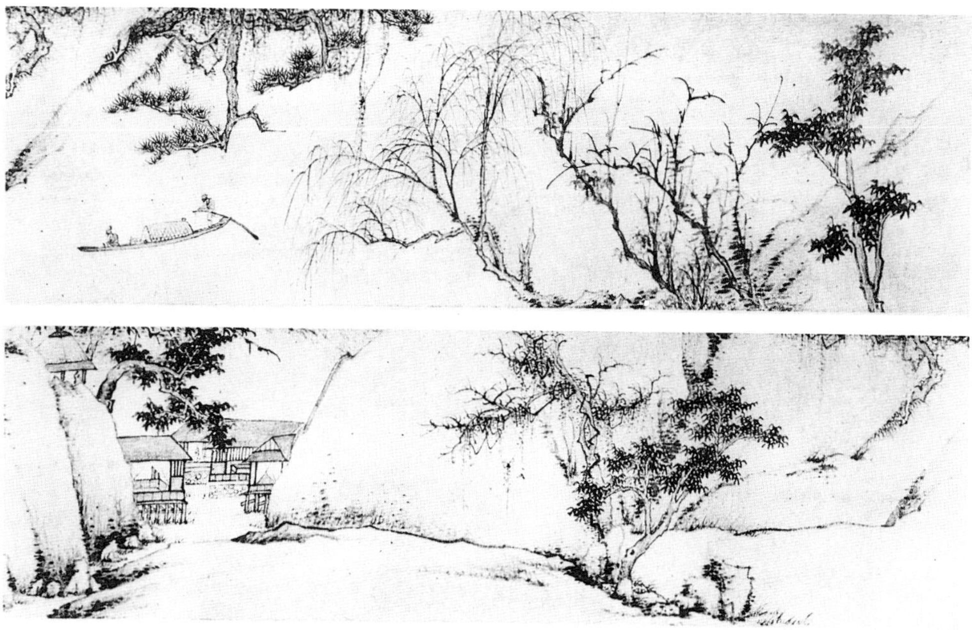

249. Chu Ch'ang: "Streams and Mountains without End," 1666. Handscroll, ink and colors on paper, 25.2×1,049.5 cm. Suchou Museum. Section 1 and Section 2.

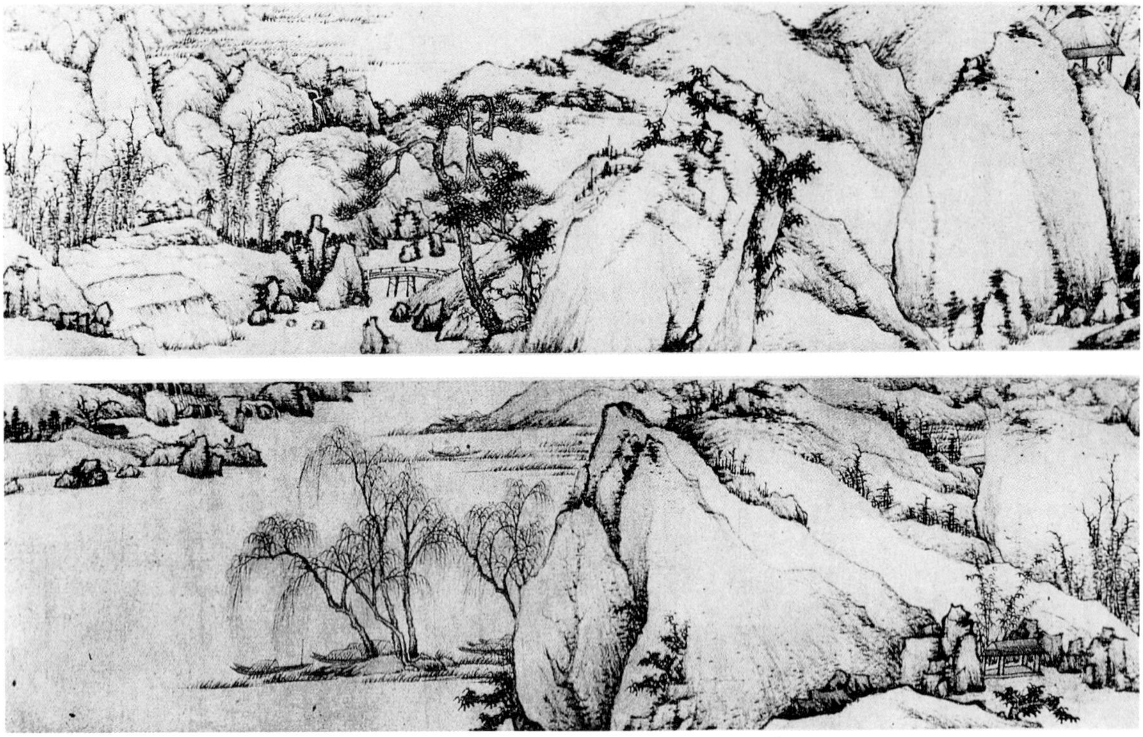

250. Chu Ch'ang: "Streams and Mountains without End." Section 3 and Section 4.

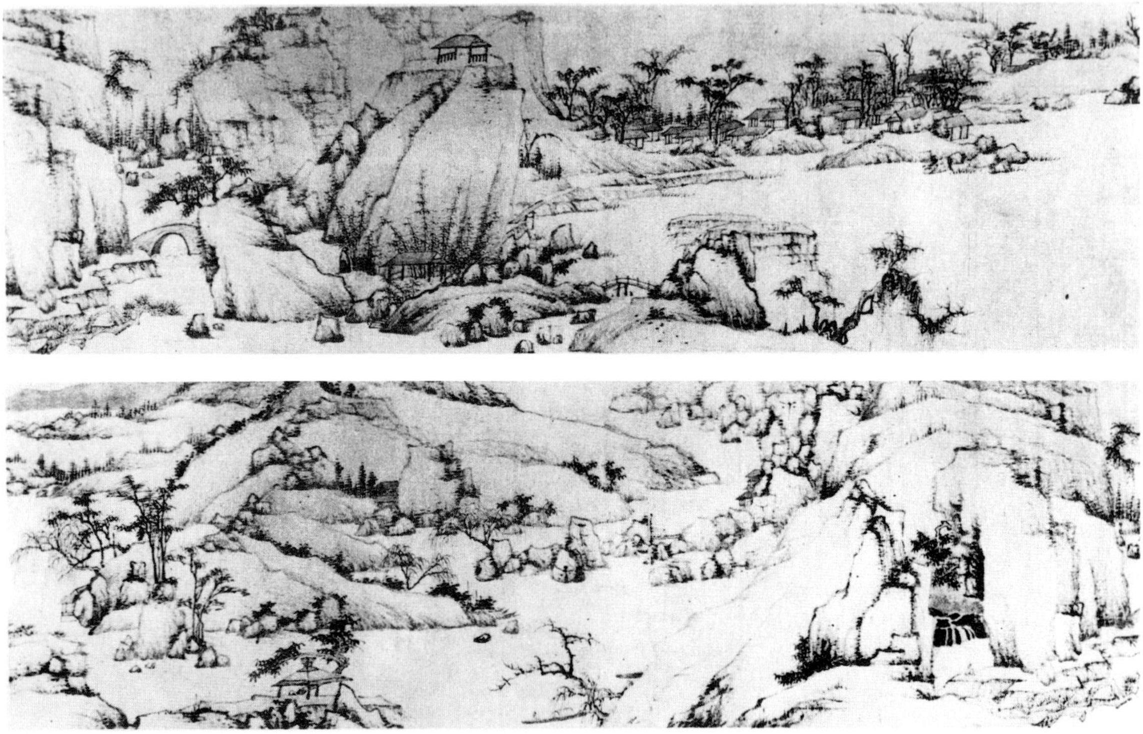

251. Chu Ch'ang: "Streams and Mountains without End." Section 5 and Section 6.

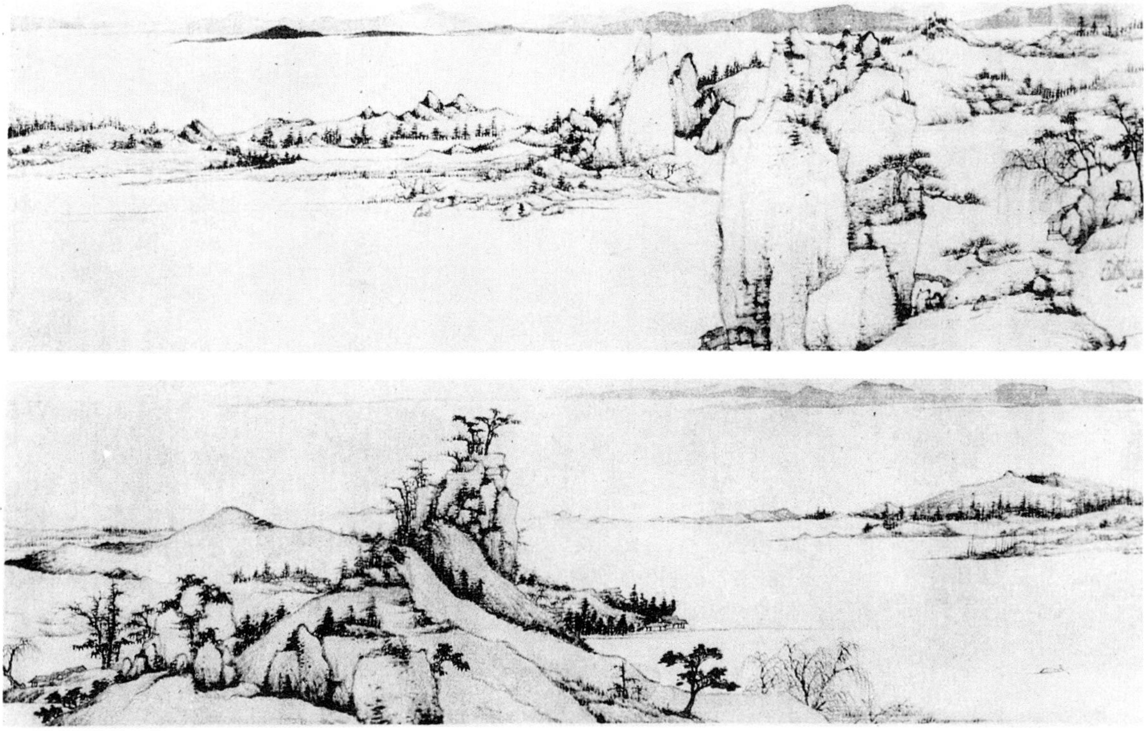

252. Chu Ch'ang: "Streams and Mountains without End." Section 7 and Section 8.

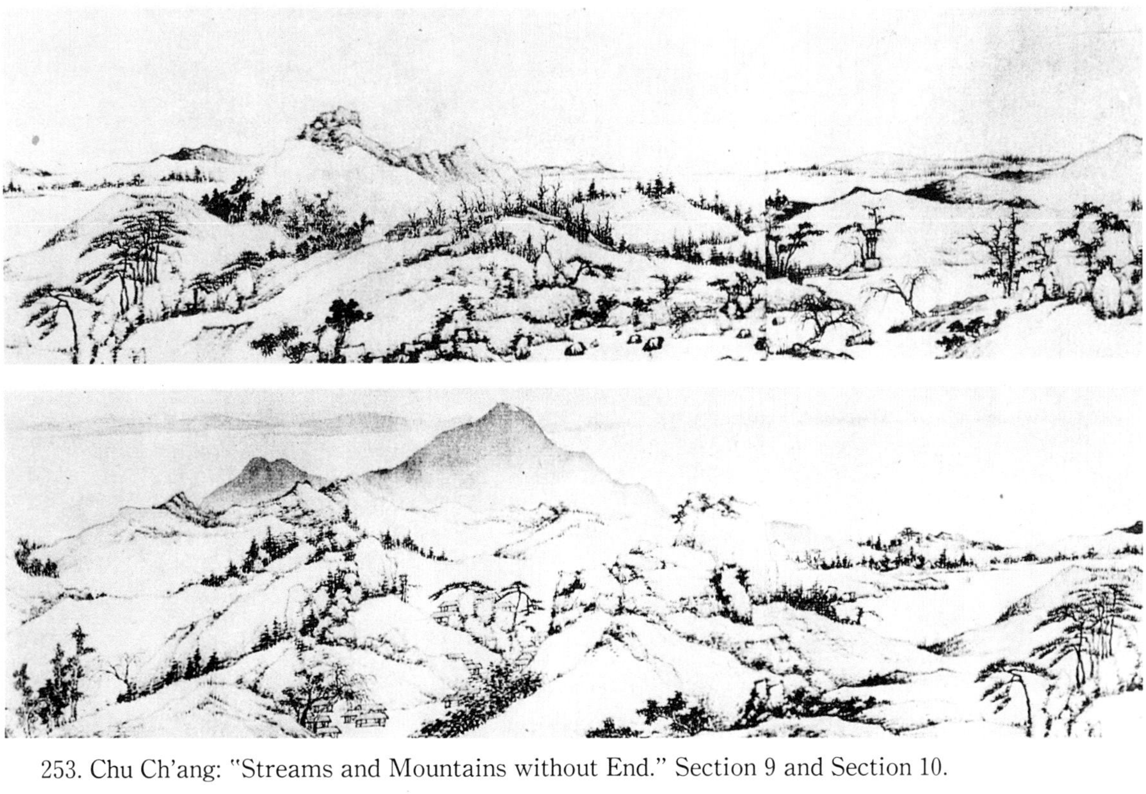

253. Chu Ch'ang: "Streams and Mountains without End." Section 9 and Section 10.

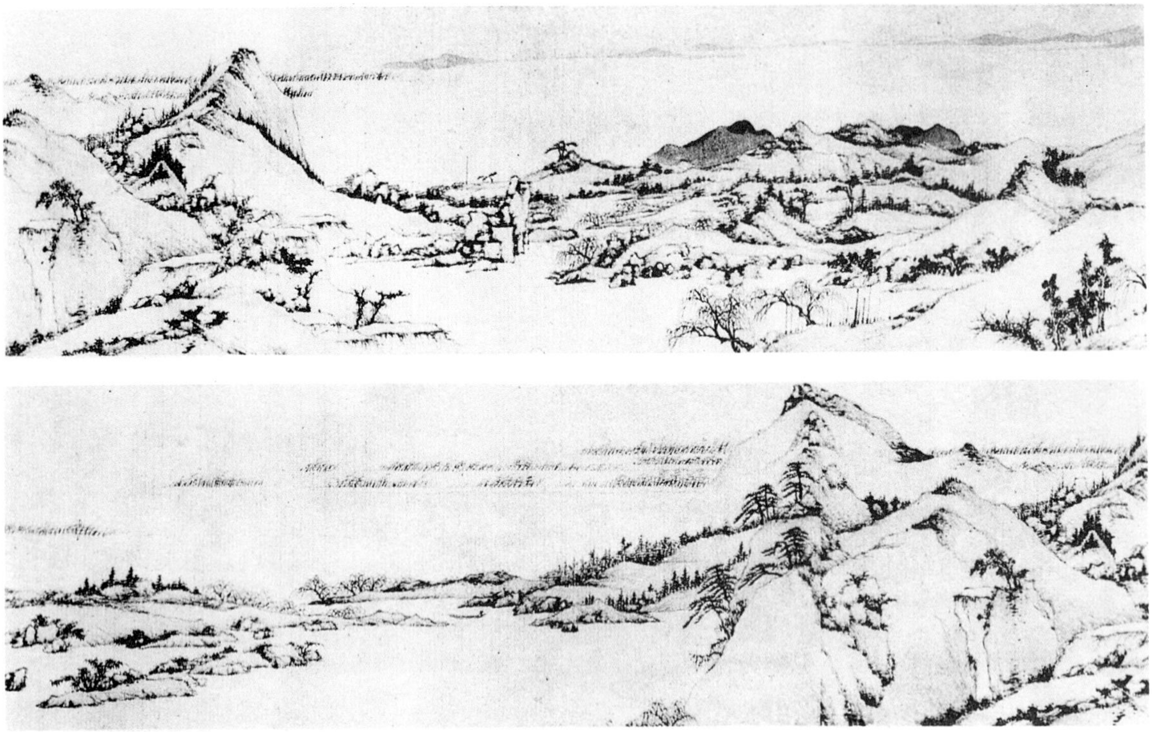

254. Chu Ch'ang: "Streams and Mountains without End." Section 11 and Section 12.

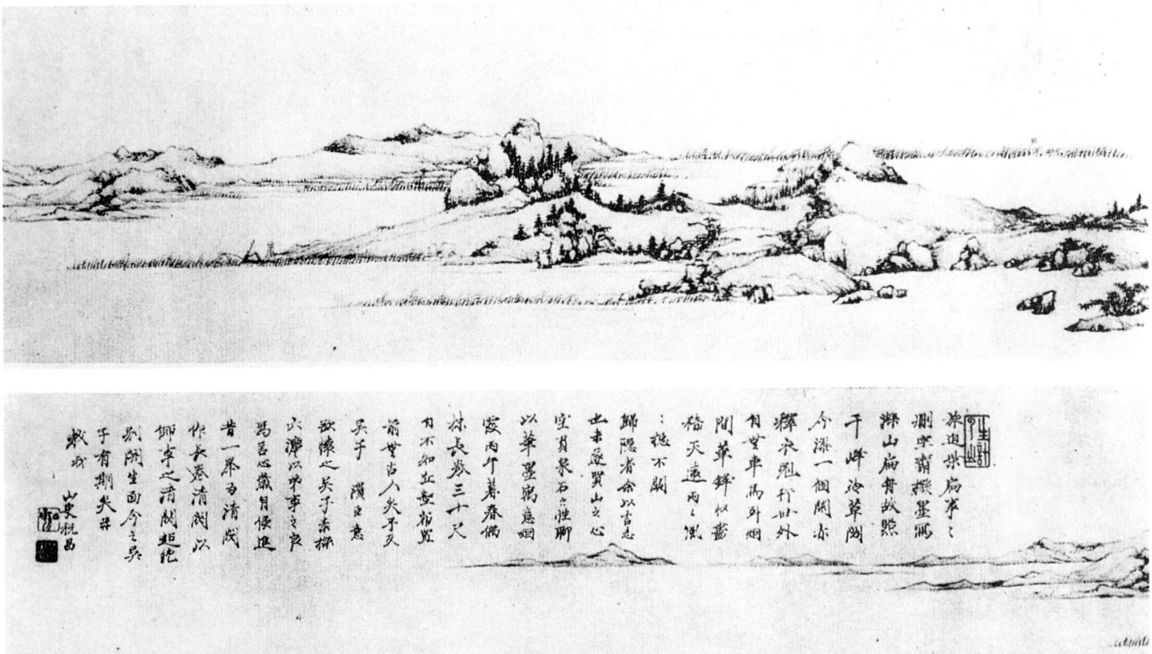

255. Chu Ch'ang: "Streams and Mountains without End." Section 13 and Section 14.

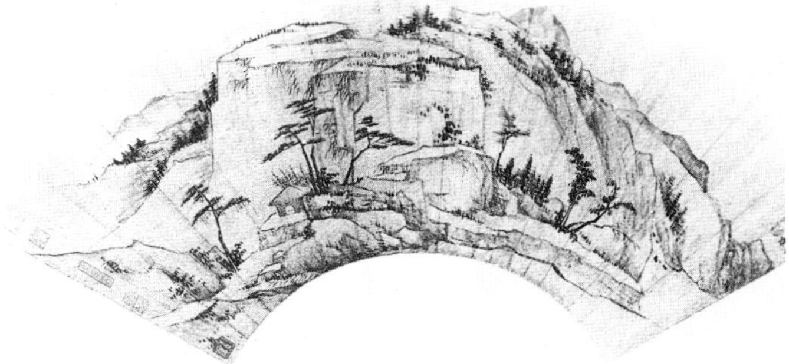

256. Cha Shih-piao: "Landscape," 1659. Fan, ink on paper. Steve Addiss collection, Lawrence, Kansas.

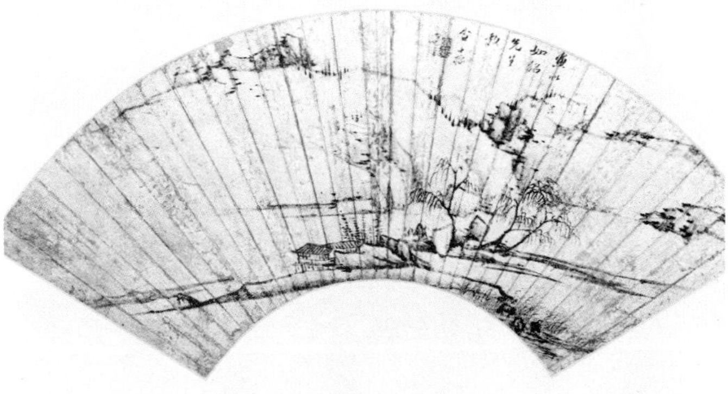

257. Cha Shih-piao: "Landscape." Fan. K. C. Ho collection, New York.

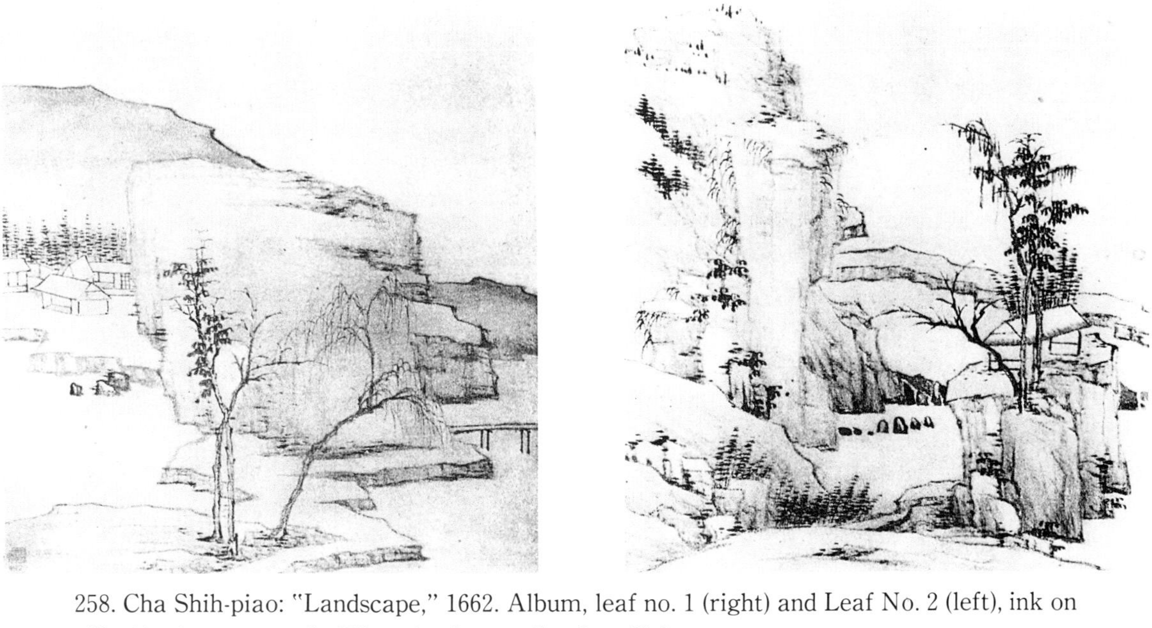

258. Cha Shih-piao: "Landscape," 1662. Album, leaf no. 1 (right) and Leaf No. 2 (left), ink on silk, 30×25.5 cm. each. T'ien-yin-t'ang collection, Tokyo.

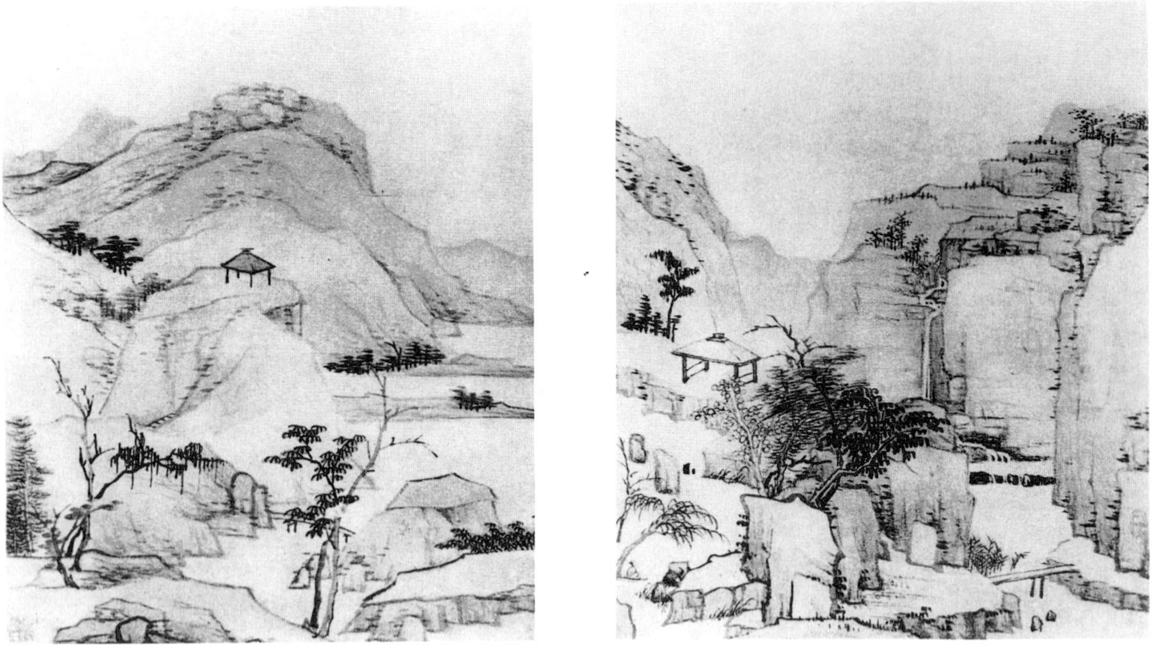

259. Cha Shih-piao: "Landscape." Leaf no. 7 (right) ana Lear no. 8 (left), ink and colors on silk, 30×25.5 cm. each.

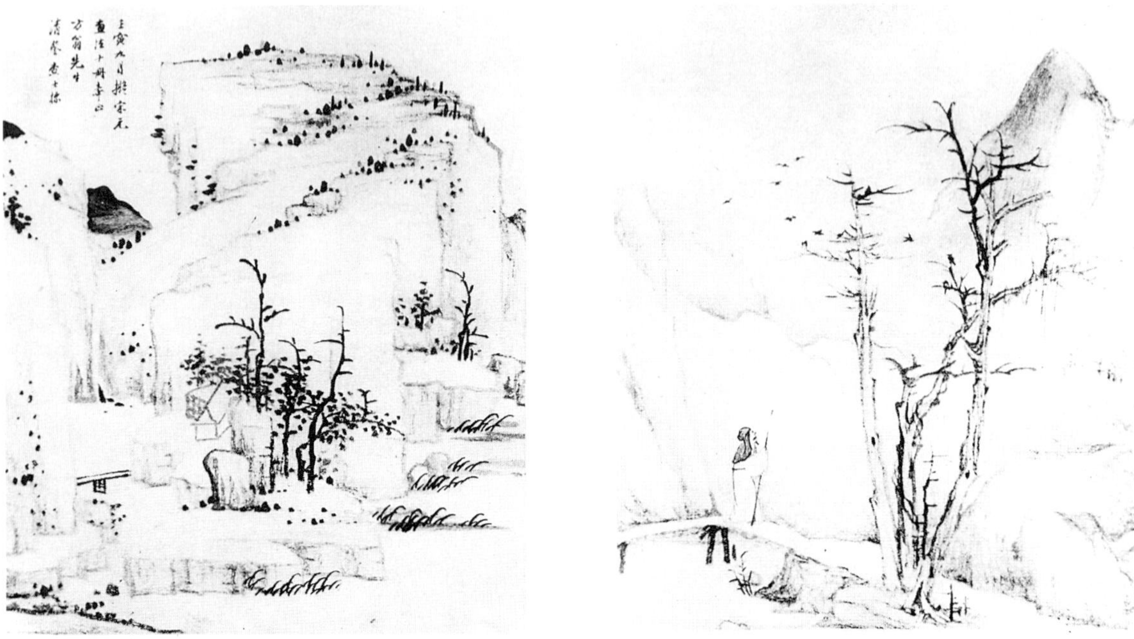

260. Cha Shih-piao: "Landscape." Leaf no. 9 (right), ink and colors on silk, and Leaf no. 10 (left), ink on silk, 30×25.5 cm. each.

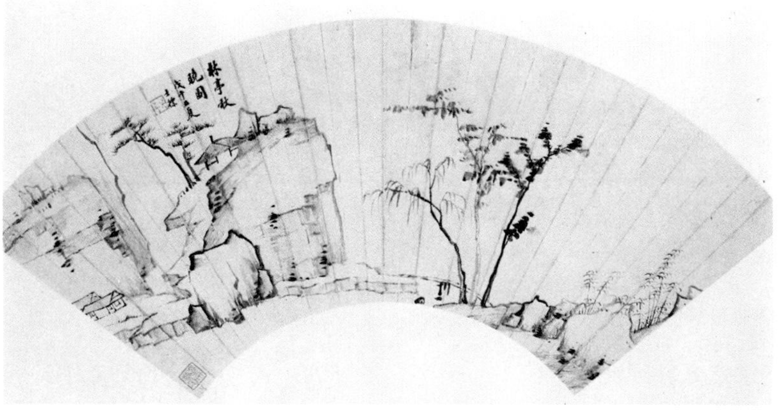

261. Cha Shih-piao: "Pavilion in Autumn Morning," 1678. Fan, ink on paper. Shanghai Museum.

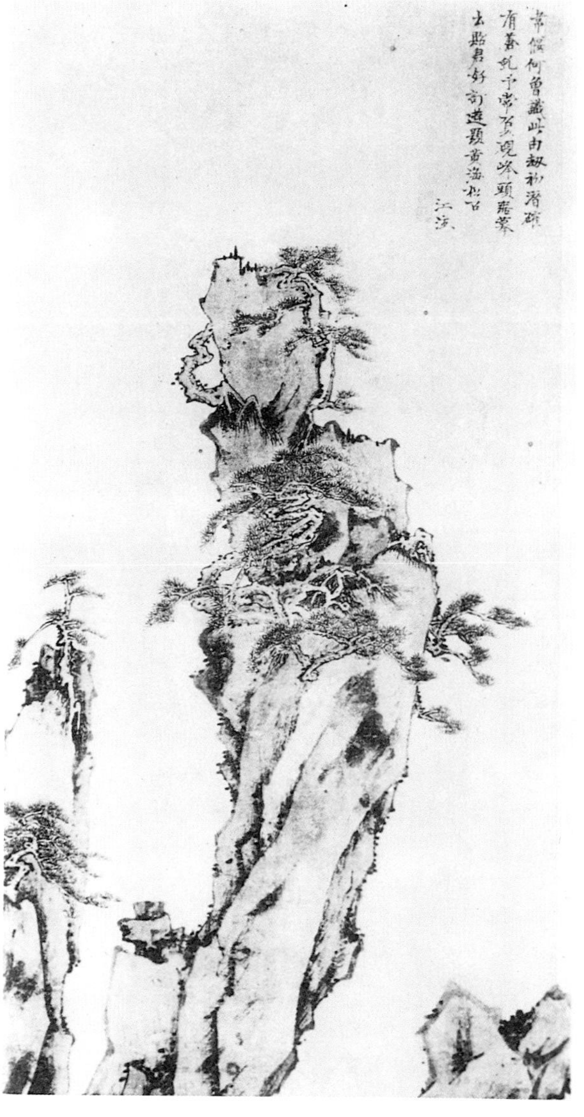

262. Chiang Chu: "Pines and Rocks at Mt. Huang." Hanging scroll. Whereabouts unknown.

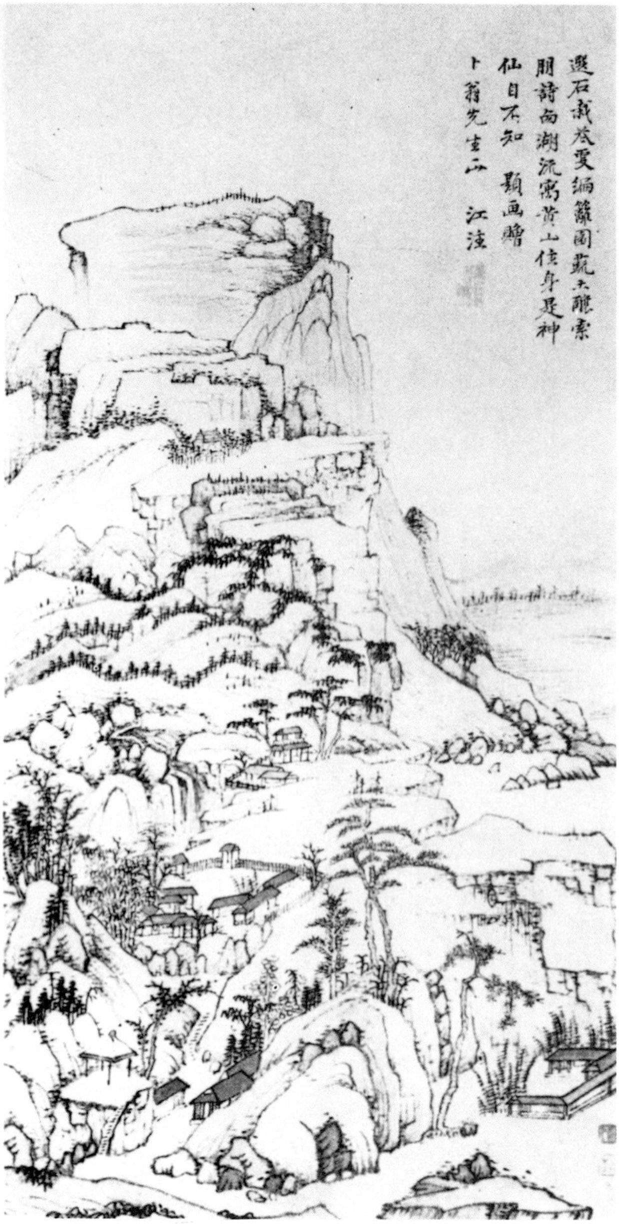

263. Chiang Chu: "Arranging Rocks and Planting Flowers." Hanging scroll, ink and colors on paper, $22\frac{1}{4} \times 10\frac{5}{8}$ in. Whereabouts unknown (formerly Richard Bryant Hobart collection).

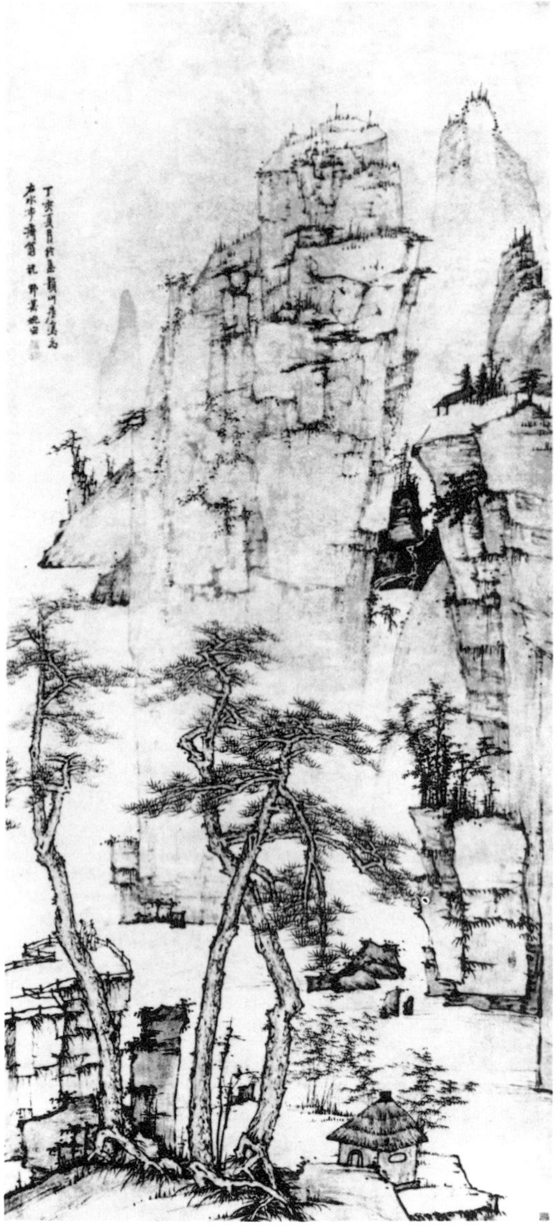

264. Yao Sung: "Landscape," 1707. Hanging scroll, ink on paper, $58 \times 23\frac{1}{2}$ in. Drenowatz collection, Zürich.

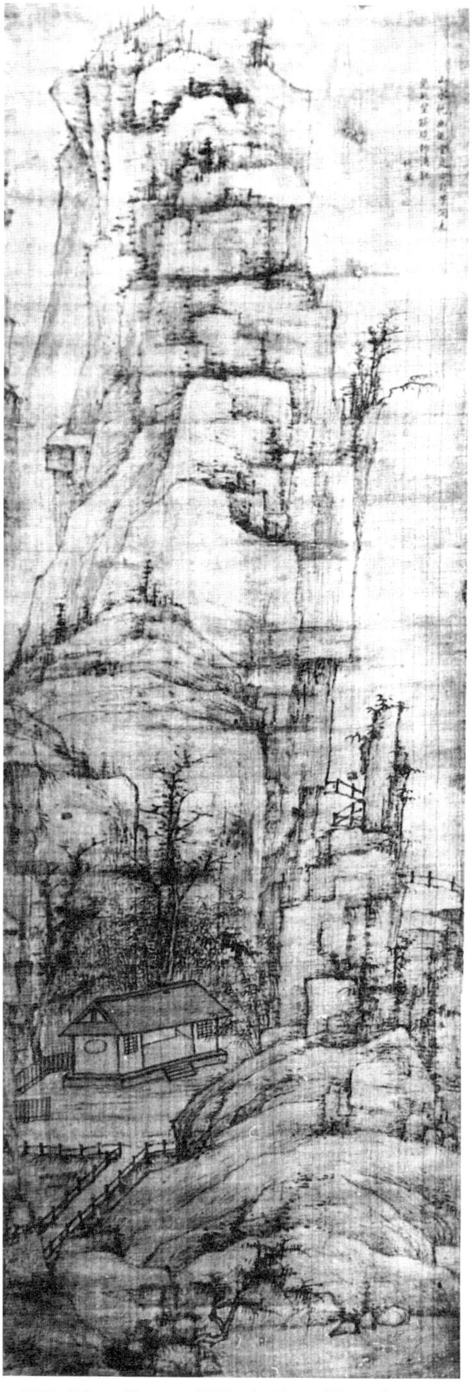

265. Yao Sung: "High Dwelling in a Quiet Valley." Hanging scroll, ink on silk, 147×50.5 cm. Hashimoto Sueyoshi collection, Japan.

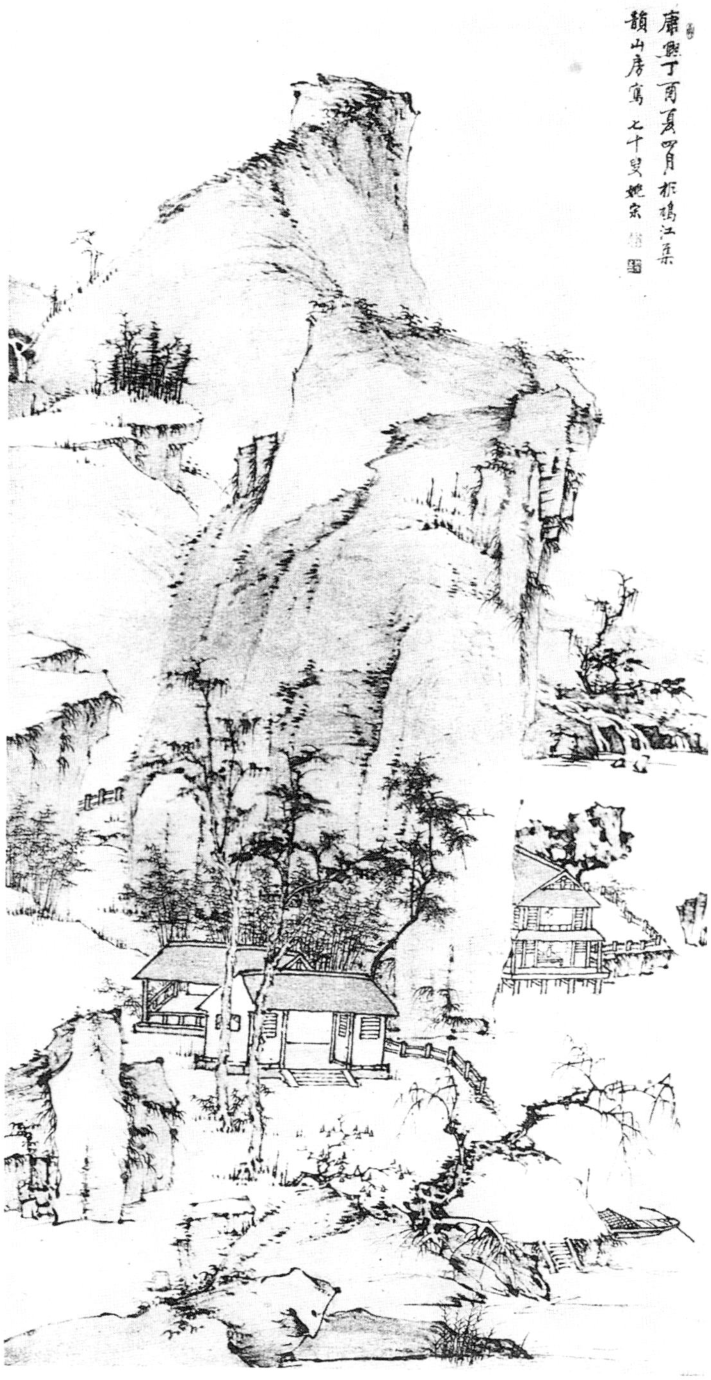

266. Yao Sung: "Leisurely View in the Autumn Mountain," 1717 . Hanging scroll, ink on paper, 202×101.5 cm. Nanking Museum .

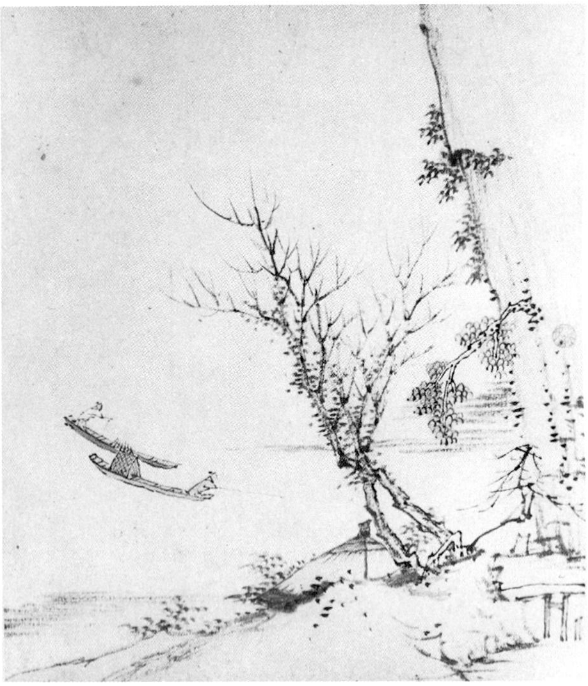

267. Yao Sung: "Landscape." Album, 21.3× 24.5 cm. each. Franco Vannotti collecton, Lugano. Leaf no. 1.

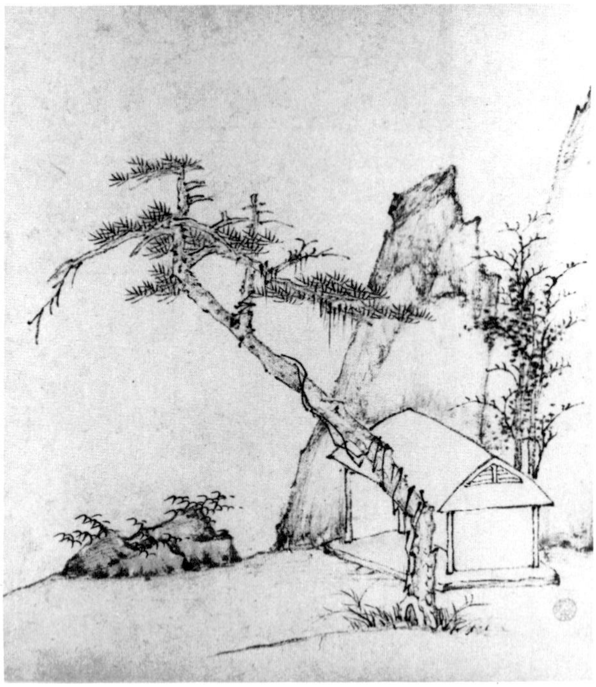

268. Yao Sung: "Landscape." Leaf no. 2.

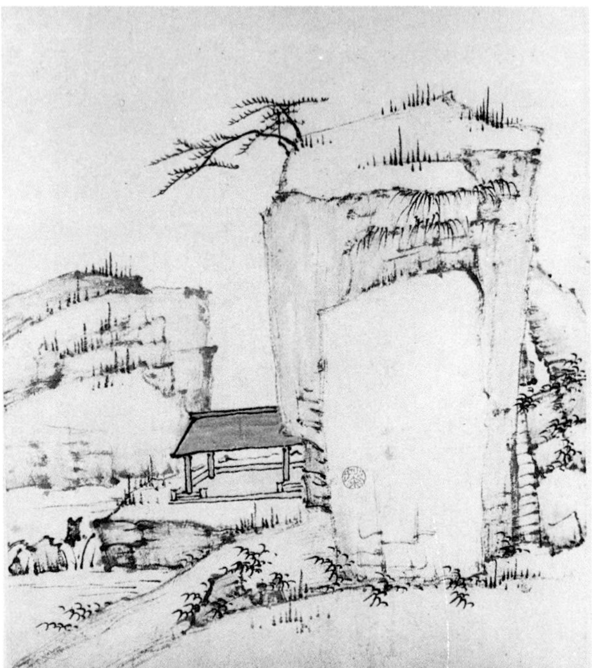

269. Yao Sung: "Landscape." Leaf no. 3.

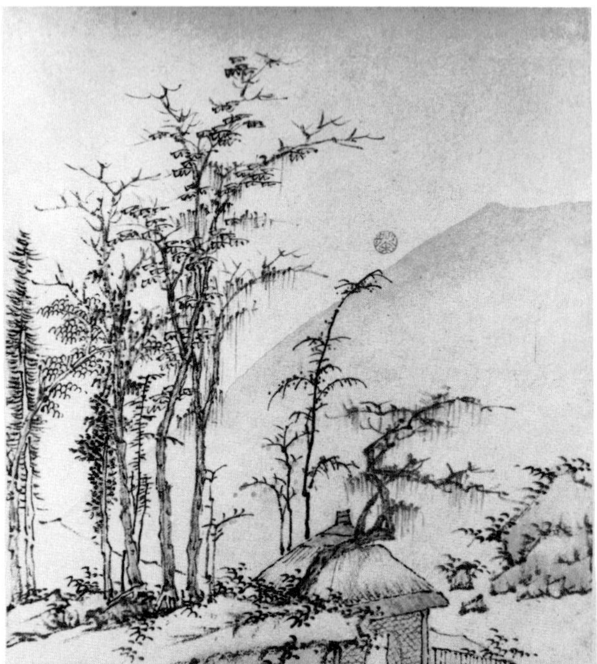

270. Yao Sung: "Landscape." Leaf no. 4.

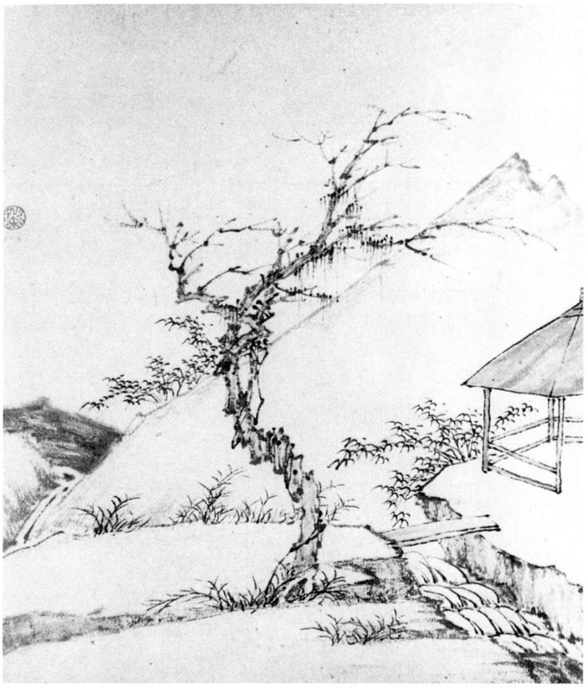

271. Yao Sung: "Landscape." Leaf no. 5.

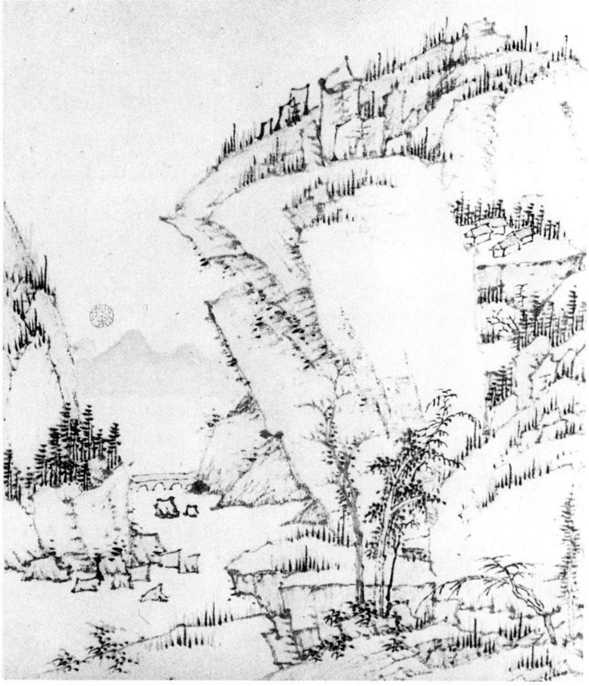

272. Yao Sung: "Landscape." Leaf no. 6.

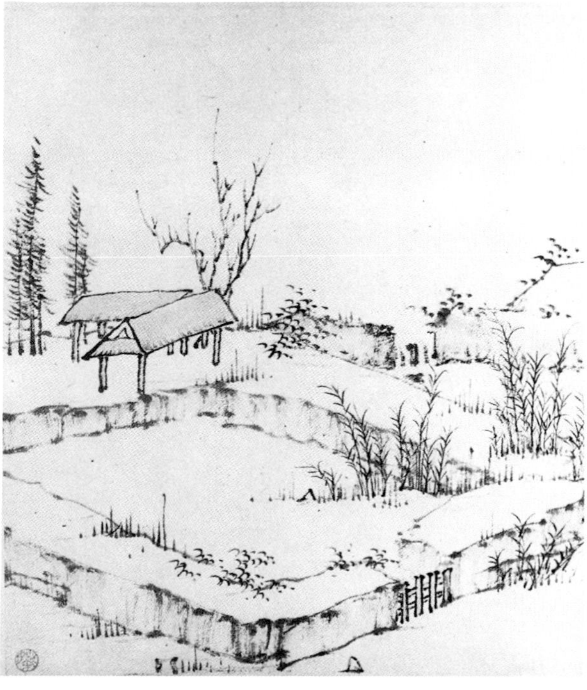

273. Yao Sung: "Landscape." Leaf no. 7.

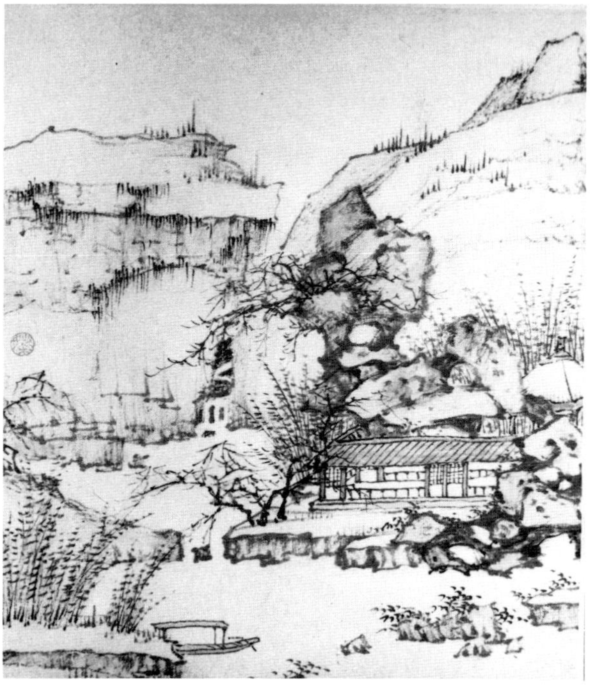

274. Yao Sung: "Landscape." Leaf no. 8.

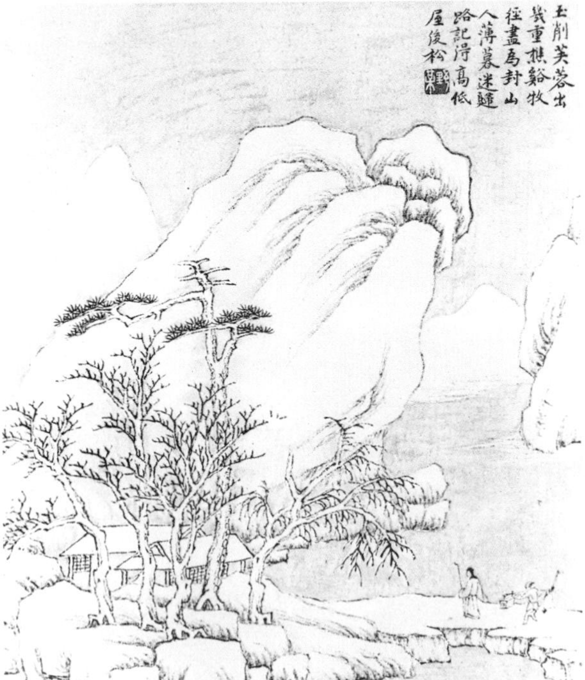

275. Yao Sung: "Landscape." Album, ink and color on paper, $9 \times 7\frac{5}{8}$ in. each. Whereabouts unknown (formerly Richard Bryant Hobart collection). Leaf no. 1.

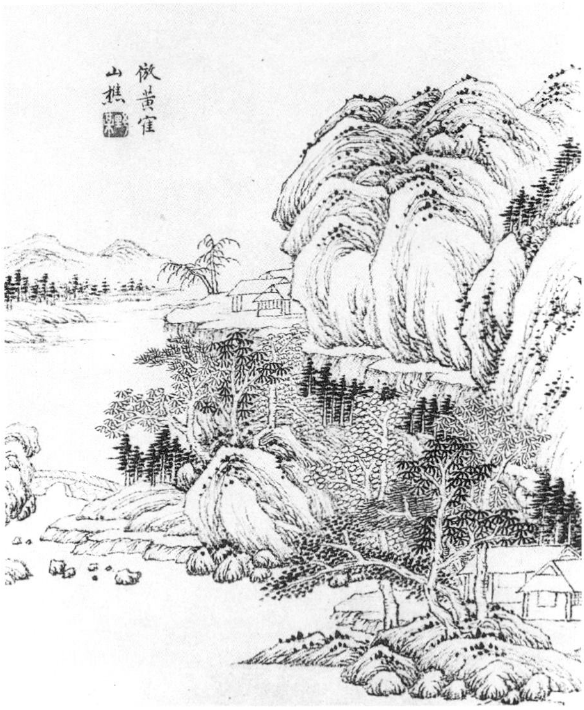

276. Yao Sung: "Landscape." Leaf no. 2.

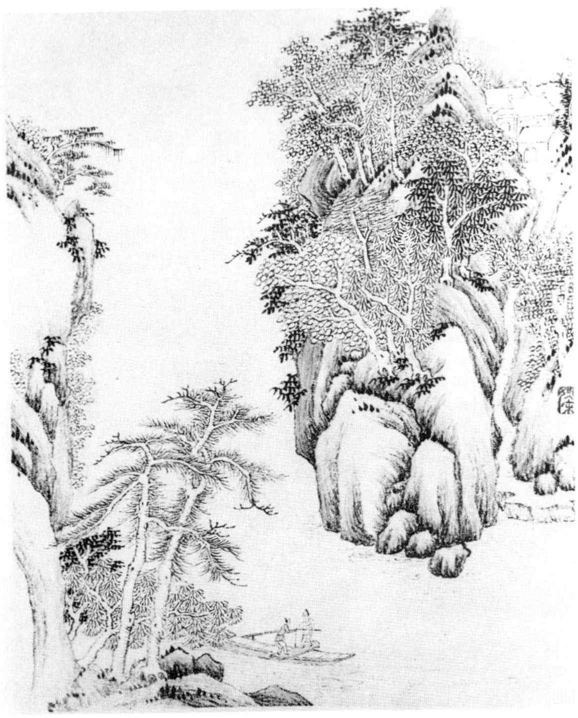

277. Yao Sung: "Landscape." Leaf no. 3.

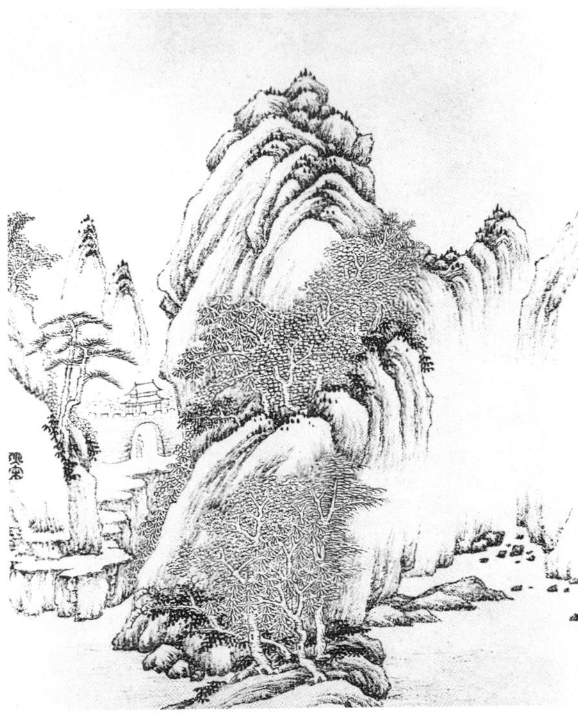

278. Yao Sung: "Landscape." Leaf no. 4.

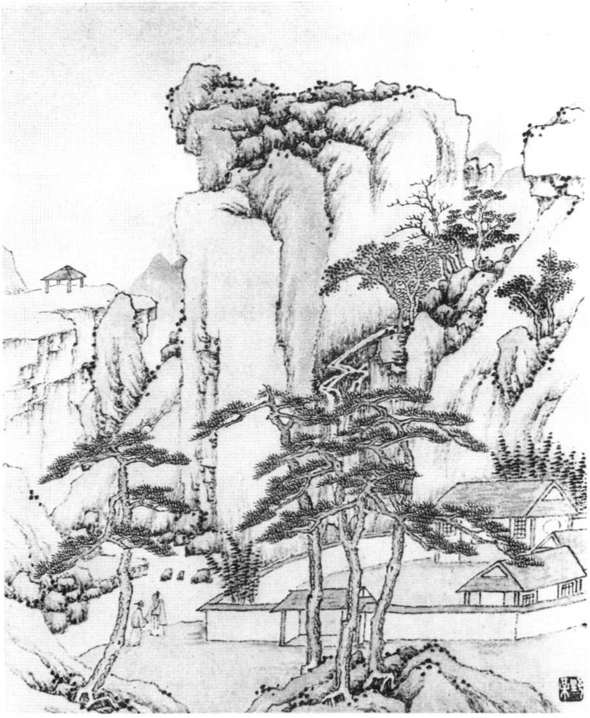

279. Yao Sung: "Landscape." Leaf no. 5.

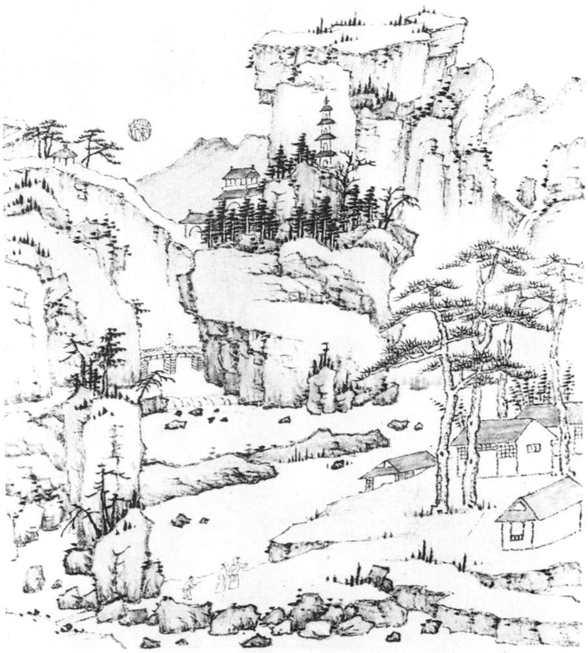

280. Yao Sung: "Landscape." Leaf no. 6.

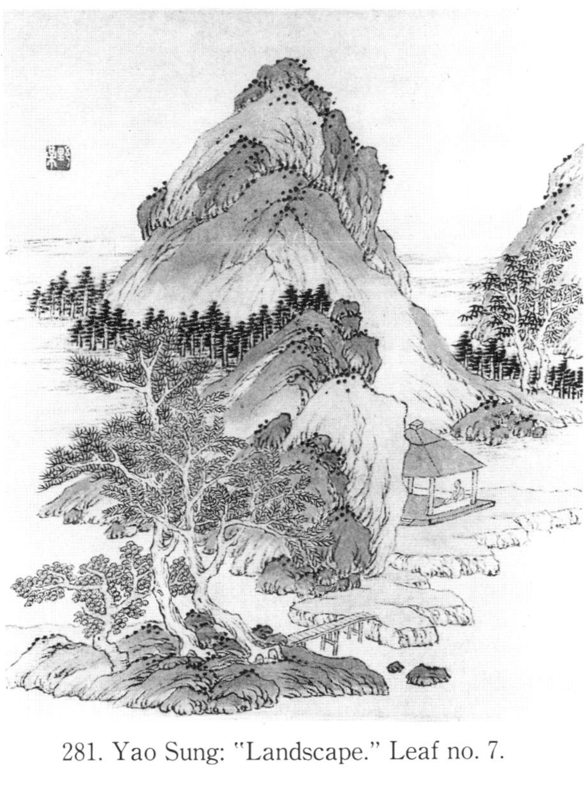

281. Yao Sung: "Landscape." Leaf no. 7.

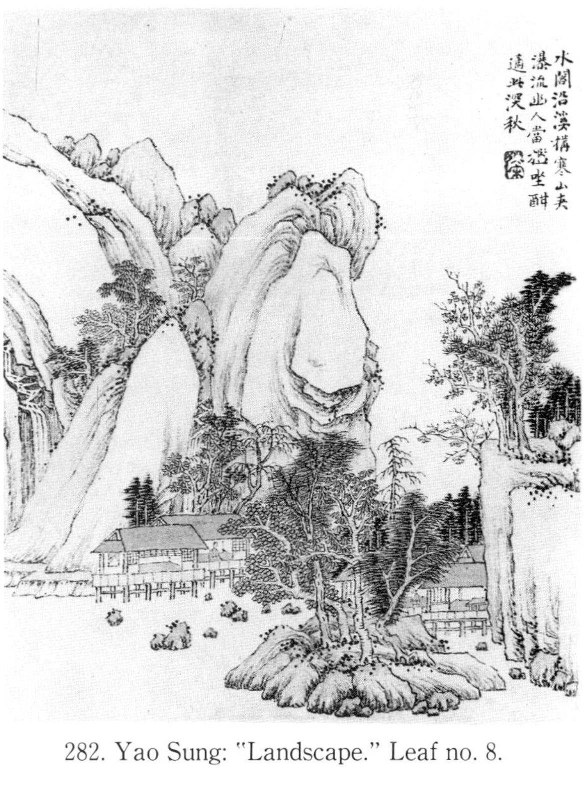

282. Yao Sung: "Landscape." Leaf no. 8.

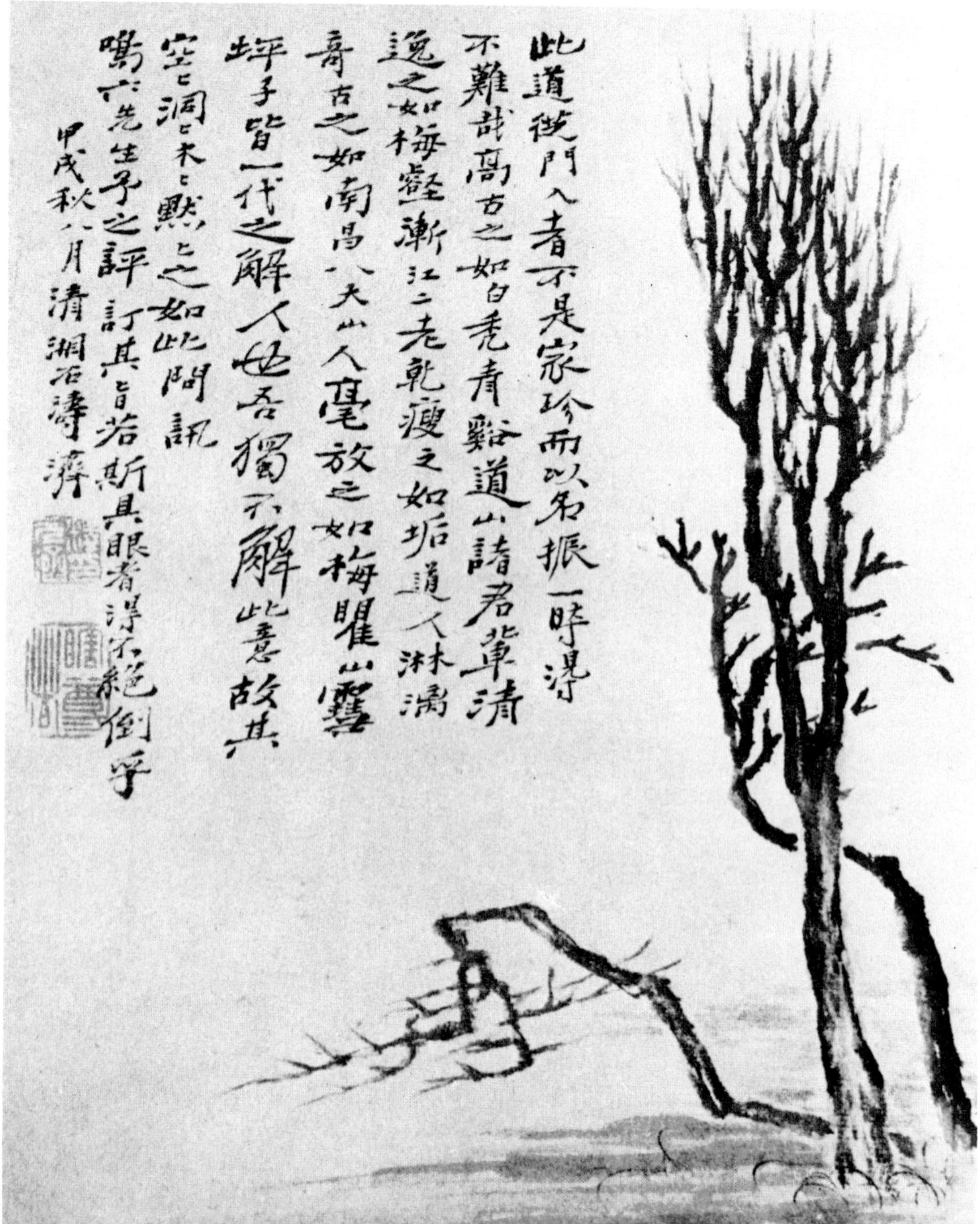

283. Tao-chi: "Landscape," 1694 Album, ink on paper, $11 \times 8\frac{3}{4}$ in. The Los Angeles County Museum of Art. Leaf no. 4.

弘仁冷峻的山水畫研究 **精裝新台幣720元**

民國七十九年三月初版發行
民國 八 十 年七月二刷發行

著　者：郭　繼　生
發行者：魏　德　文
發行所：南天書局有限公司
中華民國・台北市羅斯福路3段283巷14弄14號
☎(02)362-0190　電傳(Fax)：(02)362-3834
郵政劃撥：0108053-8號（南天書局帳戶）
登記證：局版台業字第1436號

製版廠：豪華彩色製版有限公司
☎309-1235 台北市西園路2段279之7號4樓
印刷廠：國順印刷有限公司
☎967-7226 板橋市中正路216巷2弄13號

ISBN 957-9482-11-X